GREAT EVENTS

1900-2001

GREAT EVENTS

1900-2001
REVISED EDITION

Volume 6
1984-1993

From
The Editors of Salem Press

SALEM PRESS, INC.
Pasadena, California   Hackensack, New Jersey

Editor in Chief: Dawn P. Dawson
Managing Editor: R. Kent Rasmussen *Research Supervisor:* Jeffry Jensen
Manuscript Editor: Rowena Wildin *Acquisitions Editor:* Mark Rehn
Production Editor: Joyce I. Buchea *Page Design and Graphics:* James Hutson
Photograph Editor: Philip Bader *Layout:* William Zimmerman
Assistant Editor: Andrea E. Miller Eddie Murillo

Cover Design: Moritz Design, Los Angeles, Calif.

Cover photos: Center image—Corbis, Remaining images—AP/Wide World Photos
Half title photos: Library of Congress, Digital Stock, AP/Wide World Photos

© 2002 *Great Events: 1900-2001, Revised Edition*
© 1997 *The Twentieth Century: Great Scientific Achievements, Supplement* (3 volumes)
© 1996 *The Twentieth Century: Great Events, Supplement* (3 volumes)
© 1994 *The Twentieth Century: Great Scientific Achievements* (10 volumes)
© 1992 *The Twentieth Century: Great Events* (10 volumes)

∞ The paper used in these volumes conforms to the American National Standard for Permanence of Paper for Printed Library Materials, Z39.48-1992 (R1997).

Library of Congress Cataloging-in-Publication Data
Great events : 1900-2001 / editors of Salem Press.— Rev. ed.
      v. cm.
Includes index.
    ISBN 1-58765-053-3 (set : alk. paper) — ISBN 1-58765-059-2 (vol. 6 : alk. paper)
    1. History, Modern—20th century—Chronology. 2. Twentieth century. 3. Science—History—20th century—Chronology. 4. Technology—History—20th century—Chronology.
D421 .G627 2002
909.82—dc21
                                                                          2002002008

First Printing

CONTENTS

vi

CONTENTS

vii

COMPLETE LIST OF CONTENTS

VOLUME 1

VOLUME 2

VOLUME 3

XV

xvi

VOLUME 4

VOLUME 5

xxi

VOLUME 6

xxiii

xxvii

VOLUME 8

GREAT EVENTS

1900-2001

El Salvador Has First Military-Free Election in Fifty Years

Under its new constitution, war-ravaged El Salvador elected Christian Democrat José Napoleón Duarte as president, with little interference from the Salvadoran military.

What: National politics

When: May 6, 1984

Where: El Salvador

Who:

ROBERTO D'AUBUISSON (1919-1992), leader of the right-wing ARENA party, and president of the Constituent Assembly

JOSÉ NAPOLEÓN DUARTE (1926-1990), president of El Salvador from 1980 to 1982, and 1984 to 1989

RONALD REAGAN (1911-   ), president of the United States from 1981 to 1989

The ARENA Years

In the 1982 election for a Constituent Assembly in El Salvador, Roberto D'Aubuisson and his right-wing ARENA party gained the support of smaller right-wing groups and were the clear winners. D'Aubuisson enjoyed the support of El Salvador's large landholders and businessmen and much of the military. As leader of ARENA, D'Aubuisson became president of the Assembly, and he had ambitions of serving as provisional president of El Salvador as well.

But the Salvadoran government depended on the United States for military help in its war against leftist guerrillas of the FMLN (Farabundo Martí Front for National Liberation), and many members of the U.S. Congress were quite unhappy with ARENA's victory. To them, D'Aubuisson was the brutal leader of a right-wing death squad. There was evidence that he was involved in the 1980 murder of Catholic Archbishop Oscar Romero, who had criticized Salvadoran conservatives. These members of Congress believed that El Salvador's civil war was caused by poverty and political oppression by the wealthy elite, not by FMLN Communists.

Though many military officers in El Salvador approved of D'Aubuisson's ideas, they realized that if he became provisional president, the U.S. Congress would cut off aid. Then it would be difficult for the Salvadoran Armed Forces to prevent an FMLN victory. The Reagan administration insisted that D'Aubuisson not become president, so the officers found it necessary to give in. As a result, the Assembly named Alvaro Magaña, a banker, to the presidency.

Still, the ARENA government's war against the leftists went poorly. When D'Aubuisson and his supporters won the 1982 elections, the guerrillas probably numbered about two thousand; within two years, their ranks had grown to ten thousand. The growth was partly a reaction against the nation's security forces and secret death squads, which used terrorist intimidation and assassination to punish those who criticized ARENA. Clergy, journalists, union leaders, students, and peasants fell victim of the brutality. Also, the FMLN received help from the Sandinistas in Nicaragua, whom U.S. president Ronald Reagan blamed for keeping the war in El Salvador aflame.

Meanwhile, ARENA refused to carry out land reform or to make any compromises in dealing with El Salvador's problems. D'Aubuisson, his supporters, and many military officers believed that the Christian Democrat Party (PDC) was a front for Communist guerrillas. If not held back by the United States, D'Aubuisson might have arrested or assassinated José Napoleón Duarte,

2137

AP/Wide World Photos

Thousands of voters line up to cast their ballots during the 1984 presidential election in San Salvador.

leader of the PDC and former mayor of El Salvador's capital city, San Salvador. Jubilant after his 1982 victory, D'Aubuisson had spoken of trying Duarte for treason.

A Historic Election

By the time the Constituent Assembly announced in early February, 1984, that presidential elections would be held in late March, the people were deeply frustrated with ARENA and with the government's inability to deal with the guerrillas and end the civil war. Fearful of another D'Aubuisson victory—which would drive even more Salvadorans into the FMLN—the Reagan administration gave its support to Duarte. The Central Intelligence Agency and the American Institute for Free Labor Development helped fund Duarte's campaign, as did foundations in West Germany and Venezuela.

When Salvadorans went to the polls on March 28, they chose among eight candidates. The results gave Duarte 43 percent and D'Aubuisson 30 percent. Because neither of the two had an abso-

lute majority, a second-round election was necessary. It was scheduled for May 6.

In the areas controlled by the guerrillas, the FMLN had not allowed people to vote, but these areas made up only about one-sixth of the municipalities in El Salvador. The FMLN did not attack polling places, but in a wave of bombings before and after March 28, they knocked out electricity for 80 percent of the nation. At the same time, right-wing death squads threatened to murder election commissioners, whom they accused of being "evil Salvadorans playing the game of international Communism." Yet the threat of violence remained only a threat.

In the election of May 6, Duarte won with 54 percent of the valid votes, while D'Aubuisson received 46 percent. Although the election was much freer than previous Salvadoran elections, the political terrorism of the Right and the Left kept it from being completely free. An alliance of leftist parties, the Democratic Revolutionary Front (FDR), had offered no candidates in the election. The FDR claimed that its candidates would not be allowed to campaign freely, and that they would probably be killed by security forces or death squads.

Also, because of the FMLN's threats of violence, the Salvadoran military policed polling places to protect them from terrorism, and the ballot boxes were transparent, so that election officials were able to see how many citizens had chosen to vote.

Nevertheless, probably 1.4 million of the 2.3 million eligible voters took part in the election. Afterward, D'Aubuisson and ARENA reluctantly turned over power to the Christian Democrats. D'Aubuisson claimed that the PDC had handed out false identification papers to some voters to give Duarte more votes, but his claim was not taken seriously.

Consequences

In the United States, President Reagan was delighted with Duarte's victory, for he believed that now Congress would be more willing to give economic and military aid to El Salvador. Still, Duarte faced huge difficulties as the new president. He needed to rally the armed forces against the in-

creasing strength of the guerrillas. At the same time, he needed to stop right-wing terrorism by controlling the security forces and the death squads.

Even with Reagan's support, Duarte had to work hard to convince Congress that El Salvador deserved to receive military and economic aid. Many Salvadoran business leaders, landowners, and army officers—those who had traditionally held power in the country—distrusted **Duarte**, especially after he announced that he would negotiate with the rebels.

Sadly, Duarte failed to find a solution to the civil war. His discussions with leftist leaders did not get very far; meanwhile, right-wing terrorists continued to leave their victims corpses along roads and in ditches. Still, Duarte struggled on, his health weakened by cancer. When he gave up the presidency in 1989, it was to a successor chosen through the ballot box rather than at the muzzle of a rifle. The electoral process had brought a peaceful transfer of power in El Salvador.

Kendall W. Brown

World Court Orders U.S. to Stop Blockading Nicaragua

A unanimous International Court of Justice condemned the U.S. mining of Nicaragua's harbors; the Court also declared that Nicaragua's independence should be fully respected.

What: International relations; International law

When: May 10, 1984

Where: The Hague, the Netherlands

Who:

DANIEL ORTEGA (1945-   ), leader of Nicaragua's junta from 1979 to 1984, and president of Nicaragua from 1984 to 1990

RONALD REAGAN (1911-   ), president of the United States from 1981 to 1989

TASLIM OLAWALE ELIAS (1914-1991), a Nigerian, presiding judge of the World Court

STEPHEN SCHWEBEL (1929-   ), an American judge on the World Court

U.S.-Nicaraguan Conflict

From 1912 to 1925, U.S. Marines occupied Nicaragua, and again from 1926 to 1933. The Marines left when peasant leader and revolutionary Augusto César Sandino agreed to hold talks with the U.S.-supported government of Nicaragua. A year later, Sandino was gunned down by the U.S.-trained National Guard, which was led by Anastasio Somoza. The Somoza family ruled Nicaragua from 1936 to 1979. The family owned much of the country's fertile land and most of the transportation and communication facilities. American aid was used for the rulers' favorite cause—themselves.

After Nicaragua's 1979 revolution, the Sandinista National Liberation Front gained power. The Sandinistas wanted to end poverty and inequality in Nicaragua through education programs and land reform. Nicaragua accepted aid from many sources, including the communist governments of Cuba and the Soviet Union.

The administration of U.S. president Ronald Reagan was determined to confront Sandinista leader Daniel Ortega and remove him from office. The administration accused Nicaragua of interfering in El Salvador, where a U.S.-supported government held power and was battling leftist guerrillas. The United States supported and trained rebel armies, the "Contras," made up of former members of Somoza's National Guard and others who were displeased with Sandinista rule. The Contras attacked some military targets, but they also attacked farming cooperatives, utilities, and medical clinics. The Central Intelligence Agency (CIA) offered advice to rebel leaders and also became involved directly, by placing mines in Nicaragua's harbors.

Nicaragua responded to its rebels and their U.S. supporters by boosting its armed forces. It tried to gain support from sympathetic Americans (including some members of the Congress) and from international organizations. After the harbors were mined, the Sandinistas decided to take their case to the World Court.

The Court Rules

The World Court, or International Court of Justice, is part of the United Nations. Its predecessor, the Permanent Court of International Justice, was formed by the League of Nations. In 1946 the United States agreed to allow the World Court to decide disputes between it and other countries—if the other countries agreed to abide by the Court's rulings.

The United States did not see the Court as a "supergovernment," but did find it useful for mediating international conflicts. The U.S. State Department took Iran to the Court in 1979;

when Iran complained, United States officials said that this left Iran "outside the community of civilized nations."

On April 9, 1984, Nicaragua brought a complaint to the Court. Nicaragua claimed that the United States had disrupted its economy, aided a rebel army, and committed acts of brutality. It asked the Court to take emergency action before entering into long arguments over whether the United States had broken international law.

On May 10, Presiding Judge Taslim Olawale Elias announced an "Interim Order of Protection": The United States should stop mining and blockading Nicaragua's ports, and Nicaragua's political independence should be fully respected. The Court had had to consider three questions: Should it hear Nicaragua's case? (In other words, did the Court have jurisdiction in this matter?) Should it order the United States to stop mining Nicaragua's harbors? Should it issue a general condemnation of threats to Nicaragua's independence?

The World Court is made up of fifteen judges, who are supposed to make up their own minds when deciding cases rather than obeying instructions from their governments. Those hearing Nicaragua's case reflected the world's diversity: an American, a Soviet, Africans, Asians, East and West Europeans, and Latin Americans.

The United States tried to argue that Nicaragua had never properly accepted the Court's role. Nicaragua's president had recognized the Court in 1929, and in 1935 its congress agreed, but the formal document stating this agreement (called a "Protocol of Signature") was never recorded at the League of Nations.

On April 6, only three days before Nicaragua brought its complaint, the United States had tried to escape jurisdiction by asking the Court not to consider "disputes with any Central American state or arising out of or related to events in Central America." Nicaragua's lawyers (who included an American, Harvard University professor Abram Chayes) argued that the United States could cancel its acceptance of the Court's authority, but only with six months' notice.

All fifteen judges, including the American judge, Stephen Schwebel, were convinced that the Court could and should hear Nicaragua's request for an emergency order. They would later listen to more debate over whether the Court had jurisdiction to issue a decision.

All fifteen judges agreed to order the United States to end its blockade and to stop placing explosive mines in the harbors of Nicaragua. These actions went against the United Nations Charter, the Charter of the Organization of American States, other treaties, and international custom. They affected not only Nicaragua but also the dozens of other countries that traded with Nicaragua.

By a vote of fourteen to one, the Court also decided that no nation should interfere with Nicaragua's independence. Judge Schwebel was the lone vote against this decision. He believed that the United States did have some reason to claim that it was defending the independence of El Salvador, Honduras, and Costa Rica; together these countries may have acted in "collective self-defense," which is encouraged by international law. Judges Sir Robert Jennings (from Great Britain) and Hermann Mosler (from Germany) wrote a separate opinion in which they agreed with the Court majority, but suggested that Nicaragua's actions should also be examined by the Court.

Consequences

Contra leaders announced that the Court's order would not affect their military activities, and they left open the possibility that the ports would be mined again. President Reagan also promised to continue opposing the Sandinistas. In early 1985, his administration decided not to appear at the rest of the World Court's proceedings. Because of pressure from Congress and other nations, however, the blockade and mining of harbors were stopped.

In 1986, the Court reached its final decision: U.S. sponsorship of the Contra war, the economic embargo, and the 1984 mining of Nicaragua's harbors violated international law. The United States was ordered to pay Nicaragua for the damage it had caused. In 1990, however, Nicaragua dropped its claim. The newly elected Nicaraguan president, Violeta de Chamorro, was eager to receive economic aid from the United States, and the claim against the American government was a block to that aid.

Arthur Blaser

Great Britain Promises to Leave Hong Kong in 1997

> *The British government announced that in 1997 Hong Kong would be turned over to the control of the People's Republic of China.*

What: International relations
When: September 26, 1984
Where: Hong Kong
Who:

DENG XIAOPING (1904-1997), leader of the People's Republic of China from 1978 to 1997

ZHAO ZIYANG (1919-   ), premier of the People's Republic of China from 1980

MARGARET THATCHER (1925-   ), prime minister of Great Britain from 1979 to 1990

SIR GEOFFREY HOWE (1926-   ), foreign and commonwealth secretary of Great Britain from 1983 to 1989

A Valuable Colony

In 1842, the Treaty of Nanjing (Nanking), which settled the First Opium War, gave the British control of Hong Kong. Yet the Chinese continued to resist, and after a second war (1858-1860), China was forced to give up the southern Kowloon Peninsula by the Treaty of Beijing (1860). The Hong Kong colony was completed in 1898, when China had to accept a ninety-nine-year lease of another 365 square miles called the "New Territories."

At first, British leaders did not expect the new colony, called Hong Kong, to be very important. Yet soon it became a center of trade with China—a very valuable possession.

Until the Chinese Communist Revolution of 1949, the British felt quite secure in Hong Kong. From the beginning, the new government of the People's Republic of China (PRC) insisted that the treaties creating Hong Kong were "unequal" and should not be honored. Yet they did not take action to take Hong Kong back. The colony—which was a free port, had low taxes, and was a center of capitalist business—began to bring large amounts of foreign currency into China. The PRC benefited by selling water, food, and cotton cloth to Hong Kong, and Chinese-owned banks and shops thrived.

When times were especially hard in China, such as during the Cultural Revolution of 1966-1967, large numbers of Chinese refugees fled to Hong Kong, where more laborers were always needed. Hong Kong's population grew from 2.4 million in 1951 to 5.5 million in 1984.

The Time Is Ripe

In 1972, Great Britain and the PRC began a more positive relationship, which brought more Chinese business into Hong Kong. Travel between the mainland and the colony was even encouraged. After Chinese leader Mao Zedong died in 1976, however, officials on both sides began to restrict the movement of mainland Chinese to Hong Kong. Cautiously, PRC leaders said that the Hong Kong situation should be discussed "when the time was ripe."

When the United Nations listed Hong Kong as a colonial territory that should eventually become independent, the PRC protested. They still considered Hong Kong a part of China that had been occupied by a foreign power.

By the early 1980's, the people of Hong Kong were more and more concerned about the future. The British lease was due to expire in 1997 (actually, only the New Territories were leased, but the rest of the colony was dependent on this region). What would happen to mortgages and other economic agreements if Hong Kong went back to Chinese control? Because of this uncertainty, the value of stocks, land, and the

2142

make an extra effort, and the final agreement was reached on September 26, 1984.

The Joint Declaration (in combination with several added documents) set up Hong Kong's situation after 1997. The city would become a "Special Administrative Region" of the PRC and would be allowed to govern most of its own affairs. Its government would be made up of Hong Kong residents, except for the chief executive, who would be appointed in the Chinese capital.

Hong Kong's existing economic system would not be disturbed for fifty years after the PRC takeover, and Hong Kong would continue to be involved in international trade organizations and agreements. The PRC's military forces would not intervene in Hong Kong's domestic affairs, although some troops would be stationed in the city. Rights and freedoms were assured, and China promised to draw up a basic law that would include all the provisions of the Joint Declaration.

Consequences

Hong Kong residents reacted to the Joint Declaration with relief—and suspicion that it was too good to be true. They feared that the PRC would back down on some of its promises, and many felt uncomfortable at the thought of the Chinese People's Liberation Army being present in Hong Kong.

However, the PRC had good reasons to stand by the agreement. Hong Kong's commerce could provide valuable foreign currency to the PRC. Also, if the PRC followed through on its promises about Hong Kong, it would stand a better chance of regaining Taiwan—a Chinese island that had been ruled by a separate, noncommunist government since 1949. As subsequent events were to show, the transition from British to Chinese rule in 1997 went smoothly, and in the year immediately following, China lived up to its promises.

Fred R. van Hartesveldt

Hong Kong dollar fell sharply.

In September, 1982, British prime minister Margaret Thatcher discussed the Hong Kong question with Chinese premier Zhao Ziyang. The Chinese position was firm: The government of Hong Kong must pass to China, but economic arrangements could be negotiated. Thatcher insisted that the British keep administrative control. This disagreement lasted a year, and Hong Kong's economy continued to suffer.

The next summer, Thatcher gave in. Great Britain would acknowledge Chinese sovereignty over Hong Kong as soon as an economic arrangement could be settled. Formal talks began in July, 1983, and lasted fourteen months. On August 15, 1983, the PRC increased the pressure by announcing that it would take control of Hong Kong on July 1, 1997, no matter how the negotiations turned out.

Progress in the negotiations was slow but steady. The Chinese, however, began to insist that an agreement be reached by October 1, 1984, the thirty-fifth anniversary of the founding of the PRC. The British never said openly that they would meet this deadline, but they did

China Adopts New Economic System

Chinese leaders announced major economic reforms that had the potential of changing the practice of socialism in the People's Republic of China.

What: Economics
When: October 20, 1984
Where: China
Who:

Mao Zedong (1893-1976), chairman of the Chinese Communist Party from 1949 to 1976

Deng Xiaoping (1904-1997), leader of the People's Republic of China from 1978 to 1997

Zhao Ziyang (1919-   ), premier of China from 1980 to 1987

Socialist Ideals in China

From 1949 until his death in 1976, Mao Zedong directed the course of China's socialist development. He had an unshakable faith in socialism, and he believed that the great masses of Chinese people could accomplish any task if they were motivated to work together for the good of the country.

Two of Mao's primary goals were to improve production on China's small rural farms and to organize farmers into more efficient work groups.

In the late 1950's, Mao directed the formation of communes in the countryside. These communes brought a number of villages and towns into one large community, managed in every detail by commune officials and Communist Party groups. So communes became the hub of Mao's rural economic policy; central planning by the government determined how much of what crops should be produced, and communes carried out the orders.

Communes had both strong and weak points. Because officials could organize all commune members, large-scale tasks were accomplished that aided China's economic development. Commune workers constructed dams and huge waterway projects, so that seasonal flooding was controlled and the irrigation of rice paddies and grain fields was improved.

Communes also aided Mao's goal of creating a society of equality, where all people could have decent housing, enough food, free health care, and at least a primary school education. All members of the commune shared the same standard of living—living together in dormitories, eating in common cafeterias, and working in collective fields.

In order to keep people equal, however, private ownership of property was abolished. All commune workers received approximately the same wage. Equal wages caused problems for the communes: Those who worked hard received almost the same wages as those who were lazy, so there was little motivation to work hard. As a result, productivity fell, the economy slowed, and the standard of living for the workers did not improve. By 1976, commune production could barely keep up with increases in China's population.

The Victory of Pragmatism

Two years after Mao's death, Deng Xiaoping rose to power as China's principal leader. Deng was a more practical leader than Mao; he believed that people who worked for themselves, rather than for the general good of the country, tended to work harder.

In 1978, Deng introduced the first of his economic reforms: allowing individual households in the countryside to make their own decisions about what crops to produce, how hard to work, and where to sell. As more households adopted this "responsibility system," communes began to break up.

In 1982, with the enthusiastic support of China's new premier, Zhao Ziyang, Deng introduced more reforms that officially replaced the commune system with the "responsibility system." Now farmers could lease land from the state and sell their surplus crops at newly opened "free" markets.

In cities, peddlers appeared on street corners selling everything from Tibetan shish kebab to blue jeans from Hong Kong. Family-owned restaurants opened, and small businesses flourished. Productivity increased, setting the stage for the reforms of 1984, which restructured China's socialist economy.

On October 20, 1984, Deng's government announced the latest step in the reform process: A new market system would replace China's socialist "command" economy. Instead of following a national plan designed by the government, larger businesses would be encouraged to make their own decisions on production, distribution, and pricing of goods (as is done in capitalist countries such as the United States).

As the reforms were adopted, competition developed among businesses. The result was higher efficiency and greater production—and a stronger and more adaptable national economy.

So the 1984 economic reforms were a new direction for the socialist economy in China. By adopting a market-style system, Deng liberalized Chinese socialism and made China a more active partner in the world trading community.

Consequences

Deng's bold economic reforms encouraged the Chinese people to seek reforms in other areas of China's socialist system. In 1986, and again in 1989, students and workers took to the streets in huge demonstrations for greater freedom and democracy.

In 1986, Deng refused to consider political reforms and forced the students to end the demonstrations. In 1989, however, he called on the military to end the demonstrations. On June 4, soldiers opened fire on demonstrators crowded into Beijing's Tiananmen Square, killing hundreds, perhaps thousands, of protesters.

The Tiananmen Square massacre of peaceful demonstrators marked China's return to a more conservative, repressive style of socialism. Deng announced that he was determined to preserve economic reforms, but he seemed unwilling to open China to Western-style democracy.

Daniel J. Meissner

2145

India's Prime Minister Gandhi Is Assassinated

The murder of Indian prime minister Indira Gandhi was an expression of resentment by Sikhs, a minority group in northern India.

What: Assassination
When: October 31, 1984
Where: New Delhi, India
Who:
INDIRA GANDHI (1917-1984), prime minister of India from 1966 to 1977 and from 1980 to 1984
JARNAIL SINGH BHINDRANWALE (1947-1984), a Sikh leader who wanted an independent Sikh nation to be established
HARCHAND SINGH LONGOWAL (1932-1985), leader of the Akali Dal, the largest Sikh political organization
BEANT SINGH (died 1984) and SATWANT SINGH (1941-1989), Gandhi's assassins

A Militaristic Faith

The Sikh religion comes from the teachings of ten "gurus" (teachers), the first of whom was Nānak (1469-1539) and the last Gobind Singh (1666-1708). The early Sikh leaders said that there was only one god and called their followers out of the Indian caste system into a community of equals.

The earliest gurus were pacifists, but the Sikh movement became involved in a tax revolt against the Mughals, a Muslim dynasty that ruled parts of India from the sixteenth through part of the eighteenth century. The Sikhs began to feel that they were being persecuted for their religious beliefs, and in response they began to arm themselves. Guru Gobind Singh taught that every "true" Sikh must undergo a special baptism and take the name Singh, which means "lion." Every male was supposed to carry a sword, wear war-rior's clothes, and avoid alcohol because it would keep him from being a good fighter.

In the eighteenth century, the Sikhs of the Punjab region were able to establish their own independent states, which were brought together in the empire of Ranjit Singh (1792-1839). Soon, however, the growing British empire came into conflict with the Sikhs. The Sikhs lost two wars against the British, but the British admired their courage and skill in battle and later gave them an honored place in the Indian army. As a result, though the Sikhs were only 2 percent of India's population, they made up about 25 percent of the army.

When the British decided to give up India in 1947, they reluctantly decided to divide the Indian subcontinent between a Muslim nation, Pakistan, and a mostly Hindu India. Although the Sikhs had always insisted, "We are not Hindu, we are not Muslim," the Sikh community's leaders agreed to be counted as Hindus at the time of partition, believing that it would be better to be part of India. Many Sikhs who lived in the western Punjab then had to move eastward, since the west was to be part of Pakistan.

A few Sikhs argued that the Sikh community should have joined Pakistan, while others believed that they would be better off having their own nation. Even the moderates began calling for a "Punjabi Suba"—a province in which the main language would be Punjabi.

In 1966, Indira Gandhi's government seemed to give in to that demand. The hill regions of northern Punjab, where most of the people spoke Hindi, were separated into a new state, Himachal Pradesh; the Hindi-speaking areas of southern Punjab became the state of Haryana. This made Punjab much smaller than it had been, but now it had a Sikh majority of 52 percent.

In the 1960's and 1970's, farmers in the Punjab prospered, and many roads, irrigation canals, and small factories were built in the state. The income of the average Punjabi peasant was more than twice that of farmers in other parts of India. But the boom ended in the 1980's, and the peasants began feeling the pinch. In their frustration, they began to listen to political radicals such as Jarnail Singh Bhindranwale, a holy man who wanted the Sikhs to form an independent nation.

Leaders Are Felled

In January, 1984, Bhindranwale took over parts of the Golden Temple in Amritsar, in the Indian state of Punjab. The Golden Temple was the Sikhs' most sacred shrine, and more than one hundred of Bhindranwale's followers joined him there.

In the next few months a number of moderate Sikhs and Hindus were assassinated by Sikh extremists, and the people of India began insisting that the government needed to take strong action. In May, the general staff of the Indian army began to plan how to take the Golden Temple from the Sikh militants. The plan, which was given the code name "Operation Bluestar," called for a surprise attack.

The attack began on June 4, 1984, and from the start everything went wrong. That day the Golden Temple was crowded with innocent pilgrims who were unaware of any danger. The militants fought back hard, so that the army had to bring in tanks and heavy artillery.

The battle lasted into the next day, and more than one hundred soldiers and uncounted numbers of pilgrims and militants died. The Golden Temple was seriously damaged. Bhindranwale himself died during the attack.

A number of the other militants, however, managed to escape, and in the weeks following the battle they formed an army of rebels. These radical Sikhs assassinated many of their Sikh opponents, and sometimes they opened fire on Hindus and other non-Sikhs who happened to be passing through Punjab on trains and buses.

There were Sikhs among the members of Prime Minister Gandhi's personal bodyguard, and her advisers told her to find other jobs for them, but she did not want to do anything to stir up more trouble.

On the morning of October 31, 1984, Gandhi left her house and began to walk across the walled garden that separated her house from her office. On the way, two of her Sikh guards, Beant Singh and Satwant Singh, approached her and opened fire with their automatic weapons. At least sixteen bullets pierced her body.

Beant Singh was killed immediately by other members of the bodyguard, while Satwant Singh was jailed and later hanged for his part in the assassination.

Consequences

Soon after Gandhi's death, Sikhs in Delhi began to be attacked. Buses full of young thugs were brought into Sikh neighborhoods and allowed to vandalize, rape, and kill. The police made no attempt to stop them. Although the Sikhs fought back, they were outnumbered, and many died.

The new prime minister, Rajiv Gandhi, tried to make friends with the moderate Sikhs. He made some progress in negotiating with Harchand Singh Longowal, leader of the Akali Dal (a Sikh political party). But Longowal was assassinated in 1985, and no other important Sikh leader was willing to take his place in negotiating with the government. Violence continued in Punjab and the surrounding areas, and the Indian government seemed unable to find a solution.

Gregory C. Kozlowski

Ethiopian Famine Gains International Attention

After years of drought and civil war, Ethiopia's growing famine gained the attention of world media and led to a dramatic humanitarian response.

What: Economics; International relations
When: November 22, 1984
Where: Ethiopia
Who:
HAILE SELASSIE (1892-1975), emperor of
 Ethiopia from 1930 to 1974
HAILE MENGISTU MARIAM (1937-   ),
 leader of Ethiopia from 1977 to 1991
BOB GELDOF (1954-   ), an Irish rock
 star

War and Hunger

Famines do not begin quickly; they usually develop over a period of several months or years, and they happen in poor countries, not rich ones. Droughts happen in both rich countries and poor countries, but in rich countries, food can be easily shipped to areas that are in a drought, and other services can be provided as well. In poor countries, especially in those that are suffering a civil war, communication, transportation, and services to the drought-stricken areas may be cut. This leaves drought victims without help, and the result is famine.

Ethiopia is a poor country, and it has had many droughts and famines over the years. Between 1972 and 1974, tens of thousands of people in Tigray and Wollo provinces died in a famine. The Ethiopian government, under Emperor Haile Selassie, tried to cover up this famine, hoping that other governments would not notice. This was one of the reasons the emperor was overthrown by a new socialist government.

After bloody power struggles, the new government came under the control of Haile Mengistu Mariam. Unfortunately, Mengistu was quite ruthless. Soon Ethiopia sank deeper into civil war, as opposition to Mengistu grew. The people of Eritrea, who had already been fighting for independence from Ethiopia, fought even harder, while rebel groups formed in Tigray Province to oppose Mengistu's government and claim control over their region. Meanwhile, Ethiopia's neighbor Somalia tried to take advantage of Ethiopia's troubles and recapture disputed territory. Somalia invaded Ethiopia's Ogaden region in 1977.

To win its war with Somalia, Ethiopia relied on troops from other Communist countries, especially the Soviet Union and Cuba. The Mengistu government hoped that it could also force the rebels in Eritrea and Tigray to quit fighting, with no compromises. But the victory over Somalia did not lead to victory in the civil war, even with continued help from Cuba and the Soviet Union.

The government spent more and more money on weapons to fight the civil war. Yet Ethiopians were actually no more secure, and peasant farmers throughout the country received little help from their government. When drought struck northern Ethiopia again in the early 1980's, disaster was bound to happen.

The World Gets the News

By 1984, the situation was desperate. Ethiopia needed huge amounts of food aid and medicine to prevent millions of people from dying of starvation and disease—and Ethiopia's Communist allies could not provide enough aid.

Already, governments around the world were aware that Ethiopia was in trouble. Some food was being shipped overland from Ethiopia's neighbor Sudan into the most needy areas of northern Ethiopia, including Eritrea and Tigray. This food was taken over the border and through the mountains under the supervision of the International Committee for the Red Cross (ICRC).

Because rebel forces controlled much of the countryside in Eritrea and Tigray, the ICRC turned the food over to them to be distributed to the people. The Ethiopian government did not like this program, because it seemed to help the rebels, so sometimes the food convoys were bombed. But actually, even if all the food in the ICRC program had reached its destination, there were still far too many starving people.

Eventually, the drought and famine spread. People in rebel-held territory began to flee to government-controlled areas in search of food, but famine was beginning to affect the people in government-held areas as well.

Once the famine spread beyond the rebel-held areas, the Ethiopian government decided to ask for help from the United Nations, the governments of Western Europe, and the United States. One way to get that help was to let journalists see the horrible conditions of the famine victims.

On November 22, 1984, the world was shocked by film footage from famine-stricken areas of Ethiopia. By this time, hundreds of thousands of Ethiopians had fled from their drought-ravaged homes. Many parts of northern Ethiopia had had little or no rain for the preceding two or three years. Even where rain fell, harvests had often been interrupted by fighting between rebel forces and government troops. Civil war and drought had proved to be a deadly combination.

Tens of thousands of the fleeing victims of famine died—especially the weak, the sick, the very young, and the very old. Scenes of their dis-

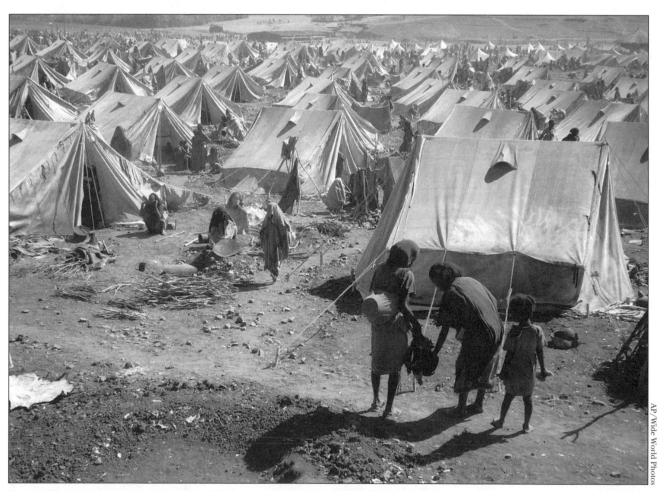

Despite the efforts of the Red Cross, in 1984, as many as 120 people a day died at this emergency feeding center on the edge of northern Ethiopia's Danakil Desert.

AP/Wide World Photos

2149

mal flight were captured on film, and these images prompted people in other countries to make donations so that food, medicine, and shelter could be provided to the desperately poor, starving victims of famine.

Consequences

The Ethiopian famine led to one of the most amazing humanitarian responses the world has ever seen. Singers in the United Kingdom held a "Band Aid" Christmas concert. Bob Geldof, an Irish rock star, sponsored a "Live Aid" Concert that was seen worldwide. Hollywood entertainers produced a song called "We Are the World" to earn money for famine relief in Ethiopia and other parts of Africa.

Governments and the United Nations provided millions of dollars worth of food aid and emergency assistance. Although the assistance was too late for many famine victims, millions more were saved.

Robert F. Gorman

Union Carbide Gas Leak Kills Hundreds in Bhopal, India

> *An accidental release of lethal gas in Bhopal killed or injured thousands of people in 1984.*

What: Environment; Medicine; Disasters
When: December 2-3, 1984
Where: Bhopal, India
Who:
WARREN ANDERSON (1921-   ),
    president and chief executive officer of
    Union Carbide
RAJIV GANDHI (1944-1991), prime minister
    of India from 1984 to 1991

Deadly Mistakes

In the 1960's, India's agriculture grew quickly during what was called the "Green Revolution": Scientists brought new breeds of rice and wheat that produced more grain and resisted drought and some kinds of plant diseases. Also, chemical fertilizers and pesticides were used to make farming more productive. India became able to produce enough food for its own people and even to export food to other countries.

The Indian government decided to begin a cooperative project with Union Carbide, a multinational corporation whose base is in the United States. Union Carbide was encouraged to build a pesticide-manufacturing plant in the city of Bhopal, which lies about three hundred miles south of New Delhi.

One of the Bhopal factory's products was methyl-isocyanide (MIC), used in pesticides that are applied to cotton and sugar cane. In its concentrated form this mixture is deadly, and it needs to be mixed and stored very carefully. Some European nations had completely banned MIC because it is so dangerous.

In its liquid state, MIC reacts violently to ordinary water: Its temperature rises quickly until it reaches boiling point—about 125 degrees Fahrenheit; then it becomes gas. On the night of December 2, 1984, water that was being used to clean a series of pipes entered one of the tanks holding MIC at the Bhopal Union Carbide plant. The temperature of the MIC rose quickly.

Because the temperature of another nearby tank was being increased deliberately, though, technicians in the control room at first thought this had caused the rising heat of the first tank. At about 11:30 P.M., several of the plant's workers began to complain that their eyes were irritated.

At midnight, some workers discovered the water running into the MIC tank and shut it off, but by then the chemical reaction could not be stopped. Several safety devices were supposed to start working if MIC vaporized: Safety valves would channel the MIC gas into a scrubbing tower, where it would be doused with caustic soda and turned back into liquid. A flame tower would then burn off any leftover gas.

Several of the valves burst, so that much of the vapor escaped. The scrubbing tower had been shut down, and the flame tower could not keep up with the flow of MIC vapor. About forty-five metric tons of MIC escaped.

All these things happened during the plant's night shift. The plant manager, the engineers, the chemists, and most of the skilled technicians were at home. At first, the men working at the plant that night did not understand what was happening. When they realized that there was some MIC gas in the air, most of them put on protective gas masks and suits; thus they did not realize how strong the gas was. Thinking that this was a minor problem that could be easily taken care of, they waited for several hours before calling their supervisors.

Bhopal Is Devastated

Most chemical factories are built away from people's homes, but the Indian government finds it almost impossible to keep settlements from growing up around such factories. Poor people are eager to take any piece of unoccupied ground near a city. To save their hard-earned rupees, they build makeshift homes from bits of wood and tin or throw old cloth over flimsy frames of sticks. Thousands of these "squatters" had moved in around the Union Carbide plant in Bhopal, and it was they who suffered the most when the great cloud of MIC gas was discharged.

If the gas had escaped during the daytime, the sun would have caused it to rise quickly into the atmosphere and to disperse. But the night of December 2-3 was cool and cloudy; smog still hung over the city, trapped by the low ceiling of clouds. The MIC stayed close to the ground, and night breezes blew it into thousands of shacks and tents where Bhopal's poorest residents were sleeping.

MIC gas affects the tissues of the eyes, nose, throat, and lungs. It causes choking, uncontrolled tears, shortness of breath, and then suffocation or heart failure. Thousands of sleeping people woke up with these symptoms. Many people began running—but, since none of them knew what the problem was, some ran toward the plant. Others tried to find buses or headed for the railroad station. Fathers and mothers lost track of their children; other families gave up, sat down, and died together.

When the first victims reached clinics and hospitals, the doctors and nurses on duty had no idea what had caused the symptoms. It took the doctors several weeks to figure out what could relieve the injured people's pain.

In the early hours of December 3, police and government officials began to suspect that the Union Carbide plant was the source of the disaster. They telephoned the plant manager, who was still at home; he insisted that his factory's safety system was foolproof. Even several months after the accident, Union Carbide executives were arguing that the incident had been caused by sabotage.

Consequences

Because so many people died so quickly, no accurate count of the dead was ever made. The Indian government estimated that twenty-five hundred had died, but some critics claimed that the dead had numbered about fifteen thousand. Similarly, the government announced that twenty thousand people (including many children) had been injured, but critics said the figure was closer to a hundred thousand.

Around the world, environmentalists pointed out how dangerous MIC-based pesticides are. Every year, they claimed, thousands of individuals die from misuse of these chemicals; Bhopal simply highlighted the problem.

Lawyers from the United States arrived in India within a few weeks of the tragedy; they offered to represent Bhopal victims in American courts. Their clients sued for a total of $100 billion. In May, 1986, however, a U.S. judge ruled that claims against Union Carbide must be filed in the courts of India.

On February 14, 1989, Union Carbide offered to settle out of court with the victims: All criminal charges against Union Carbide employees (including War-

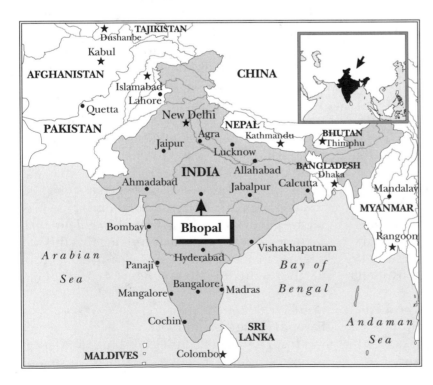

Indians in Bhopal walk past the bloated bodies of animals killed by a poisonous gas leak, which also caused the deaths of an estimated twenty-five hundred people.

ren Anderson, the company's head) would be dropped in exchange for payments totaling $470 million. In 1990, however, the Indian government asked the Indian Supreme Court to stop the settlement. Until negotiations could reopen, the government agreed to pay survivors and the injured two hundred rupees (about twenty dollars) per month.

In October, 1991, the Indian court upheld the basic settlement terms, while adding several new requirements, including the establishment of a health insurance fund for people who might later show symptoms of gas poisoning and the funding of a hospital in Bhopal. Union Carbide accepted the new terms. Afterward, the Union Carbide Company sold off its interest in its Indian operations and turned the proceeds over to the fund set up to build the new hospital, which was completed in 2000.

Gregory C. Kozlowski

Buckminsterfullerene Is Discovered

> *Buckminsterfullerene, which is produced when carbon is vaporized in a helium atmosphere, may always have been present in candle flames.*

What: Chemistry
When: 1985
Where: Houston, Texas
Who:

ROBERT F. CURL, an American chemist
HARRY W. KROTO (1939-   ), an
    English chemist
RICHARD E. SMALLEY (1943-   ), an
    American physical chemist

The Buckyball Mystery 60

In 1985, while astronomers peered into their telescopes and noted unusual features in the light that came from distant stars, physical chemists Robert F. Curl, Harry W. Kroto, and Richard E. Smalley stumbled on a previously unknown form of carbon that had been created in a chamber filled with helium gas in which graphite had been vaporized with a laser. This experiment was intended to re-create the conditions that exist in the outer reaches of the universe. The soot that remained in the chamber was found to contain a molecule of sixty carbon atoms that were arranged in a perfectly spherical shape resembling that of a soccer ball. This molecule of carbon, carbon 60, also resembles a geodesic sphere, and for that reason it has been named buckminsterfullerene for R. Buckminster Fuller, the creator of the geodesic dome. Scientists affectionately call the buckminsterfullerene molecule the "buckyball."

Buckminsterfullerene may be one of the most abundant and oldest molecules in the universe; it can be traced to the first generation of stars, which were formed some ten billion years ago. It also promises to be the favorite plaything of chemists for years to come. Many scientists believe that fullerene chemistry (fullerene molecules consist of clusters of carbon atoms) will be-

come a field of chemistry with a wide array of applications, including superconducting materials, lubricants, and spherical "cage compounds" that can be used to carry medications or radio-isotopes to patients. Some chemists believe that the structure of the buckyball is so symmetrical and beautiful (chemically speaking) that it must have many applications that have not yet been discovered.

The Buckyball's Magic Peak

Mountain climbers are thrilled when they sight a snow-capped mountain rising out of the surrounding countryside. Chemists feel that same kind of excitement when they come across their own kind of "magic peak." Such a peak occurs occasionally for chemists as an upward-pointing blip on the readout of a "mass spectrometer," an instrument that sorts molecules by mass (a molecule's mass is the amount of material that the

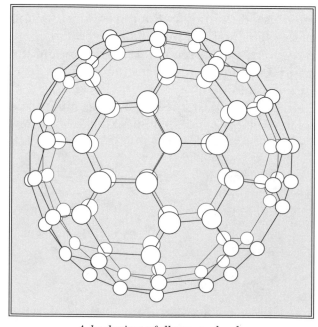

A buckminsterfullerene molecule.

2154

molecule contains). Those doing research find such peaks baffling and fascinating, because they indicate that an experiment has yielded a large number of molecules of one particular mass.

Precisely that kind of peak led to the discovery of buckminsterfullerene, a cagelike molecule made up of sixty carbon atoms elegantly arranged in the shape of a soccer ball. The discovery of this form of carbon fascinated the chemistry and materials science communities.

As is often the case with scientific discoveries, the buckyball was discovered by accident rather than design. As early as 1983, scientists were studying the ultraviolet light given off by graphite smoke, hoping to learn something about the composition of interstellar dust. It was at that time that they noticed some unusual absorption patterns (referred to earlier as "magic peaks"). These patterns were not understood until 1985, when it was found that they had been produced by buckminsterfullerene. Although scientists have not yet discovered buckminsterfullerene in space (they had expected to find it in abundance around carbon-rich stars), they now know exactly what to look for.

The procedure used to produce the first crystals of carbon 60 was quite inelegant: It involved studying the soot produced in a chamber. The hardware of the experiment was a cylindrical steel vacuum chamber three feet across. Within this cylinder was a steel block with a hollow tube that held a one-inch disk of a sample material—in this case, graphite. A laser was fired through a one-millimeter hole in the side of the block, blowing a plume of carbon atoms off the graphite. At the same time, a pressurized blast of helium (an inert gas) was sent through the end of the block and over the graphite. Because helium is inert, it did not interact with the carbon; instead, it captured and carried along the freed carbon atoms, allowing them to collide and cluster into groupings. Then, the high-pressure gas, now crowded with carbon clusters, rushed through

a hole in the wall of the block and into the vacuum of the spacious main chamber. There, the atomic collisions diminished, their temperature dropped to just a few degrees above 0 Kelvin, and the clusters that were formed were preserved and recorded by the mass spectrometer. In the output recording of the mass spectrometer, chemists Smalley and Kroto observed the magic peaks of carbon 60, which caused great excitement in the scientific community.

Consequences

It is now believed that fullerenes may be among the most common of molecules; they are thought to be formed along with soot and to be present in candle flames. To most people, soot is not very interesting, but combustion engineers find it fascinating. These highly specialized technologists spend much time and energy trying to control soot output—to maximize or minimize it. In a furnace, soot is needed to transfer heat outward from the flame to the water or air that is being heated. In a jet engine, however, soot must be eliminated, since it can damage or destroy delicate components.

Now that chemists know how to create carbon 60, they can modify it to create new chemicals—by encapsulating atoms or molecules inside its spherical "cage," for example. Scientists are also enjoying naming new derivatives of the buckyball molecule. Smalley has been able to "dope" the buckyball by placing boron atoms inside its "cage." He calls the resulting structure the "dopeyball."

Among the proposed uses of buckminsterfullerene are the creation of such novel products as specialized automotive lubricants; strong, light materials to be used for airplane wings; and rechargeable batteries. It remains to be seen how many derivatives of buckminsterfullerene will be created and what their uses will be.

Jane A. Slezak

Construction of Keck Telescope Begins on Hawaii

Using a new technology of mirror construction and computer alignment, the Keck Telescope is the most sophisticated ground-based optical telescope in the world.

What: Astronomy
When: 1985
Where: Mauna Kea, Hawaii
Who:

JAMES ROGER PRIOR ANGEL (1941-   ), a British-born engineer, astronomer, and major innovator in telescope mirror fabrication

Pushing the Limits

In 1985, a newly organized consortium, the California Association for Research in Astronomy (CARA), announced innovative plans for the design of a large new telescope facility to be located on the slopes of the Mauna Kea volcano in Hawaii. The site offered some of the best observing conditions in the world. A gift of $70 million from the W. M. Keck Foundation provided major funding for construction.

Since the mid-1970's, when Soviet engineers decided to build a reflecting telescope with a mirror diameter of almost 600 centimeters, the prevailing opinion had been that ground-based optical telescopes had reached their practical size limits. The Soviet mirror was not very successful. Conventional methods of mirror preparation, which use fused silicates for material and precision machinery to grind the parabolic surface of the mirror, create problems that compound themselves as mirror diameter increases. These problems became overwhelming for the ponderous Soviet telescope.

The construction of the previous record holder for size, the 508-centimeter Hale reflector at Mount Palomar Observatory in California, which had been completed in 1948, had demanded many technical innovations, including the use of patterns of wafflelike indentations that reduced the enormous weight of the single piece of Pyrex glass that functions as the mirror. It required several attempts to manufacture a usable blank mirror disk for Hale and months of grinding and polishing of the parabolic surface, resulting in tons of expensive Pyrex being wasted.

A Radical Design

The engineers of the Keck Telescope promised to find ways to overcome the physical and financial obstacles to building large mirrors. Instead of using a single mirror, the Keck instrument would align thirty-six hexagonal mirrors, each approximately 180 centimeters in diameter, into a single array with a light-gathering power roughly four times that of the Hale Telescope. The mirrors would be arranged in a mosaic pattern reminiscent of a honeycomb. The positioning of each mirror segment would be controlled by computers throughout observing sessions, so that they would focus their reflected light on a single, secondary mirror as a composite image. Computers would also continuously adjust thirty-six support points under each mirror segment and align the telescope support structure to minimize changes caused by gravity and temperature.

The major competition to this Multi-Mirror Telescope (MMT) design came from British-born engineer James Roger Prior Angel and his team at the Steward Observatory Mirror Laboratory at the University of Arizona. They were convinced that rigid mirrors much larger than 500 centimeters could be successfully manufactured. In the long run, they reasoned, a simple single mirror would be far more desirable in isolated observatories than a segmented mirror system with the extremely complex computer programs and other devices that it required.

Their process involves the use of a spinning furnace that melts the borosilicate mirror material (which is similar to Pyrex) over a honeycombed support structure. Spinning the furnace as it holds molten material allows the team to use centrifugal force to create the parabolic surface of the mirror. This method makes it possible to create a deeper parabolic surface than traditional grinding methods can achieve, thereby forming a more compact mirror. Barring unforeseen difficulties with the process, mirrors 800 centimeters in diameter would be ready for installation by the mid-1990's.

Other major projects, such as the European Southern Observatory's Very Large Telescope (VLT) at the Paranal Observatory in Atacama, Chile, incorporated new technologies. When the VLT began operating in 2001, it was the world's largest and most advanced optical telescope. It had four 8.2-meter reflecting unit telescopes, as well as several moving 1.8-meter auxiliary telescopes, whose light beams are combined in the VLT interferometer. With its unrivaled optical resolution and surface area, the VLT produces exceptionally sharp images and can pick up light from the most remote objects in the universe.

Consequences

In 1975, ground-based visual astronomy was widely perceived as a technological dead end and an increasingly marginal pursuit. However, in the last quarter of the twentieth century, it embarked upon a period of unprecedented technological progress and daring engineering.

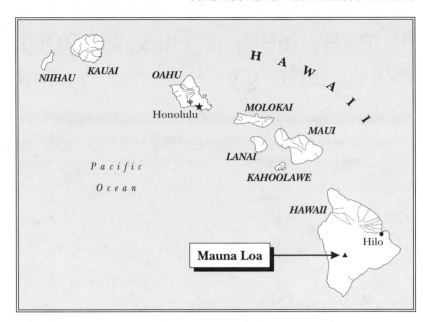

As the Keck Observatory became fully operational, as larger mirrors were produced by the Angel Team, and as work progressed on the European VLT, the public became aware of astronomical breakthroughs unlike any since early in the twentieth century. Some of the new technologies learned in ground-based telescopes were also applied to orbiting platforms, such as the Hubble Space Telescope.

The stakes could not be higher in terms of knowledge to be gained. The largest telescopes of mid-century design can detect objects some eight to ten billion light-years distant from Earth. Instruments such as the Keck Telescope have detected phenomena as far as fifteen billion light-years distant from Earth. They have thus brought astronomers to the very frontiers of the universe, in the sense that light observed from these immensely distant sources must have been emitted at around the time the universe began. The new telescopes are not only instruments of observation but also potential keys to the most basic questions about the nature and history of the universe.

Ronald W. Davis

Norway Begins First Reactor to Use Wave Energy

> *The wave-driven electrical power facility at Toftestallen, Norway, had the potential to provide cheap electrical power while creating minimal environmental pollution.*

What: Energy
When: 1985
Where: Toftestallen, Norway
Who:
KVÆRNER BRUG, a Norwegian
   hydroelectric company

Energy Without Pollution

Humanity produces electrical energy by converting other forms of energy, such as mechanical energy, into electricity. Engineers strove for many years to develop methods of producing electricity that would neither pollute the environment nor require the use of nonrenewable resources such as coal. For a time, hydroelectricity seemed to meet those goals.

Hydroelectricity involves the conversion of the mechanical energy of moving water into electrical energy and is, especially when compared to systems that use fossil fuels, a very clean technology. Conventional hydroelectric power systems, however, are subject to criticism for disrupting the environment in other ways. A large hydroelectric plant located on a river may require a high dam and an extensive reservoir that significantly alter the habitat for fish and other wildlife and plant species. Dams can prevent fish from traveling upstream to spawn, and the hydroelectric turbines can change the oxygen content and temperature of the water. In addition, conventional hydroelectric plants are vulnerable to drought and seasonal fluctuations in the water supply.

Engineers recognized these and other weaknesses almost simultaneously with the early development of hydroelectricity in the 1880's, and they began thinking about harnessing ocean wave energy. Engineering journals published the first plans for a proposed wave-energy plant before the end of the nineteenth century.

Ocean Waves Captured

It took almost one hundred years for these theoretical proposals to be translated into a working prototype wave-energy power plant. In 1985, a Norwegian company that specializes in hydroelectric development, Kværner Brug, began construction of the world's first electrical power generating station that would convert wave energy into electrical energy.

The proposed facility differed from many previously proposed wave-energy plants in that it would not utilize the wave energy directly. Previously proposed plants used water wheels or turbines mounted on breakwaters, or floats on the ocean surface that would move with the waves. The Kværner Brug facility, however, employed an air-driven turbine that transmitted mechanical energy to a generator.

This plant was similar to an air-driven plant proposed by the British a few years earlier. Both plants would use wave energy to create a column of air that would drive a wind turbine that would, in turn, transmit energy to an electrical generator. The British plan differed from the Norwegian plan in one crucial area. Whereas the British planned to erect their power station in a breakwater, the Norwegian engineers designed their facility to be built on the coast. This would reduce both initial construction costs and maintenance and repair costs, and it may be one reason that the Norwegians completed the Toftestallen plant, while the British plant was never built.

Seen from the ocean, the facility resembled a lighthouse hugging the cliffs. A metal cylinder 64 feet (19.5 meters) tall rose from a concrete caisson (chamber) on the ocean floor. The caisson was designed to channel breaking waves into an 11.5-foot-diameter (3.5-meter-diameter) steel pipe. As the waves rushed in, they served as a water piston and forced a column of air up through the pipe to the blades of a 6.5-foot-diameter (2-meter-diameter) wind turbine. As the water flowed back out into the ocean, it created suction that drew the air back down through the turbine blades, creating mechanical energy from the downdraft. The turbine blades had a symmetrical cross section that permitted them to rotate in a single direction despite reversals in air flow. This design eliminated the need for pivoting blades or air ducts and flap valves. The turbine, in turn, drove an electrical generator mounted above it.

Kværner Brug officials predicted that the prototype plant would generate 500 kilowatts of energy for every 82 feet (25 meters) of wavefront. Engineers designed the opening in the caisson and tower to match the frequency of the ocean waves, thereby increasing the waves' oscillation and doubling the energy intake. The steel tube was meant to be an energy multiplier. A Kværner Brug engineer compared the effect of the tube to that of pushing a child's swing at just the right moment. The simplicity of the design and the projected low maintenance costs contributed to company predictions of energy production costs of approximately four cents per kilowatt-hour.

Consequences

The wave-energy generating station at Toftestallen had the potential to serve as a model for low-cost, environmentally sound electrical power generating stations. Unfortunately, however, waves created by two exceptionally violent storms in the North Sea destroyed the plant in December of 1988. Engineers with Kværner Brug believe that waves from the first storm loosened the bolts anchoring the metal tower to the cliff. The second storm three days later washed the plant into the sea.

Although the Toftestallen plant was destroyed, it provided important information for advocates of renewable energy. The plant operated for several years prior to being lost and demonstrated effectively that electrical generation from ocean wave energy is possible. It also served as a prototype for electrical generating systems that Kværner Brug designed for underdeveloped nations and remote sites. Since the construction of the Toftestallen plant, other ocean wave-energy electrical generating stations have been built that employ the same principle of sucking a pocket or column of air into and out of a turbine. Each such turbine reduces reliance on nonrenewable fossil fuels and helps to eliminate air pollution. As the first operating wave-energy plant, Toftestallen showed that renewable electrical energy can be cost effective and can compete with other methods of power generation.

Nancy Farm Mannikko

Scientists Discover Hole in Ozone Layer

Scientists discovered a seasonal loss of ozone in the atmosphere above Antarctica.

What: Earth science; Environment
When: 1985
Where: Halley Bay, Antarctica
Who:

JOSEPH C. FARMAN, a researcher at the British Antarctic Survey at Halley Bay

SUSAN SOLOMON (1956-　　), an atmospheric scientist

MARIO JOSÉ MOLINA (1943-　　), a chemist who shared the 1995 Nobel Prize in Chemistry

F. SHERWOOD ROWLAND (1927-　　), a chemist who shared the 1995 Nobel Prize in Chemistry

A Threat to the Earth

Ozone is a chemical substance consisting of three oxygen atoms per molecule. In the early 1920's, scientists discovered that a relatively high concentration of ozone can be found at altitudes between 15 and 35 kilometers above sea level, a region of the atmosphere called the lower stratosphere. Because ozone molecules absorb ultraviolet (UV) light, this ozone layer prevents most high-energy UV light given off by the sun from reaching the earth's surface. Since high levels of UV light cause skin cancers and other illnesses in humans, decreased yields in UV sensitive food crops, and damage to microorganisms and plants, the presence of ozone in the stratosphere is of critical importance in maintaining the health and vitality of life on the earth.

In 1974, chemists F. Sherwood Rowland and Mario J. Molina proposed that changes in the composition of the earth's atmosphere as a result of human activity could lead to decreases in the levels of stratospheric ozone. Their theory concerned a class of compounds called chlorofluorocarbons (CFCs) that were commonly used in refrigerators and air conditioners and as aerosol propellants. Rowland and Molina suggested that CFCs released in the lower atmosphere would over time migrate into the stratosphere, where they would chemically react and remove ozone, leading to increased levels of UV light at the surface of the earth.

Rowland and Molina's theory began a debate among scientists about possible damage to the ozone layer from synthetic chemicals. By 1978, the use of CFCs as aerosol propellants had been banned in the United States, Sweden, and Canada; however, conclusive evidence that CFCs would damage the ozone layer proved difficult to obtain. Many of the chemical reactions involved in the proposed mechanism for ozone destruction were poorly understood. In addition, natural processes cause short-term fluctuations in the concentration of stratospheric ozone, making it difficult to demonstrate ozone depletion by direct observation. By the early 1980's, efforts to limit the use of CFCs had stalled as a result of the uncertainty concerning their effects on the atmosphere.

Finding the Hole

In 1957, as part of an international effort to study the earth and its atmosphere, the British established a research base in Halley Bay, Antarctica. At that time, a team under the supervision of Joseph Farman began a series of measurements on ozone in the atmosphere above Antarctica. By the early 1980's, twenty-five years of continuous observations on ozone had been recorded.

Beginning in the late 1970's, Farman and his coworkers began to see a seasonal decrease in the concentration of ozone above Antarctica. The decrease appeared each year during early spring and lasted one or two months. As years passed, the magnitude of ozone depletion grew larger, so that by 1981, ozone losses of 20 percent were found.

At first, the British researchers worried that the apparent decrease in ozone might have resulted from a malfunction in the instrument used to make the experimental observations; however, a new instrument brought in to check the observations confirmed the original findings. By 1984, even larger decreases in ozone had been observed. In addition, a second British base, located a thousand miles away at Argentine Island, had also begun to observe decreased ozone levels in early spring.

Convinced that the observed ozone depletion was real, Farman decided to publish his results. On May 16, 1985, a summary of the observations on seasonal ozone loss above Antarctica appeared in the British research journal *Nature*.

Consequences

The report of springtime ozone loss in the atmosphere above Antarctica came as a shock to the scientific community. Farman's observations were soon confirmed by examination of measurements made by the Nimbus 7 satellite orbiting Earth.

While the reality of the "ozone hole" was quickly established, it was at first not clear whether the observed ozone loss was a natural phenomena or was caused by CFCs. Different theories to explain the appearance of the ozone hole were put forth by various scientists, but the lack of detailed information on the chemical and physical properties of the atmosphere above Antarctica made it difficult to decide which theory, if any, was correct.

In August, 1986, Susan Solomon, a research scientist who had collaborated with Rowland, led an expedition of scientists to the American research base at McMurdo Station, Antarctica, to conduct a series of experimental observations on the atmosphere. The following year, a second expedition was sent to Antarctica for further observations. The results of these studies provided conclusive evidence that the formation of the ozone hole was linked to the release of CFCs in the atmosphere.

The publicity surrounding the discovery of the ozone hole revitalized international efforts to limit the production and use of CFCs. On September 16, 1987, representatives from twenty-four nations, including the United States, Japan, and the nations of the European Economic Community, signed the Montreal Protocol, an agreement restricting the production of CFCs. A modified agreement approved by ninety-three countries in 1990 provided for a total ban on the manufacture of CFCs and related compounds by industrial nations by the year 2000, with a ten year extension in the manufacturing ban granted to less-developed countries. In the United States, even more stringent limitations on the use and manufacture of CFCs were enacted as part of the 1990 Clean Air Act Amendments.

While springtime ozone depletion above Antarctica continued, by 1995, there was evidence that the restrictions on the manufacture and release of CFCs were slowing the rate of ozone destruction over Antarctica and in the stratosphere in general. In recognition of the importance of their theory, Rowland and Molina shared the 1995 Nobel Prize in Chemistry.

Jeffrey A. Joens

Iran-Contra Scandal Taints Reagan Administration

> *In a complicated series of events, American weapons were sold to Iran and funds were secretly provided to the Contra fighters of Nicaragua; an investigation raised questions of the balance of power between the executive and legislative branches of government in the United States.*

What: National politics; International relations

When: 1985-1987

Where: Washington, D.C.

Who:

RONALD REAGAN (1911-), president of the United States from 1981 to 1989

JOHN M. POINDEXTER (1936-), national security adviser to the president from 1985 to 1986

OLIVER L. NORTH (1943-), a lieutenant colonel who served as an aide to the National Security Council

Arms for Hostages

The Iran-Contra scandal involved the United States' dealings with two countries: Iran and Nicaragua. The United States considered the Iranian government unreliable and dangerous. During the regime of Ayatollah Ruhollah Khomeini, who came to power in Iran in 1979, sixty-six employees of the American embassy were seized in Tehran, Iran's capital. Fourteen were later released, but the rest were held hostage for fourteen months. The release came just after Ronald Reagan was inaugurated as president in 1981. Under "Operation Staunch," begun in 1983, the United States had convinced other governments not to sell weapons to Iran.

In a nationally televised speech on November 13, 1986, however, President Reagan said that U.S. arms had been sold in Iran in an attempt to create better relations with certain Iranian moderates. The president denied that the weapons had been sold in exchange for the release of American hostages who were being held by pro-Iranian terrorists in Lebanon. In fact, Reagan had often stated his firm policy against bargaining with terrorists in any way. By March, 1987, however, the president admitted that "what began as a strategic opening to Iran deteriorated in its implementation into trading arms for hostages."

The second nation involved in the scandal was Nicaragua—though not Nicaragua's government, but a group of antigovernment rebels, called the Contras. On November 25, 1986, the Reagan administration announced that the Justice Department had found evidence that some of the money earned by the arms sale had been passed along to the Contras. As a result of their involvement in this scheme, National Security Adviser John Poindexter resigned, and National Security Council aide Oliver North was fired.

Since 1981, Reagan had insisted that the Sandinistas who governed Nicaragua were Marxist extremists who would form close ties with Cuba and the Soviet Union. Some members of Congress disagreed, arguing that the Sandinistas were more committed to nationalism than to Marxist ideas. Trying to end the Reagan administration's support for the Contras, in October, 1984, Congress had passed the Boland Amendment, which prohibited aid to the Contras by any executive-branch agency that was involved in intelligence activities. Congress wanted the United States to provide only humanitarian aid to the Contras—not military aid. So the Iran-Contra affair raised an important question: Had Reagan's staff broken the law as stated in the Boland Amendment by supplying money for the Contras' military operations?

AP/Wide World Photos

Lieutenant Colonel Oliver L. North is sworn in prior to giving testimony on the Iran-Contra scandal.

Investigations and Findings

Events of the Iran-Contra scandal were uncovered in a series of investigations that involved both the executive and legislative branches. After the Justice Department discovered a memo from North stating that funds from the Iran arms sales had been passed along to the Contras, President Reagan appointed a commission, headed by Senator John Tower, to investigate the matter further.

The Tower Commission report, released February 26, 1987, criticized President Reagan for his management style. According to the report, Reagan had given considerable power to his staff members, including Poindexter and North—who carried out the weapons sale and support to the Contras without reporting back to the presi-

dent. A special prosecutor, retired judge Lawrence Walsh, was appointed to investigate whether there had been criminal wrongdoing.

In congressional hearings between May 5 and August 3, 1987, the Select Committees of the House of Representatives and the Senate investigating the affair held dramatic joint hearings. Thirty-two witnesses appeared before the committees, including North. In forceful, effective testimony, North defended both the sale of weapons to Iran and his efforts to help the Contras. Admitting that he had made earlier statements to Congress that were misleading, North explained that because Congress had been unreliable in providing support for the Contras, the executive branch needed to take secret action to keep the Contras afloat. North claimed that the

Boland Amendment did not apply to the National Security Council, only to intelligence agencies.

In his testimony, Poindexter insisted that he had not informed President Reagan that some funds from the weapons sales had been used to help the Contras. Poindexter believed that this information was a detail that the president did not need to know, since Reagan had made it clear that he did wish to support the Contras. In the end, the committees found no evidence that Reagan had known about the diversion of funds to the Contras.

Consequences

The majority report issued by the select committees stated that the secret Iran-Contra initiatives had been marked by "secrecy, deception, and disdain for the rule of law." The committees said that the aid to the Nicaraguan Contras did violate the Boland Amendment, and that it had disregarded Congress's authority to make decisions about how government money should be spent.

According to the report, several officials of the executive branch had misled and lied to Congress. It suggested that if President Reagan had not been aware of the use of funds for the Contras and other aspects of the affair, he should have been.

The Iran-Contra scandal highlights the need for cooperation between the executive branch and Congress in foreign policy. Although the executive branch often takes the lead in foreign affairs, the U.S. Constitution also assigns certain powers in this area to Congress. The Iran-Contra affair also shows that the president must stay in control over his staff; otherwise they may make independent decisions that can actually work against the best interests of the president and the nation.

Mary A. Hendrickson

Jeffreys Produces Genetic "Fingerprints"

Alec Jeffreys produced "fingerprints" of human DNA that are completely specific to an individual and can be used to establish family relationships and to identify criminals.

What: Biology; Genetics
When: March 6, 1985
Where: Leicester, England
Who:

ALEC JEFFREYS (1950-   ) and
VICTORIA WILSON (1950-   ), English
  geneticists
SWEE LAY THEIN (1951-   ), a
  biochemical geneticist

Microscopic Fingerprints

In 1985, Alec Jeffreys, a geneticist at the University of Leicester in England, developed a method of deoxyribonucleic acid (DNA) analysis that provides a visual representation of the human genetic structure. Jeffreys' discovery had an immediate, revolutionary impact on problems of human identification, especially the identification of criminals. Whereas earlier techniques, such as conventional blood typing, provide evidence that is merely exclusionary (indicating only whether a suspect could or could not be the perpetrator of a crime), DNA fingerprinting provides positive identification.

For example, under favorable conditions, the technique can establish with virtual certainty whether a given individual is a murderer or rapist. The applications are not limited to forensic science; DNA fingerprinting can also establish definitive proof of parenthood (paternity or maternity), and it is invaluable in providing markers for mapping disease-causing genes on chromosomes. In addition, the technique is utilized by animal geneticists to establish paternity and to detect genetic relatedness between social groups.

DNA fingerprinting (also referred to as "genetic fingerprinting") is a sophisticated technique that must be executed carefully to produce valid results. The technical difficulties arise partly from the complex nature of DNA. DNA, the genetic material responsible for heredity in all higher forms of life, is an enormously long, double-stranded molecule composed of four different units called "bases." The bases on one strand of DNA pair with complementary bases on the other strand. A human being contains twenty-three pairs of chromosomes; one member of each chromosome pair is inherited from the mother, the other from the father. The order, or sequence, of bases forms the genetic message, which is called the "genome." Scientists did not know the sequence of bases in any sizable stretch of DNA prior to the 1970's because they lacked the molecular tools to split DNA into fragments that could be analyzed. This situation changed with the advent of biotechnology in the mid-1970's.

The door to DNA analysis was opened with the discovery of bacterial enzymes called "DNA restriction enzymes." A restriction enzyme binds to DNA whenever it finds a specific short sequence of base pairs (analogous to a code word), and it splits the DNA at a defined site within that sequence. A single enzyme finds millions of cutting sites in human DNA, and the resulting fragments range in size from tens of base pairs to hundreds or thousands. The fragments are exposed to a radioactive DNA probe, which can bind to specific complementary DNA sequences in the fragments. X-ray film detects the radioactive pattern. The developed film, called an "autoradiograph," shows a pattern of DNA fragments, which is similar to a bar code and can be compared with patterns from known subjects.

The Presence of Minisatellites

The uniqueness of a DNA fingerprint depends on the fact that, with the exception of identical twins, no two human beings have identical DNA sequences. Of the three billion base pairs in human DNA, many will differ from one person to another.

In 1985, Jeffreys and his coworkers, Victoria Wilson at the University of Leicester and Swee Lay Thein at the John Radcliffe Hospital in Oxford, discovered a way to produce a DNA fingerprint. Jeffreys had found previously that human DNA contains many repeated minisequences called "minisatellites." Minisatellites consist of sequences of base pairs repeated in tandem, and the number of repeated units varies widely from one individual to another. Every person, with the exception of identical twins, has a different number of tandem repeats and, hence, different lengths of minisatellite DNA. By using two la-

beled DNA probes to detect two different minisatellite sequences, Jeffreys obtained a unique fragment band pattern that was completely specific for an individual.

The power of the technique derives from the law of chance, which indicates that the probability (chance) that two or more unrelated events will occur simultaneously is calculated as the multiplication product of the two separate probabilities. As Jeffreys discovered, the likelihood of two unrelated people having completely identical DNA fingerprints is extremely small—less than one in ten trillion. Given the population of the world, it is clear that the technique can distinguish any one person from everyone else. Jeffreys called his band patterns "DNA fingerprints" because of their ability to individualize. As he stated in his landmark research paper, published in the English scientific journal *Nature* in 1985, probes to minisatellite regions of human DNA produce "DNA 'fingerprints' which are completely specific to an individual (or to his or her identical twin) and can be applied directly to problems of human identification, including parenthood testing."

Consequences

In addition to being used in human identification, DNA fingerprinting has found applications in medical genetics. In the search for a cause, a diagnostic test for, and ultimately the treatment of an inherited disease, it is necessary to locate the defective gene on a human chromosome. Gene location is accomplished by a technique called "linkage analysis," in which geneticists use marker sections of DNA as reference points to pinpoint the position of a defective gene on a chromosome. The minisatellite DNA probes developed by Jeffreys provide a potent and valuable set of markers that are of great value in locating disease-causing genes. Soon after its discovery, DNA fingerprinting was used to locate the

David Parker/Science Photo Library

Alec Jeffreys.

defective genes responsible for several diseases, including fetal hemoglobin abnormality and Huntington's disease.

Genetic fingerprinting also has had a major impact on genetic studies of higher animals. Because DNA sequences are conserved in evolution, humans and other vertebrates have many sequences in common. This commonality enabled Jeffreys to use his probes to human minisatellites to bind to the DNA of many different vertebrates, ranging from mammals to birds, reptiles, amphibians, and fish; this made it possible for him to produce DNA fingerprints of these vertebrates. In addition, the technique has been used to discern the mating behavior of birds, to determine paternity in zoo primates, and to detect inbreeding in imperiled wildlife. DNA fingerprinting can also be applied to animal breeding problems, such as the identification of stolen animals, the verification of semen samples for artificial insemination, and the determination of pedigree.

The technique is not foolproof, however, and results may be far from ideal. Especially in the area of forensic science, there was a rush to use the tremendous power of DNA fingerprinting to identify a purported murderer or rapist, and the need for scientific standards was often neglected. Some problems arose because forensic DNA fingerprinting in the United States is generally conducted in private, unregulated laboratories. In the absence of rigorous scientific controls, the DNA fingerprint bands of two completely unknown samples cannot be matched precisely, and the results may be unreliable.

Maureen S. May

Gorbachev Becomes General Secretary of Soviet Communist Party

> *With his election as general secretary of the Soviet Communist Party, Mikhail Gorbachev began programs of reform that would bring enormous changes in the Soviet Union.*

What: National politics
When: March 11, 1985
Where: Moscow
Who:

LEONID ILICH BREZHNEV (1906-1982), first secretary of the Soviet Communist Party from 1964 to 1982

YURI ANDROPOV (1914-1984), first secretary of the Soviet Communist Party from 1982 to 1984

KONSTANTIN CHERNENKO (1911-1985), first secretary of the Soviet Communist Party from 1984 to 1985

MIKHAIL GORBACHEV (1931-   ), first secretary of the Soviet Communist Party from 1985 to 1991

Rapid Rise

Mikhail Gorbachev was born on March 2, 1931, in the Northern Caucasus region of the Soviet Union. His parents were peasants who worked on a collective farm. Gorbachev studied law at Moscow University; after he was graduated in 1955, he studied for a second degree in agricultural economy at the Stavropol Agricultural Institute. Like many other young Soviets, Gorbachev became a member of Komsomol (Young Communist League); he became a top leader of Komsomol during his time in Moscow.

In 1955, Gorbachev returned to his home district, Stavropol, where he worked as an agricultural economist. He worked closely with Fyodor Kulakov, a high-ranking Communist Party member who had developed methods for increasing productivity on farms. When Kulakov was called to Moscow to become a member of the Politburo (the party's executive committee), he rewarded

Gorbachev by making him first secretary of the city of Stavropol.

By 1970, Gorbachev was party chief of the entire district, and by 1971, at the age of thirty-nine, he had become a member of the party's Central Committee. When Kulakov died suddenly in 1978, Gorbachev was called to Moscow to take his place. He was given responsibility for farm policy and became part of the inner circle of powerful party leaders.

When Gorbachev moved to Moscow, a power struggle was going on within the Politburo. Yuri Andropov, head of the security forces (Komitet Gosudarstvennoi Bezopaznosti, or KGB), was trying to gain influence at the expense of Leonid Brezhnev, who had been first secretary (or general secretary) of the party for many years. Andropov was a reformer who wanted to end corruption within the party, increase discipline, and make changes in the Soviet economy.

Gorbachev quickly joined forces with Andropov and a group of younger, progressive-minded party members who wanted to see reforms in their nation. With Andropov's support, Gorbachev became a "candidate member" of the Politburo in 1979 and a full member in 1980.

Gorbachev Is Elected

When Brezhnev died in 1982, Andropov became general secretary, and within eight months he had also become president of the Supreme Soviet. Yet Brezhnev's allies, the "old guard," still held some important positions in the Politburo and on the Central Committee; one of these men was Konstantin Chernenko.

With Andropov in power, Gorbachev gained considerable control over the Soviet economy and over party cadres (small working groups). Even so, when Andropov died on February 9,

1984, it was the older, more conservative Chernenko who was elected as the new general secretary. The top leaders of the Politburo were not yet ready for the changes Gorbachev was sure to bring.

In exchange for supporting Chernenko, Gorbachev gained even more power. He was made chairman of the Foreign Affairs Commission of the Supreme Soviet and took on other important responsibilities as well. Gorbachev's trip to Great Britain in December, 1984, marked his arrival on the world stage. Former British defense minister Dennis Healy described him as "a man of exceptional charm . . . frank and flexible with a composure full of inner strength."

Chernenko was elderly, and soon his health began to fail. When he was ill, Gorbachev took his place in leading Politburo meetings; beginning in December, 1984, Gorbachev actually was responsible for running the country. Chernenko died in March, 1985.

That same month, Andrei Gromyko, a senior Politburo member, nominated Gorbachev to be the new general secretary. No others were nominated, and Gorbachev was elected by the Central Committee Plenum as the new first secretary of the Communist Party and the seventh leader of the Soviet Union.

Between 1980 and 1985, under the leadership of three elderly, ailing men—Brezhnev, Andropov, and Chernenko—the Soviet Union had been like a large ship without a rudder. Now Gorbachev was ready to take over and begin changing things.

He realized that the government needed to stop keeping Soviet citizens under tight control, and also that the Soviet Union's centrally planned economy was not working and needed to be reformed. So he proposed a program of major changes which he called *glasnost* (openness) and *perestroika* (restructuring).

Consequences

Gorbachev brought many changes to the Soviet Union. He began reforming the economy, bringing free-market practices into the Communist system. Many political prisoners were released, and the Soviet people gained new freedom to express their political views. Gorbachev was also responsible for ending the long, expensive war the Soviets had been fighting in Afghanistan. He met several times with United States president Ronald Reagan, and the two men became friends and began planning to reduce their nations' stores of nuclear weapons.

On August 19, 1991, hard-line party members tried to seize power so that communism in the Soviet Union could be saved. Gorbachev, who was on vacation in the Crimea, was put under house arrest. But the coup was not supported by most party members or by the Soviet people, and it soon failed; the Soviet Communist Party itself began to crumble. On August 24, Gorbachev resigned as first secretary of the party; he ordered the government to confiscate the party's property and asked the Central Committee to disband.

Shlomo Lambroza

Japan's Seikan Tunnel Is Completed

> *The building of Japan's Seikan Tunnel, the world's longest tunnel, required the development of revolutionary new tunneling methods.*

What: Engineering; Transportation
When: March 16, 1985
Where: Japan
Who:
YASUHIRO NAKASONE (1918-   ), prime minister of Japan from 1982 to 1987

From Drawing Board to Reality

In 1985, the Seikan Tunnel, which connects the Japanese islands of Honshu and Hokkaido, was completed. This tremendous project took more than forty-five years to move from the planning stage to completion. The tunnel, the world's longest, had to include a 23-kilometer central section beneath the treacherous and highly earthquake-prone Tsugaru Strait. Its completion was viewed as essential, because it would open up living space and provide essential natural resources—on rich but sparsely settled Hokkaido—to citizens living on crowded Honshu.

The building of Seikan, a railroad tunnel for Japanese high-speed bullet trains, involved many enormous technical problems. Two of the biggest were that the faults beneath the strait had a great potential for earthquakes and that digging through the soft seabed for 23 kilometers was extremely difficult. In spite of these problems, however, a combination of careful long-term planning, excellent engineering work, sophisticated computer technology, and innovative tunneling made the Seikan Tunnel a reality. The techniques developed to construct the tunnel are exceptionally valuable and will be utilized at other tunnel locations throughout the world.

Building the Tunnel

In 1939, the first engineering study for the planned tunnel between Honshu and Hokkaido was carried out. The Seikan Tunnel was to be twice as long as any undersea passage then in operation, so it was expected to be much more difficult to construct than any other tunnel. The tunnel was required because the island of Honshu, Japan's chief island and home of Tokyo, was overcrowded and also required the use of more natural resources. The tunnel project was delayed for the first time when Japan entered World War II and set aside most civilian projects.

After the war, more studies were carried out, over a period of twelve years, and it was decided that the tunnel would require ten years and $168 million to complete. With this plan in mind, digging began on Hokkaido in 1963. The digging actually took twenty-two years and cost $2 billion. Constructed by Japan's national railway system engineers, the Seikan Tunnel is 54 kilometers long, including its entrances on Honshu and Hokkaido. These entrances slope downward, dropping 91 meters below the surface of the Tsugaru Strait. At its deepest undersea point, the tunnel is nearly 244 meters below sea level. The tunnel passes through nine earthquake fault zones that make it extremely earthquake-prone. In fact, several seismic episodes during its construction let ocean water into the tunnel, delaying the tunnel's completion and contributing to the deaths of thirty-four construction workers.

The Seikan Tunnel was constructed from both sides of the Tsugaru Strait, and sophisticated computer monitoring was used to make sure that the two halves of the tunnel met exactly. The tunnel actually consists of three interconnected tunnels: a pilot tunnel, a service tunnel, and a railroad tunnel. The pilot tunnel was designed to be completed first, and its main function was to allow engineers to identify and solve any problems that they might encounter in digging the other tunnels. It was completed in January, 1983, when Yasuhiro Nakasone, then prime minister of Japan, detonated a dynamite charge

2170

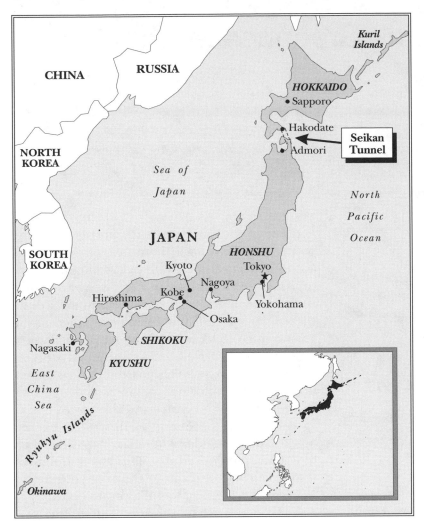

personnel to work for more than a few hours per day. This meant that automation and speedy drilling would be essential, and this led to the development of new high-speed, automated rock drilling technology.

Consequences

It is known that underwater tunnels have been built since 2000 B.C.E., and when the Seikan Tunnel was being planned, many tunnels existed in Europe, Asia, and North America. The Seikan Tunnel, however, had a tremendous impact on tunneling because it produced valuable technological advances that were desired by many nations. These advances came about because of the tremendous length of the tunnel—twice that of any other tunnel—and the problems of tunneling in the Tsugaru Strait, which involved the likelihood of earthquakes and the nature of the seabed itself.

Because the 23-kilometer undersea section of the tunnel was constructed from both sides of the strait, it had to be dug exactly straight so that the two halves of the tunnel would match exactly when they met. This was accomplished by means of a computer-monitored laser guidance system developed in Japan.

Another tunneling innovation, which had to be developed because of the softness of the seabed and the area's proneness to earthquakes, involved the redesign of tunnel-digging machinery and the injection of a chemical that solidified on contact with water into the areas being dug. This technology, based in part on earlier Western work, was utilized by other nations in the 1990's.

Finally, difficult working conditions in the tunnel led to the development of automated and rapid drilling machinery that minimized use of human workers. This new machinery, hailed as the greatest advance made in modern tunneling technology, has also been widely adopted.

Sanford S. Singer

that joined its two halves. Two years later, the service tunnel was completed, and on March 16, 1985, the main tunnel was completed and was ready for construction of the rail line, which was estimated as costing $1 billion.

One of the engineering problems in the construction of the Seikan Tunnel had to do with the softness of the seabed, which at first allowed the ocean to breach the tunnel during construction. Japanese engineers solved this problem by devising a method for injecting into the rock a chemical that solidified when it came into contact with water. The mixture was injected into the areas to be tunneled before the process of boring was begun.

Another related problem was the fact that a combination of high pressure, a working temperature of 26.6 degrees Celsius (80 degrees Fahrenheit), and a level of humidity higher than 90 percent made it impossible for construction

2171

French Agents Sink Greenpeace Ship
Rainbow Warrior

> *When the* Rainbow Warrior *was sunk, people around the world became more aware that nuclear tests were being conducted in the Pacific.*

What: Environment; International relations
When: July 10, 1985
Where: Auckland Harbor, New Zealand
Who:

CHARLES HERNU (1923-1990), French minister of defense from 1981 to 1986

FRANÇOIS MITTERRAND (1916-1996), president of France from 1981 to 1995

LAURENT FABIUS (1946-   ), prime minister of France from 1984 to 1986

Nuclear Controversy

Greenpeace, an organization founded in 1969 by Canadian environmentalists, takes non-violent action to call attention to environmental problems around the world. In 1977, the organization used money supplied by the Dutch branch of the World Wildlife Fund to buy a thirty-year-old boat. Renamed *Rainbow Warrior,* this trawler became a protest ship that sailed to Iceland, Spain, and Peru to protest whale hunting. The *Rainbow Warrior* also sailed into Siberian waters so that its crew could gather information about illegal whaling there. The crew also called attention to Canada's slaughter of harp seal pups, the dumping of hazardous chemicals by American companies, and the disposal of nuclear waste in the North Atlantic.

Sometimes the *Rainbow Warrior* crew's protests helped persuade governments to change their policies. For example, after the ship's campaign off the coast of Peru, the Peruvian government closed down its whaling operations. But the *Rainbow Warrior*'s crew members were quite aggressive, and at times they were taken into custody

and had to find a way to escape or persuade the government to release them.

Greenpeace International decided that its priority for 1985 would be the Pacific Ocean—especially the nuclear testing that was being done there. Nuclear tests had been performed in the Pacific since 1948, when three large nuclear explosions were carried out at Eniwetok Atoll. At first government officials insisted that the radiation from the explosions was not very dangerous. In 1958 the United States and the Soviet Union agreed to stop these tests; both nations backed out of the agreement in 1961, but a test-ban treaty was signed in 1963. This treaty outlawed testing nuclear weapons in the atmosphere, but underground testing was allowed to continue.

In 1968, the two superpowers signed the Nonproliferation Treaty, agreeing that they would not help other nations gain nuclear weapons. As part of this treaty, nations that did not have nuclear weapons promised not to try to develop them.

France refused to sign the Nonproliferation Treaty, because President Charles de Gaulle was sure that only a French nuclear force could guarantee that France would survive and would remain strong.

Rainbow Warrior Sinks

At first, France tested its nuclear weapons in the Sahara Desert, but after the Algerian Revolution it was forced to seek another location. In 1966 de Gaulle set up a French test facility in the Pacific, at Mururoa Atoll. Nuclear tests were conducted there in 1972 and 1973.

In 1985, Greenpeace made plans to disrupt France's activities in Mururoa, where a nuclear test was planned for October. The *Rainbow Warrior* sailed to Auckland, New Zealand, to prepare

2172

for the protest. The ship was anchored in Auckland Harbor on July 10 when two bombs ripped through its hull at midnight.

The first blast struck near the engine room and made a hole big enough to drive a truck through. Water poured into the *Rainbow Warrior* at the rate of six tons per second. The second blast came a minute later, near the propeller; it destroyed the ship's propulsion system. One member of the ship's fourteen-person crew, a photographer named Fernando Pereira, was killed.

Who had placed the bombs? Suspicion quickly fell on the French, who had had conflicts with Greenpeace and its protest ships in years past. For example, in 1966 a ship was crippled on its way to Mururoa when sugar was poured into its tanks. The next year, a mysterious illness struck a member of that same ship's crew. In 1972, another protest ship was sabotaged, and yet another was found to be damaged after French officials had boarded the ship to inspect it.

The French government quickly denied that it was responsible for bombing the *Rainbow Warrior,* but after three days of investigations, New Zealand officials arrested a French-speaking couple who held false Swiss passports. Soon the authorities were able to identify those who had actually done the bombing and the ship they had used. Both the ship and these suspects were able to flee, however, before they could be arrested.

AP/Wide World Photos

The Greenpeace vessel Rainbow Warrior *founders in Auckland Harbor, New Zealand, after it was sunk by French agents to prevent its setting sail to protest French nuclear testing in the South Pacific.*

2173

Because French newspapers published information suggesting that the French government was involved in the bombing, in August Prime Minister Laurent Fabius announced that Bernard Tricot, a former adviser to de Gaulle, would lead an investigation into the attack on the *Rainbow Warrior*. Tricot examined the evidence New Zealand officials had gathered and announced that it was disturbing but that it did not prove anything. He said that the suspects New Zealand had identified were French agents, but that they were innocent of the bombing.

Fabius then asked Defense Minister Charles Hernu to begin a more detailed study. To show his support for the nuclear program—and to try to show that his government was innocent—French president François Mitterrand went to Mururoa in September to witness a nuclear test.

Soon after Mitterrand returned, the newspaper *Le Monde* suggested that high officials in the French government had ordered the bombing, or at least had known about it ahead of time. Among these officials were Defense Minister Hernu, General Jean Saulnier, and Pierre Lacoste, head of the French intelligence service. In late September, Fabius finally admitted that French agents had sunk the ship and blamed Hernu and Lacoste, both of whom were forced to resign.

In July, 1986, United Nations secretary general Javier Pérez de Cuellar, who had been asked to help settle the dispute, announced his decision: France should officially apologize to New Zealand and pay seven million dollars in damages. New Zealand would turn the two convicted French agents over to France, where they would spend three years in jail.

Consequences

The bombing of the *Rainbow Warrior* actually accomplished Greenpeace's goal: calling attention to nuclear tests in the Pacific. Earlier in 1985, New Zealand had refused to allow a U.S. destroyer into its waters because the United States would not reveal whether it carried nuclear weapons. The people of New Zealand were already very opposed to nuclear weapons, and now people around the world began asking whether it was fair to conduct nuclear tests in other small island nations of the South Pacific.

Though New Zealand received a payment from France as part of the settlement, little was done to compensate Greenpeace or the family of Fernando Pereira. France continued its program of nuclear testing, and the United States provided secret help for its efforts to build a strong nuclear arsenal.

Glenn Hastedt

Tevatron Accelerator Begins Operation

The Tevatron particle accelerator at Fermilab generated collisions between beams of protons and antiprotons at the highest energies ever recorded.

What: Physics
When: October, 1985
Where: Batavia, Illinois
Who:
ROBERT RATHBUN WILSON (1914-2000), an American physicist and director of Fermilab from 1967 to 1978
JOHN PEOPLES (1933-   ), an American physicist and deputy director of Fermilab from 1987

Putting Supermagnets to Use

The Tevatron is a particle accelerator, a large electromagnetic device used by high-energy physicists to generate subatomic particles at sufficiently high energies to explore the basic structure of matter. The Tevatron is a circular, tube-like track 6.4 kilometers in circumference that employs a series of superconducting magnets to accelerate beams of protons, which carry a positive charge in the atom, and antiprotons, the proton's negatively charged equivalent, at energies up to 1 trillion electronvolts (equal to 1 teraelectronvolt, or 1 TeV; hence the name Tevatron). An electronvolt is the unit of energy that an electron gains through an electrical potential of 1 volt.

The Tevatron is located at the Fermi National Accelerator Laboratory, which is also known as Fermilab. The laboratory was one of several built in the United States during the 1960's.

The heart of the original Fermilab was the 6.4-kilometer main accelerator ring. This main ring was capable of accelerating protons to energies approaching 500 billion electronvolts, or 0.5 teraelectronvolt. The idea to build the Tevatron grew out of a concern for the millions of dollars spent annually on electricity to power the main ring, the need for higher energies to explore the

inner depths of the atom and the consequences of new theories of both matter and energy, and the growth of superconductor technology. Planning for a second accelerator ring, the Tevatron, to be installed beneath the main ring began in 1972.

Robert Rathbun Wilson, the director of Fermilab at that time, realized that the only way the laboratory could achieve the higher energies needed for future experiments without incurring intolerable electricity costs was to design a second accelerator ring that employed magnets made of superconducting material. Extremely powerful magnets are the heart of any particle accelerator; charged particles such as protons are given a "push" as they pass through an electromagnetic field. Each successive push along the path of the circular accelerator track gives the particle more and more energy. The enormous magnetic fields required to accelerate massive particles such as protons to energies approaching 1 trillion electronvolts would require electricity expenditures far beyond Fermilab's operating budget. Wilson estimated that using superconducting materials, however, which have virtually no resistance to electrical current, would make it possible for the Tevatron to achieve double the main ring's magnetic field strength, doubling energy output without significantly increasing energy costs.

Tevatron to the Rescue

The Tevatron was conceived in three phases. Most important, however, were Tevatron I and Tevatron II, where the highest energies were to be generated and where it was hoped new experimental findings would emerge. Tevatron II experiments were designed to be very similar to other proton beam experiments, except that in this case, the protons would be accelerated to an energy of 1 trillion electronvolts. More important still are the proton-antiproton colliding beam experiments of Tevatron I. In this phase, beams of protons and antiprotons rotating in opposite directions are caused to collide in the Tevatron, producing a combined, or center-of-mass, energy approaching 2 trillion electronvolts, nearly three times the energy achievable at the largest accelerator at Centre Européen de Recherche Nucléaire (the European Center for Nuclear Research, or CERN).

John Peoples was faced with the problem of generating a beam of antiprotons of sufficient intensity to collide efficiently with a beam of protons. Knowing that he had the use of a large proton accelerator—the old main ring—Peoples employed the two-ring mode in which 120 billion electronvolt protons from the main ring are aimed at a fixed tungsten target, generating antiprotons, which scatter from the target. These particles were extracted and accumulated in a smaller storage ring. These particles could be accelerated to relatively low energies. After sufficient numbers of antiprotons were collected, they were injected into the Tevatron, along with a beam of protons for the colliding beam experiments. On October 13, 1985, Fermilab scientists reported a proton-antiproton collision with a center-of-mass energy measured at 1.6 trillion electronvolts, the highest energy ever recorded.

Consequences

The Tevatron's success at generating high-energy proton-antiproton collisions affected future plans for accelerator development in the United States and offered the potential for important discoveries in high-energy physics at energy levels that no other accelerator could achieve.

Physics recognized four forces in nature: the electromagnetic force, the gravitational force, the strong nuclear force, and the weak nuclear force. A major goal of the physics community is to formulate a theory that will explain all these forces: the so-called grand unification theory. In 1967, one of the first of the so-called gauge theories was developed that unified the weak nuclear force and the electromagnetic force. One consequence of this theory was that the weak force was carried by massive particles known as "bosons." The search for three of these particles—the intermediate vector bosons W^+, W^-, and Z^0—led to the rush to conduct colliding beam experiments to the early 1970's. Because the Tevatron was in the planning phase at this time, these particles were discovered by a team of international scientists based in Europe. In 1989, Tevatron physicists reported the most accurate measure to date of the Z^0 mass.

When the Tevatron was built, it was the only particle accelerator in the world with sufficient power to conduct further searches for the elusive Higgs boson, a particle attributed to weak interactions by University of Edinburgh physicist Peter Higgs in order to account for the large masses of the intermediate vector bosons. In addition, the Tevatron had the ability to search for the so-called top quark. Quarks are believed to be the constituent particles of protons and neutrons. Evidence had been gathered of five of the six quarks believed to exist. Physicists then had yet to detect evidence of the most massive quark, the top quark. However, evidence for its existence was finally found by rival research teams working at Fermilab in 1994.

During the late 1990's, the Tevatron accelerator was revamped to work at even faster speeds, and plans were made for even greater advances in the decade that followed.

William J. McKinney

Palestinians Seize *Achille Lauro* Cruise Ship

Four Palestinian terrorists, wishing to force Israel to release some of their comrades from prison, hijacked an Italian ship and murdered one of their hostages.

What: International relations
When: October 7, 1985
Where: The eastern Mediterranean Sea
Who:
MOHAMMED ABU ABBAS, head of the Palestine Liberation Front
ABDULRAHIM KHALED, Abu Abbas's "right-hand man"
LEON KLINGHOFFER (1916-1985), an American passenger on the *Achille Lauro*

Murder on the High Seas

On October 7, 1985, the Italian cruise ship *Achille Lauro* left the Egyptian port of Alexandria with 350 crew members and 97 passengers. The ship was headed toward Port Said, Egypt, where it was to pick up 750 more passengers. Once the passengers were all on board, the vessel was scheduled to visit Ashdod, Israel.

The ship, which had begun its Mediterranean cruise at Genoa, Italy, had originally carried five members of the terrorist organization known as the Palestine Liberation Front (PLF), which is part of Yasir Arafat's wing of the Palestine Liberation Organization (PLO). At Alexandria, one of the five, Abdulrahim Khaled, had left the vessel. He is thought to have been the right-hand man of Mohammed Abu Abbas, who planned how the ship would be hijacked.

The four sea pirates' original plan was to seize control of the vessel when it reached Israel, and to hold it until Israeli authorities released Palestinian terrorists who were being held in prison. But an Italian steward aboard the *Achille Lauro* accidentally discovered the hijackers' weapons and alerted the captain. Realizing that they could no longer hope to coerce Israel, the four terrorists decided to seize the ship immediately and force it to head for Syria.

Unfortunately for the hijackers, however, they belonged to the wrong faction of the PLO: Arafat and Abu Abbas had been hostile to Syria since 1983. So the government of Hafez al-Assad in Syria refused to give sanctuary to the hijackers.

The four hijackers then seem to have panicked, taking a step that had not been part of the original plan: They seized an American passenger, Leon Klinghoffer of New York. Klinghoffer was sixty-nine years old and had suffered a stroke, so he was confined to a wheelchair. The terrorists shot him, and his body and wheelchair were dumped into the sea.

The International Response

After holding the ship at gunpoint for forty-eight hours, the four hijackers were exhausted. Having lost the advantage of surprise, they could no longer go to Israel. They decided to let the *Achille Lauro* head for Port Said, where they surrendered to the Egyptian authorities.

Meanwhile, Abu Abbas had flown from Tunisia to Cairo, where he negotiated a deal with the Egyptians: The four prisoners would be allowed to leave Egypt for Tunisia, and Arafat himself would put them on trial and decide on their punishment. Even though no one really trusted this promise, it would take Egypt out of an embarrassing dilemma. Egyptian president Hosni Mubarak and his government did not want to make other Arabs angry by acting against the interests of the PLO; yet Egypt also did not want to give open offense to the United States and Israel, the countries most concerned about punishing the terrorists for Klinghoffer's murder.

President Mubarak quickly announced that the four prisoners, in the custody of Abu Abbas, had already left for Tunisia. In fact, they were still

2177

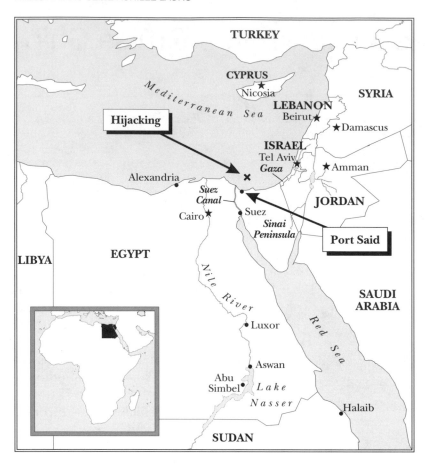

slavia, Abu Abbas denied that any of his group had murdered Klinghoffer and insisted that the American had died of a heart attack. This claim was repeated by Arafat and the PLO's observer at the United Nations.

A week after the murder, however, the body of the dead man washed ashore near Tartus, Syria, and was turned over to the U.S. embassy to be delivered to his widow, Marilyn Klinghoffer. The Syrians were quite pleased to embarrass Arafat, Abu Abbas, and their branch of the PLO by revealing that a bullet hole through the head plainly revealed that Leon Klinghoffer had been murdered.

Consequences

It was not until September, 1991, that Yasir Arafat finally broke ties with Mohammed Abu Abbas and had him dismissed from the Executive Committee of the PLO. It is possible that Arafat took this step only to please the United States, for the American government was involved in planning an Israeli-Arab peace conference that was scheduled to take place in Madrid in October and November. The PLO had angered many of its former friends by its support for Iraq during the Gulf War in early 1991.

Abdulrahim Khaled, who had been aboard the *Achille Lauro* until just before the hijacking, and who had designed the aborted plot, escaped justice until March, 1991. He was then arrested in Greece, on unrelated charges. The Greek courts found him guilty of illegal possession of explosives, drug dealing, and attempted escape from jail, and sentenced him to ten years in prison. The Italian government requested his extradition to Italy so that he could serve a life term to which he had already been sentenced, in absentia, for his role in the hijacking. The Greek minister of justice, Michael Papakostanlinou, ruled that he could be extradited to Italy only after he had served his ten-year term in Greece.

Arnold Blumberg

in Egypt. The matter was confused even more when Tunisia announced that it did not plan to grant asylum to the hijackers. Nevertheless, on October 10, an Egyptian plane left for Tunisia, carrying the four hijackers and their chief, Abu Abbas.

Both U.S. and Israeli intelligence, however, had been able to track the whereabouts of Abu Abbas and the other four terrorists. Fighter planes from the American aircraft carrier *Saratoga* managed to intercept the Egyptian plane, forcing it to land at Sigonella, Italy. The Italian authorities took Abu Abbas and the four hijackers into custody.

The Italians quickly released Abu Abbas, because he was not actually a part of the group of hijackers. Because Italy had a close, friendly relationship with the Arafat wing of the PLO, it did not wish to face the embarrassment of putting Abu Abbas on trial for his role in planning the hijacking. The four hijackers, however, were tried and imprisoned in Italy.

Having made his way to safety through Yugo-

Great Britain Approves Ulster Pact

Great Britain and the Republic of Ireland signed an agreement that gave Ireland a "consultative" role in the government of Northern Ireland.

What: International relations; Terrorism
When: November 15, 1985
Where: Hillsborough Castle, Northern Ireland
Who:
MARGARET THATCHER (1925-   ), prime minister of Great Britain from 1979 to 1990
GARRET FITZGERALD (1926-   ), prime minister of the Republic of Ireland from 1981 to 1987

Eight Centuries of Strife

Anglo-Irish strife has been going on for eight centuries. Ever since England invaded Ireland in the twelfth century, there have been many Irish revolts followed by brutal repression.

The "Irish question" has been a problem for British governments since the nineteenth century. In the early part of the twentieth century, a rising nationalism in Ireland led to the Easter Rebellion of 1916. The rebellion failed, but Great Britain granted independence to the Republic of Ireland in 1922. Ireland was partitioned, and the six counties that constitute Northern Ireland—in which Protestants outnumbered Catholics by approximately two to one—chose to remain part of the United Kingdom.

New troubles flared up in 1969, when the Catholic minority in Northern Ireland began to protest insistently against discrimination in voting rights, jobs, and housing. There were about 600,000 Catholics in Northern Ireland and more than one million Protestants. When clashes between the two sides in the capital city, Belfast, became violent, Great Britain sent in troops to keep the peace.

During the 1970's, however, the two groups fought even more violently. The Irish Republican Army (IRA), supported by many Catholics, began a terrorist campaign with the goal of forcing the British to withdraw their troops and give up control of Northern Ireland. The IRA's bombing campaign spread to England. Protestant paramilitary groups were formed as well, and they also committed many acts of terrorism. In 1972, when the conflict was at its height, 476 people died in terrorist-related violence.

The Sunningdale Agreement in 1974 attempted to establish a way for Catholics and Protestants in Northern Ireland to share power. After a two-week general strike by Protestant Unionists, however, the agreement collapsed.

By the mid-1980's, the time seemed ripe for a new agreement. In 1985, there were sixteen thousand British troops in Northern Ireland. Since 1969, twenty-five hundred people had been killed and twenty-six thousand injured as a result of the conflict. Yet during 1985 only fifty-four people died in terrorist-related incidents.

Still, it did not seem possible that the conflict would end anytime soon. Old battles remained vivid in the minds of people on both sides. Many of the Protestants in Northern Ireland are descendants of English settlers who arrived in Ireland several centuries ago. These Protestants still celebrated the Battle of the Boyne in 1690, when the forces of the Protestant English king, William III, defeated the Catholic army of James II. Like the Protestants, the Catholics were proud of their own culture and traditions. The two communities kept mostly to themselves and viewed each other with great suspicion.

Working Together

Trying to bring an end to the long conflict, the leaders of Great Britain and the Republic of Ireland met at Hillsborough Castle in Northern Ireland. There they signed an agreement that gave the Republic of Ireland a consultative responsibility in the government of Northern Ireland.

The Hillsborough agreement, called the Ulster Pact, came after eighteen months of intense and sometimes secret negotiations. It set up an Intergovernmental Conference, chaired by representatives of both countries, to take care of concerns such as human rights, administering justice and security, and cooperation between the United Kingdom and the Republic of Ireland.

In 1972, the Northern Irish parliament had been dissolved, and the six Northern Irish counties had come under the direct rule of the British government. The Hillsborough agreement supported the idea of setting up a new local government for Northern Ireland—one that would be accepted by both Protestants and Catholics.

The Ulster Pact worked toward two main goals. The first was to act more effectively against the IRA and its terrorism. Giving the Republic of Ireland a share in the government of the northern counties was Great Britain's way of admitting that the nationalist feelings of many people in Northern Ireland—especially Catholics—needed to be recognized. It was the first time since Ireland gained independence in 1922 that the British had made this kind of acknowledgment. British leaders hoped that this action would give Catholics in Northern Ireland a sense that they were finally being included in the political process. Then they would be less likely to support Sinn Féin, the political wing of the IRA, in local elections.

The Ulster Pact's second goal was to reassure the Protestant majority in Northern Ireland that the British government would keep Northern Ireland as part of the United Kingdom. Under the pact, the Republic of Ireland acknowledged that Northern Ireland was still in the United Kingdom. The agreement also stated that any change in the status of Northern Ireland could come about only with a consent of the majority of its people. Both Great Britain and the Republic of Ireland promised to work toward a united Ireland if a majority in Northern Ireland ever expressed such a desire.

Consequences

The Protestant community reacted angrily to the Ulster Pact and threatened mass protests. Calling the agreement treachery, they claimed that it was the first step on the road to a united Ireland—a possibility that they dreaded and had often vowed to resist, with violence if necessary. Fifteen Protestant Unionist members of the British parliament resigned their seats in protest.

In the year following the agreement, terrorist attacks and civilian casualties increased in Northern Ireland, although the death toll, sixty-two, was not much higher than that of 1985. In spite of this backlash, the Ulster Pact survived. British prime minister Margaret Thatcher, renowned for her determination, refused to give in to Protestant pressure.

Also, there were some positive signs. There were not so many complaints from Catholics about police and military harassment. The moderate Social Democratic and Labor Party (SDLP) gained some support among Catholics, while the IRA's Sinn Féin was not quite as popular as it had been.

Bryan Aubrey

U.S.-Soviet Summit Begins in Geneva

In the first summit conference between leaders of the United States and the Soviet Union since 1979, Ronald Reagan and Mikhail Gorbachev met in Geneva to establish better communication between the two superpowers.

What: International relations
When: November 19, 1985
Where: Geneva, Switzerland
Who:
RONALD REAGAN (1911-   ), president of the United States from 1981 to 1989
GEORGE SHULTZ (1920-   ), U.S. secretary of state from 1982 to 1989
MIKHAIL GORBACHEV (1931-   ), first secretary of the Communist Party of the Soviet Union from 1985 to 1991
EDUARD SHEVARDNADZE (1928-   ), Soviet foreign minister from 1985 to 1990

Hostile Superpowers

Ronald Reagan was elected president of the United States in 1980 with promises to stand firm against the Soviet Union and to restore the United States to greatness. This message was welcomed by Americans who were weary of painful, embarrassing crises—the Vietnam War, the Watergate scandal, Iran's taking of hostages at the U.S. embassy in Tehran.

During the late 1970's, the Soviet Union and Cuba had built up influence in Africa, the Soviet Union had invaded Afghanistan, and the Soviets had given considerable help to the Sandinistas in Nicaragua. The Soviet Union had also been building up its stockpile of nuclear weapons, and Americans were concerned. The 1980's had begun on an ominous note: In response to the Soviet invasion of Afghanistan in December, 1979, President Jimmy Carter withdrew the Strategic Arms Limitation Talks II (SALT II) treaty from consideration by the Senate. The two superpow-

ers were definitely at odds with each other, and the atmosphere was not good for arms-control talks.

Under Reagan and Soviet leader Leonid Brezhnev, the United States and the Soviet Union held completely opposite positions on how best to control nuclear forces. Believing that the Soviet arms buildup in the 1970's had given the Soviets important advantages, the Reagan administration was reluctant to begin new negotiations until U.S. defenses were improved. So arms-control talks took a back seat to the building up of American nuclear weapons.

Many Americans, however, were concerned about the threat of nuclear war, and they wanted the government to return to negotiations with the Soviets. By the fall of 1981, the Reagan administration decided to reopen talks about reducing strategic and intermediate nuclear force (INF) weapons. INF talks began in November, 1981, and the Strategic Arms Reduction Talks (START) came the following year.

Soon it was clear that the Americans and the Soviets held very different positions, and that progress would not come quickly. The Soviet delegates walked out of the START and INF negotiations in late 1983, after the United States deployed ground-launched cruise missiles in Great Britain and Pershing II ballistic missiles in West Germany.

Peace Initiatives

With the Soviet Union boycotting all arms-control talks with the United States, the initiative to start up negotiations again came from American leaders. In January, 1984, Reagan stated that "1984 finds the United States in the strongest position in years to establish a constructive and realistic working relationship with the Soviet Union."

The United States and the Soviet Union returned to the negotiating table in January, 1985. When the Soviet Union gained a young, energetic leader—Mikhail Gorbachev—in March, 1985, the arms-control process was accelerated. Soon Reagan and Gorbachev began considering a summit—a meeting between the two heads of state of the superpowers. Such a summit had not taken place in six years.

Reagan and Gorbachev met in Geneva, Switzerland, on November 19, 1985. After years of mutual criticism and foot-dragging on arms control, the leaders of the superpowers were meeting face to face. Reporting to the U.S. Congress on the meetings, Reagan said, "We met as we had to meet. I had called for a fresh start—and we made that start. I can't claim that we had a meeting of the minds on such fundamentals as ideology or national purpose—but we understand each other better. And that's a key to peace."

The Geneva summit did not bring any agreements on arms control, but it gave momentum for future talks. The two sides agreed to work together on a temporary agreement that would cover medium-range missiles in Europe; they also promised to work to cut certain kinds of offensive strategic nuclear weapons by 50 percent.

Both Reagan and Gorbachev recognized that their governments had serious differences on key issues: the American Strategic Defense Initiative program (SDI, sometimes called "Star Wars"), Soviet involvement in Afghanistan and Nicaragua, and human rights. Yet they agreed to speed up negotiations in Geneva and to plan two more summit meetings, as well as continuing talks between other high-ranking American and Soviet representatives.

The summit ended with the signing of six accords. The United States and the Soviet Union promised to cooperate on cultural and scientific exchanges, to restore civil airline links between the two nations, to improve air safety in the northern Pacific region, to set up consulates in New York City and Kiev, to cooperate in research on magnetic fusion, and to work together to protect the environment.

Arms-control experts in the United States were disappointed that Reagan had not made specific progress on arms control during the summit. His commitment to SDI was a major obstacle for the

AP/Wide World Photos

U.S. president Ronald Reagan (left) and Soviet leader Mikhail Gorbachev shake hands before the beginning of the summit held in Switzerland.

Soviets, and some Americans blamed him for being unwilling to negotiate about a defense program that they considered impractical anyway. Still, nearly everyone agreed that the Geneva summit was still important because it reopened dialogue between the superpowers and brought the possibility of better relations between them.

Consequences

The months and years after the Geneva summit, real progress was made toward controlling and reducing nuclear stockpiles. The Soviets became willing to compromise on both INF and SDI. An INF treaty was signed in 1987 during a Reagan-Gorbachev summit in Washington, D.C., and the START treaty was completed in 1991.

The 1985 Geneva summit was a significant turning point in Soviet-American relations. It turned the superpowers away from confrontation and toward cooperation. It allowed the United States and the Soviet Union to work toward arms control, ending interference in the affairs of other countries, and lessening tensions between the two leading nations of the world.

Vidya Nadkarni

Argentine Leaders Are Convicted of Human Rights Violations

> *The trial of Argentine military leaders revealed their involvement in government-led terrorism.*

What: Human rights; Military
When: December 9, 1985
Where: Buenos Aires, Argentina
Who:

JUAN PERÓN (1895-1974), president of Argentina from 1946 to 1955 and from 1973 to 1974

JORGE RAFAEL VIDELA (1925-   ), commander of the Argentine army, and head of the first junta from 1976 to 1981

ROBERTO EDUARDO VIOLA (1924-1994), commander of the Argentine army, and head of the second junta in 1981

LEOPOLDO FORTUNATO GALTIERI (1926-   ), commander of the Argentine army, and head of the third junta from 1981 to 1982

RAÚL ALFONSÍN (1926-   ), president of Argentina from 1983 to 1989

The Generals Rule

From the time Juan Perón rose to power in Argentina in 1943, he billed himself as a champion of the working classes and the poor. This made him very popular among the lower classes, but other groups were not so pleased with his programs. The armed forces, with the support of the Catholic Church, overthrew Perón and sent him into exile in 1955. But the idea of "Peronism" did not die.

After the Cuban Revolution (1959), some Argentine radicals announced that Peronism meant revolution, and guerrilla groups began using kidnapping and terrorism to try to change Argentine society. These groups included the Montoneros, founded in 1969, and the People's Revolutionary Army (ERP), founded in 1970.

The Montoneros succeeded in kidnapping and executing General Pedro Eugenio Aramburu, who had led the coup against Perón in 1955.

In June, 1973, Perón returned to Argentina, and four months later he was elected president. Soon he began criticizing the radical leftists. With this encouragement, a secret right-wing group called the Argentine Anti-Communist Alliance (AAA) began its own campaign of terrorism against people suspected of being leftists.

In July, 1974, Perón died, and the presidency passed to his wife, Eva Perón. As political violence spread, the new president gave the armed forces new powers to fight terrorism. But the armed forces were not willing to stop at fighting terrorism; they decided to take over the government itself. On March 24, 1976, a junta of commanders of each of the armed services, led by army commander Lieutenant General Jorge Rafael Videla, seized control of the government.

The generals saw themselves as taking part in a worldwide battle against left-wing terrorism. Between 1976 and 1983, a series of four juntas ruled Argentina as dictators. Nearly 350 secret detention camps were established, and somewhere between nine thousand and thirty thousand Argentine civilians "disappeared" during these years. Military task forces raided homes and kidnapped, tortured, and murdered suspects. When relatives tried to find out what had happened to their loved ones, the government insisted that it knew nothing of these people.

In 1975, some political and religious leaders banded together to form Argentina's Permanent Assembly for Human Rights and began to investigate more than two thousand disappearances. Two years later, mothers and other relatives of the "disappeared" began regular protest marches in the Plaza de Mayo, a public square in Buenos Aires.

Downfall and Trial

In April, 1982, the Argentine armed forces invaded the Falkland Islands. In a short war with Great Britain, Argentine forces were defeated, and the junta was humiliated. Facing growing opposition, some military officers organized a new junta in June, 1982, and prepared the way for national elections so that Argentina could have a civilian president once again.

In October, 1983, Raúl Alfonsín, candidate of the Radical Party, won the presidential election. After taking office, Alfonsín announced that all members of the first three juntas must be brought to trial. He also formed the National Commission on Disappeared Persons (CONADEP) to look into the disappearances and report to the government.

In a report published in 1984 and titled *Nunca más* (Never Again), CONADEP listed 8,960 disappearances. The detention centers were identified, and the commission accused the juntas of using the government to violate the human rights of Argentineans.

On December 18, 1983, the members of the three juntas were brought to trial. State Prosecutor Julio Strassera stated that they had used power illegally against Argentine citizens in what was being called "the dirty war."

The first part of the trial was held before the Supreme Council of the Armed Forces. In April, 1985, the trial was transferred to the Buenos Aires Federal Court of Criminal Appeals. Strassera brought 711 charges against the nine defendants, who included Videla, Roberto Eduardo Viola, and Leopoldo Fortunato Galtieri. A twenty-thousand-page report containing three thousand statements was presented, along with the words of more than eight hundred witnesses who told of torture, rape, murder, and illegal detention.

Strassera emphasized that the trial was not political. The accused were not being tried for organizing a coup or overthrowing the constitution, but for criminal violations of the Argentine Penal Code. The defense argued that the armed forces had been fighting a real war against armed rebels, and that the generals had been obeying the orders of an earlier civilian government (the Peróns) to destroy the enemy. Perhaps some government employees had gotten carried away in this war, but that was not the generals' fault.

On December 8, 1985, the federal court announced its verdict. Five of the nine defendants were found guilty, and two, including Videla, were sentenced to life in prison. Viola received seventeen years, while Galtieri was not found guilty. All those found guilty were disqualified from ever holding public office, stripped of their military rank, discharged from the armed forces, and ordered to pay court costs.

Consequences

The trial of the generals was the first time in Latin American history that government leaders were brought before a court of law to be tried for human rights violations. The court's decision showed that even if the generals had not committed rapes and murders themselves, they were responsible for planning acts of terrorism. Chile, a neighboring country, returned to civilian rule in 1990, and its new president, Patricio Aylwin, followed Argentina's lead in setting up a Truth and Reconciliation Committee to investigate human rights abuses that occurred during the dictatorship of Augusto Pinochet (1973-1990).

The people of Argentina were horrified by the testimonies presented during the trial. A number of politicians were anxious to put the horrors behind them, and in 1986 and 1987 the Argentine congress passed laws to stop any new prosecutions and to keep military persons under the rank of lieutenant colonel from going on trial.

In 1989 and 1990, President Carlos Menem granted pardons to the generals who had been convicted. Polls showed that more than 80 percent of Argentineans opposed the pardons. The terrible years of the "dirty war" could not be put away that easily.

Tully Discovers Pisces-Cetus Supercluster Complex

R. Brent Tully mapped a complex of galaxy superclusters more than 1 billion light-years in diameter, possibly the largest structure in the observable universe.

What: Astronomy
When: 1986-1987
Where: Hawaii
Who:
R. BRENT TULLY (1943-   ), the optical astronomer who proposed the existence of galactic supercluster complexes
J. RICHARD FISHER (1943-   ), the radio astronomer who collaborated with Tully in mapping clusters and superclusters
GÉRARD HENRI DE VAUCOULEURS (1918-1995), a specialist in large-scale structure in the universe who compiled catalogs of galaxies

Ever-Increasing Perspective

R. Brent Tully announced in December, 1987, that he had found a plane of galaxies occupying about one-tenth of the visible universe, a structure larger than any previously thought to exist. This Pisces-Cetus supercluster complex posed a major problem for the current theories concerning the origin and development of the cosmos, which suggested that the distribution of galaxies in the universe should be fairly uniform. Tully's analyses showed otherwise; the universe, he said, is lumpy and uneven.

In the 1920's, the American astronomer Edwin Powell Hubble had surprised cosmologists by proving that the universe is much larger than thought at the time and also that it is expanding. Previously, it had been assumed that the universe was more or less static. In 1927, the Belgian astrophysicist Georges Lemaître formulated the big bang theory, which seemed to account for this expansion. Logically, astronomers agreed, the explosion that occurred would send matter into the universe evenly in all directions. Tully's work surprised cosmologists again.

Astronomers widely believed that galaxies gather in clusters of about 30 million light-years in diameter. (A light-year is the distance light travels in one year through a vacuum.) Clusters, in turn, often form superclusters a hundred million light-years in diameter, or line up in long chains. The Milky Way, for example, is part of a modest group of about twenty nearby galaxies known as the "local supercluster." Furthermore, in 1981, scientists announced the discovery of pockets of empty space as large as superclusters between groups of galaxies.

Tully was a leading advocate for the existence of the local supercluster, along with Gérard Henri de Vaucouleurs, despite general resistance to the idea. Identification of such structures depended upon mapping the vast region beyond the Milky Way. Tully's interest in the project began in 1972, as he and J. Richard Fisher were completing their doctoral work at the University of Maryland.

Mapping the Universe

Tully prepared the first prototypes of ten maps while working at the Observatoire de Marseille in France, and revised prototypes were made at the National Radio Astronomy Observatory's drafting department in Charlottesville, Virginia. Meanwhile, Tully and Fisher continued to map the skies, using earlier surveys of galaxy groups by George O. Abell and de Vaucouleurs and drawing data from radio observatories in both the Northern and the Southern Hemispheres. Tully moved to the Institute for Astronomy at the University of Hawaii in 1982 and enlisted the help of cartographer Jane Eckelman of

2186

R. Brent Tully.

Manoa Mapworks in Honolulu in compiling accurate visual representations of the data that were accumulating. This was difficult, Tully and Fisher later wrote, because they still were not sure what they were trying to map; additionally, they wanted to produce complicated three-dimensional graphic illustrations to correspond to sectors of the sky.

In 1987, they published their work in the *Nearby Galaxies Catalog* and *Nearby Galaxies Atlas*. While preparing the maps for publication, Tully searched for the edge of the local supercluster. It became increasingly apparent that the supercluster was far larger than had been suspected.

Using a supercomputer to plot the distribution of densely populated (or "rich") clusters in the Northern Hemisphere, their motion relative to one another, and their distances from one another, he found that the clusters appear to lie in a plane. He also discovered that this extension of the local supercluster corresponds to a plane of rich clusters in the southern galactic hemisphere

centered about 650 million light-years from the Milky Way in the Pisces-Cetus region of the sky. The chance that this correlation of flattened distribution was an accident of random motion, his analyses suggested, was statistically small. Tully concluded that the planes of clusters must be elements of the same structure.

In the *Nearby Galaxies Atlas*, Tully and Fisher call this structure the Pisces-Cetus supercluster complex. The entire complex is about 1 billion light-years in length and 150 million light-years in width, making it the largest single structure in the observable universe. Additionally, Tully and Fisher also list four supercluster complexes that are found in other sectors of the universe.

The evidence for these supercluster complexes is suggestive rather than conclusive. The center of the Pisces-Cetus supercluster, for example, lies in the southern galactic hemisphere, and knowledge of clusters in that area is still relatively sketchy. Furthermore, the center of the Milky Way blocks direct observation of a section of the structure. Therefore, Tully's map of the Pisces-Cetus plane is, as he put it, full of holes. Still, he insisted that the alignment looked too exact to be dismissed as chance.

Consequences

By the mid-1980's, the big bang theory had changed drastically, but a central feature of the theory still predicted that matter in the modern universe should be evenly distributed because matter in the early universe expanded at a rapid, uniform rate. Cosmologists were able to accept structures the size of clusters in the big bang theory, but when voids and superclusters were found, these scientists had increasing difficulty explaining why matter was clumping in large, intricate structures. The most widely accepted theory was that soon after the big bang, the universe underwent a sudden, temporary inflation, often called the "cosmic burp." This sent gravity rippling through space that created density fluctuations in the universe's matter and fostered the formation of structures. Another theory suggested that the light-producing and light-reflecting matter that is observed accounts for only about 10 percent of the total matter in the universe. The "missing matter" is cold and dark—invisible to contemporary instruments—and spread evenly through the cosmos: Visible matter is like the froth on top of dark beer, one astrophysicist remarked.

The discovery of increasingly large structures in the universe—Tully's supercluster complexes, among them—led some observers to speculate that science may be on the verge of a new basic understanding of the environment as revolutionary as the American physicist Albert Einstein's general theory of relativity (1916) or Hubble's proof that the universe is expanding (1929).

Roger Smith

Müller and Bednorz Discover Superconductivity at High Temperatures

K. A. Müller and J. G. Bednorz found a ceramic material that "superconducted" (conducted electricity without resistance) at a temperature much higher than those at which other materials could act as superconductors.

What: Physics
When: January 27, 1986
Where: Rüschlikon, Switzerland
Who:
KARL ALEXANDER MÜLLER (1927-   ), a
  Swiss physicist
JOHANNES GEORG BEDNORZ (1950-   ), a
  German physicist
HEIKE KAMERLINGH ONNES (1853-1926), a
  Dutch physicist

The Discovery of Superconductivity

At the end of the nineteenth and the beginning of the twentieth century, when modern physics was being created, scientists found that the world did not behave in extreme conditions in the same way that it behaved in ordinary conditions. For example, Albert Einstein discovered that objects moving at speeds close to the speed of light contracted in length. Similarly, Heike Kamerlingh Onnes found that matter behaved in unusual ways at extremely low temperatures. By the first decade of the twentieth century, all gases except helium had been liquefied, and in 1908, using an elaborate device that cooled helium gas by evaporating liquid hydrogen, Onnes succeeded in liquefying helium. He was then able to use this liquid helium to cool various materials down to temperatures near absolute zero. In 1911, he discovered, to his surprise, that when mercury was immersed in liquid helium and cooled to 4 Kelvins (four degrees above absolute zero), its electrical resistance disappeared. When its temperature rose above 4 Kelvins, mercury lost this property of "superconductivity."

During the decades after Onnes's discovery, scientists found that other metals and many metal alloys were superconductors when they were cooled to temperatures near absolute zero, but they found no superconductor with a "transition temperature" (the temperature at which superconductivity occurs) higher than 23 Kelvins. They were, however, able to deepen their understanding of superconductivity. For example, in 1933, Karl Wilhelm Meissner, a German physicist, discovered that superconductors expel magnetic fields when cooled below their transition temperatures. This property and the property of resistanceless current flow became the defining characteristics of superconductivity.

Despite these and other discoveries, it was not until 1957 that a satisfactory theory of superconductivity was published. In that year, John Bardeen, Leon N. Cooper, and J. Robert Schrieffer, working at the University of Illinois, used the idea of bound pairs of electrons (Cooper pairs) to explain superconductivity. They showed how the interaction of these electrons with the vibrations of ions in the crystalline lattice of the metal caused the electrons to attract rather than repel one another, and because the movements of neighboring Cooper pairs are coordinated, the electrons could travel unimpeded. Despite these theoretical and experimental advances, however, scientists had been able to achieve only a transition temperature of 23 Kelvins for a niobium-germanium alloy.

The Discovery of High-Temperature Superconductivity

Great discoveries are often made by taking risks, and Alex Müller and Georg Bednorz, working at the International Business Machines

International Business Machines Corporation, AIP Emilio Segrè Visual Archives

Karl Alexander Müller (left) and Johannes Georg Bednorz.

(IBM) Corporation's laboratory near Zurich, Switzerland, chose to investigate complex metal oxides for superconductivity rather than the usual metals and alloys. Although other scientists had shown that some metal oxides superconducted at very low temperatures, most metal oxides turned out to be insulators (nonconductors). From 1983 to 1985, Bednorz and Müller combined metal oxides to create new compounds to test for superconductivity. More than a hundred compounds turned out to be insulators before Bednorz and Müller heard about a ceramic compound of lanthanum, barium, copper, and oxygen that French chemists had made but had failed to test for superconductivity. Bednorz discovered that this compound was indeed a superconductor, and they were able to shift its superconductivity to temperatures as high as 35 Kelvins, twelve degrees above the previous record.

Bednorz and Müller had established that their ceramic material carried electrical current without resistance, but they realized that for a mate-

rial to be a genuine superconductor, it also had to exhibit the "Meissner effect"—that is, it had to prevent magnetic fields from entering its interior. In 1986, their tests showed that their material exhibited the Meissner effect. Their results were soon confirmed at several other laboratories, which led to a frantic search among physicists for materials with even higher transition temperatures. When this search led to amazing successes, it became clear that Bednorz and Müller had made a revolutionary breakthrough. In 1987, a little more than a year after their discovery, they received the Nobel Prize in Physics for their work.

Consequences

After Bednorz and Müller's discovery, other researchers made compounds that superconducted at even higher temperatures, the most famous of which was yttrium-barium-copper oxide, a material that superconducted at almost 100 Kelvins, a temperature higher than that of liquid nitrogen (77 Kelvins). The discovery of this compound's transition temperature led to a flurry of activity among physicists, since it meant that liquid nitrogen—instead of the inconvenient and much more expensive liquid helium—could now be used to study superconductivity. About a year later, researchers at IBM's Almaden Research Center in San Jose, California, announced that they had found a thallium-calcium-barium-copper oxide with a transition temperature of 125 Kelvins.

Because substances with no electrical resistance could have many profitable applications, businessmen, engineers, and government officials were fascinated by these high-temperature superconductors. Corporations and governments invested heavily in their development (about $450 million worldwide in 1990). Politicians and businessmen were convinced of the potential of superconductors in electronics (especially high-speed computers), transportation (especially levitating trains), and power generation and distribution. Several difficulties quickly arose, however,

that tempered the initial promise of superconductors. Complex metal oxides are not easily formed into wire, for example, which would be required for many applications. Researchers have tried various ways to solve this problem, but so far they have been unsuccessful.

By the early 1990's, the commercial development of superconductors still seemed far in the future, but worldwide research efforts continued unabated. Entrepreneurs saw great market potential for superconductors, and undoubtedly much commercial development would follow when scientists solved various fabrication problems and the problems caused by the destruction of the superconducting state when superconductors were placed in magnetic fields. Physicists warned that the transition from laboratory to marketplace would not be easy, since many questions remained to be answered. Nevertheless, it seemed almost certain that high-temperature superconductors would eventually transform the way many people live and work.

Robert J. Paradowski

Space Shuttle *Challenger* Explodes on Launch

> *About seventy-four seconds after liftoff, the space shuttle* Challenger *exploded because of a malfunction, killing all seven persons on board.*

What: Technology; Economics; Disasters
When: January 28, 1986
Where: Cape Canaveral, Florida
Who:
FRANCIS R. SCOBEE (1939-1986), the commander
MICHAEL J. SMITH (1945-1986), the pilot
ELLISON S. ONIZUKA (1946-1986), mission specialist one
JUDITH A. RESNIK (1949-1986), mission specialist two
RONALD E. McNAIR (1950-1986), mission specialist three
S. CHRISTA McAULIFFE (1948-1986), payload specialist one
GREGORY J. JARVIS (1944-1986), payload specialist two

Countdown to Launch

The National Aeronautics and Space Administration (NASA) scheduled fifteen flights for 1986: communication satellites, probes of the solar system, three secret military payloads, and a variety of earth-monitoring experiments. It was an ambitious, expensive program, but most people did not suspect that it would lead to lowered flight-safety standards.

The flight of the space shuttle *Challenger* was originally set for January 23, but unexpected rain caused a delay. There was some question about *Challenger*'s thermal protection system, the thousands of tiles that were laid over its skin, protecting it from the heat of reentry. At top speed, raindrops could shred the tiles.

As the launch date approached, *Challenger* was missing some equipment. Certain parts were taken from older shuttles and installed in *Chal-*

lenger. A few people warned that this mixing of parts was dangerous, especially since the process was sped up to meet the takeoff deadline.

Congress had already been criticizing NASA in 1984 and 1985 because the space agency had been unable to meet all of its announced deadlines. There had been so many delays and cancellations that Congress had voted a freeze on the agency's budget. The 1970's had been a time of great optimism and generous budgets for NASA, but in the mid-1980's Congress was demanding that the shuttle program start paying its own way.

To save money and time, NASA designers took shortcuts. One of the most serious shortcuts was that the "launch-and-abort" rocket escape system in the original plan was removed when the shuttle was built. This saved weight but took away an escape hatch for the crew. Another cost-saving step had to do with the design and use of rubber O-rings to seal joints and prevent hot gas leaks. Some engineers reported that tests showed the O-rings were unsatisfactory, but no redesign was ordered.

During the countdown before *Challenger*'s launch, there was a question about how sensitive the booster joints' O-rings were to cold temperatures. Some engineers were concerned that below-freezing temperatures would erode the O-rings. A recommendation was made that NASA delay the launch, but NASA officials refused, stating that it was not clear that there was any problem. Pressured by the desire to meet the long-set deadline for the flight, managers overrode the engineers.

A Fireball in the Sky

In the hours before the launch, ice spread across the launch tower and on the shuttle. Some eighteen-inch icicles were found the morning of the flight. Inspection showed that ice might be drawn into the boosters. But because of the countdown pressure, no one made much of this possibility. The go-ahead for launch was given.

On January 28, 1986, at 11:38 A.M., *Challenger* lifted off its pad at the Kennedy Space Center on Cape Canaveral, Florida. Seventy-four seconds into the flight, a malfunction caused the shuttle to explode. All seven crew members were killed.

NASA officials determined that a violent wind shear had shattered the glassy oxides that sealed the joint of the right-hand solid rocket booster. The crumbling of the glassy oxides allowed the booster's flame to escape its steel casing. The flame struck the external tank near an attachment connecting the aft end of the booster to the tank, which contained pressurized liquid hydrogen.

The flame heated and blasted through the tank's insulation, and the fuel poured out. When the external tank crumpled, tons of hydrogen fuel burst into flame. The attach-struts ripped loose and the booster swiveled, crashing into the liquid oxygen tank and breaking it open.

Challenger's engines shut down as its fuel and oxidizer were used up. It was subjected to gravity forces beyond its limits. In turn, the orbiter was ripped apart and the *Challenger* broke into its various parts. The two wings flipped in different directions, the payload bay was not much damaged, and the crew compartment tumbled free.

Mission 51-L was only minutes away from orbit when the accident occurred. Onlookers were stunned, not sure exactly what had happened. Space flights had become almost routine, so that the nation had forgotten how risky they actually were. Some pieces of *Challenger* were recovered by ships that same day, but it was not until March that the crew compartment was found.

Consequences

President Ronald Reagan and Congress immediately began investigations into the causes of the accident. Witnesses testifying before the House

The space shuttle Challenger *explodes shortly after liftoff.*

Science and Technology Committee described the communication failures that had affected the launch decision. It became clear that poor judgment and a lack of careful analysis were major causes of the disaster.

The twelve-member Presidential Commission handed in a detailed report on June 6, 1986. The report included not only findings about the causes of the tragic accident but also recommendations for future NASA space programs. Tracing the chain of events that had led up to the accident, the commission found a series of mistakes, oversights, and bad decisions.

Some of the recommendations included a new design or redesign of the joint and seal, setting up a safety advisory panel to report to the space transportation system program manager, and forming an independent solid rocket design oversight committee to consider safety standards and how to certify rocket technology.

In spite of the *Challenger* tragedy, the commission praised NASA for its past achievements and declared that the United States' space program was valuable despite all of its risks. In the years after the disaster, safety problems were corrected, and Congress and NASA gained a new willingness to accept practical limits to future missions.

S. Carol Berg

Haiti's Duvalier Era Ends

After years of tyrannical rule, Haitians celebrated the forced departure of Jean-Claude "Baby Doc" Duvalier, the son of, and successor to, "Papa Doc" Duvalier.

What: Political reform; Coups
When: February 7, 1986
Where: Port-au-Prince, Haiti
Who:
JEAN-CLAUDE "BABY DOC" DUVALIER
    (1952-   ), president of Haiti from
    1971 to 1986
HENRI NAMPHY (1932-   ), president of
    Haiti from 1986 to 1987, and in 1988

How to Rear a Dictator

François "Papa Doc" Duvalier became president of Haiti in 1957, through honest elections. He had begun his career as a doctor. While working in the country, Duvalier had realized that the power of voodoo could be used for political purposes in his country. When he became president in 1957, he made voodoo a part of his image.

Because of his use of voodoo as a political tool, many Haitian peasants believed Papa Doc to be a flesh-and-blood appearance of the voodoo god Baron Samedi—the Spirit of Death. To encourage this belief, Papa Doc did several things. He often dressed in black suits, and he changed the colors of the Haitian flag from red and blue to red and black (red and black are the colors of voodoo secret societies). He slept in a tomb once a year to commune with spirits, and he kept the head of an enemy on his desk.

Papa Doc kept strengthening his position by using a security force known as Tonton Macoute. A Haitian legend claims that Tonton Macoute is the bogeyman who comes to houses at Christmas to carry naughty children away in his knapsack, never to be seen again.

Papa Doc reigned for fourteen years. During this time, the Tonton Macoute took away thousands of Haitians. The security forces infested the cities and the countryside, slitting the throats of political opponents and collecting "taxes" of lands, chickens, or anything else they wanted. Under President Duvalier's personal orders, more than three hundred people were killed during his first year as president. There are even rumors that he did some of the dirty work himself, torturing political prisoners in the basement of the presidential palace.

In 1971, Papa Doc died, and his son Jean-Claude inherited the position of president-for-life of Haiti. Before his death, Papa Doc had rewritten the country's constitution to lower the minimum age for a president to twenty. When Papa Doc died, Jean-Claude was only nineteen. But that did not stop him; he simply forced the legislature to say that he was twenty-one. The new president became known as "Baby Doc."

The Voodoo President Departs

In response to pressure from some human rights groups, Baby Doc changed the name of the Tonton Macoute to "Volunteers for National Security." He tried to break his ties to voodoo, but he could not. Baby Doc continued in his father's footsteps—taking large amounts of foreign aid for himself and increasing the family fortune with land, sports cars, yachts, and racehorses. By January, 1986, it was said that he had a fortune of between $200 and $500 million.

Yet Baby Doc's grip on the nation had begun to falter in November, 1985, when his security forces opened fire on student demonstrators in the coastal town of Gonaives. Three people had been killed. Protests followed. Many young Haitians believed that they had little chance of improving their lives so long as Duvalier was president. The opposition movement was supported by the Roman Catholic Church. Since Pope John Paul's visit to Haiti in 1983, Church leaders had been speaking out against Duvalier because the

2195

dictator did not seem to care about the poverty and suffering of the Haitian people.

At a demonstration in Gonaives, where a crowd burned down a customs house on January 27, 1986, the protesters meant to kill an army captain and two militiamen who were believed to be responsible for shooting the three students in November. The next day, thousands of protesters broke into a relief warehouse owned by a Catholic charity. In the turmoil, five people were crushed to death. These incidents started the chain of events that led up Duvalier's departure.

American leaders were concerned about the problems in Haiti. Many Haitians had been setting out for the United States in the hope of escaping the violence and poverty of their country. The United States put pressure on Duvalier to improve human rights in Haiti.

Duvalier's actual departure was to take place at 2:00 A.M. on February 7, 1986, but the dictator threw a quick champagne party for his close friends before he left, and the celebration delayed his flight. At 3:30 A.M. Duvalier and his family, along with seventeen close friends, arrived at their plane, which then took off toward France. France was the only country that was willing to grant asylum to Haiti's dictator.

Three hours after the plane left Haiti, a videotaped message from Duvalier was broadcast on Haitian television, confirming that he had departed and was no longer the head of the government of Haiti. The people of Haiti were delighted. At the Leogane Traffic Circle, south of the capital, Port-au-Prince, hundreds of Haitians pushed down a structure of ironwork that celebrated the Duvaliers' "accomplishments." At the national cemetery in Port-au-Prince, a mob tore

A French police officer guards the hotel in the French Alps in which the former Haitian president Jean-Claude "Baby Doc" Duvalier has taken refuge.

apart Papa Doc's marble and granite mausoleum. There were many wild parties of celebration throughout Haiti and in the Haitian communities of Miami and New York.

Consequences

Shortly after Duvalier's escape, a junta of six men arrived on the National Palace steps. Their leader, Lieutenant General Henri Namphy, announced in a radio broadcast that the military forces had decided to form a "National Council of Government." Namphy insisted that these new leaders should not become politically ambitious, for he hinted that elections might be held soon.

In March, 1987, Haitian voters adopted a new constitution, which set up a presidency and a national assembly elected by the people. Many Haitians protested, however, because the government tried to give control of the elections to the army instead of a civilian council.

The presidential election was to be held on November 29, 1987, but it was canceled because of terrorist violence. New elections were held in January, 1988, but a few months later Namphy overthrew the new civilian government and declared himself president of a military regime. In September, 1988, officers of the Presidential Guard seized power from Namphy.

In 1990, a popular Roman Catholic priest named Jean-Bertrand Aristide was elected to the presidency but was overthrown in a military coup within months. Thousands of Haitians fled the new regime, seeking refuge in Cuba and the United States. An executive order, however, went out from U.S. president Bush instructing the Coast Guard to turn back Haitian refugees intercepted at sea. Post Duvalier Haiti remained adrift, a society divided by class, without a strong economy or a stable government.

Rodney D. Keller

Mir Space Station Is Launched

Building on the long-duration flights of the Salyut missions, the Soviet Union demonstrated the ability to maintain astronauts on the Mir space station permanently.

What: Space and aviation
When: February 20, 1986
Where: Tyuratam, Soviet Union, and Earth orbit
Who:
Leonid Kizim (1941-   ), a veteran cosmonaut who commanded the Soyuz T-15 mission to occupy the Mir space station
Vladimir Solovyev (1946-   ), a Soyuz T-15 cosmonaut who was instrumental in setting up initial habitation of Mir
Yuri Romanenko (1944-   ) and
Alexander Laveikin (1951-   ), Soyuz TM-2 cosmonauts
Vladimir Titov (1947-   ),
Musa Manarov, and
Anatoli Levchenko (died 1988), Soyuz TM-4 cosmonauts

Interior Decorating

While the American space shuttle was grounded in the wake of the 1986 *Challenger* accident (the spacecraft had exploded about seventy-four seconds after launch, killing all aboard), the Soviet Union moved to a new plateau in space station development. The Mir (meaning "peace" or "commune") space station was launched from the Baikonur Cosmodrome on February 20, 1986. Mir was cylindrical and had a docking port at each end, as well as four extra docking ports circling the forward port. The Mir was meant to serve as the core for a larger space station constructed by the attachment of specialized modules.

The heavy multiple-docking adapter at the front of Mir forced Soviet mission planners to launch the space station with almost no scientific or long-duration life-support equipment. These supplies would be sent up later with a combination of manned flights, unmanned Progress freighters, and added modules.

The crew compartment represented a vast improvement over the earlier designs for the Salyut (salute) space station. For privacy, there were two individual cabins, each with a chair, table/desk, sleeping bag, and window for a soothing view of Earth. Mir had a galley complete with a refrigerator and a pair of food warmers, as well as a folding table and removable chairs. For personal hygiene, there was a zero-gravity toilet, makeshift shower unit, and wash station. Stowed beneath the crew compartment floor was a bicycle ergometer and treadmill for exercise and relaxation. The interior of the space station was organized in such a way as to preserve a natural sense of up and down and was painted in pleasing, soft pastel colors.

Living in Space

Cosmonauts Leonid Kizim and Vladimir Solovyev were launched on March 13, 1986, in the Soyuz (union) T-15 spacecraft and rendezvoused and docked at Mir at the forward port. Soyuz T-15 carried some experimental equipment that the cosmonauts installed and tested. Further equipment, in addition to propellant, food, water, and air supplies, was delivered by the unmanned Progress 25 freighter that docked at the aft port on March 21. A second Progress vehicle was sent to Mir a month later to complete the refueling of Mir's propulsion system and change the space station's orbit, making it possible for the Soyuz T-15 to take a trip to Salyut 7, an earlier space station that had been abandoned in orbit.

Kizim and Solovyev left Mir on May 5 and docked at Salyut 7 the next day. The cosmonauts reactivated the station, left the spacecraft to retrieve space-exposed specimens, and removed several pieces of working scientific equipment

2198

The Mir orbital station during preflight testing.

for use in Mir. On June 25, Soyuz T-15 left Salyut 7 and returned to Mir late the next day. After some initial scientific work, they left Mir on "automatic pilot" and then returned to Earth on July 16. They had spent 125 days in space and lived on two different space stations.

Mir was next inhabited by cosmonauts Yuri Romanenko and Alexander Laveikin, when Soyuz TM-2 docked at Mir on February 8, 1987. During their stay, Romanenko and Laveikin were visited by seven unmanned Progress freighters, the Soyuz TM-3 international cosmonaut team (Alexander Viktorenko, Alexander Alexandrov, and Mohammed Faris) that remained on Mir for six days, and the first expansion module, called "Kvant."

Kvant (quantum) was a combination astrophysical research laboratory, work area, and space station support module. Kvant carried a total of 1,500 kilograms of support equipment and 2,500 kilograms of consumables and other

cargo. Inside the expansion module was an extra 40 cubic meters of working space. Kvant docked at the aft port of Mir after some initial difficulty caused by debris preventing the docking mechanism from latching. Romanenko and Laveikin had to perform an emergency space walk in order to dock Kvant to Mir successfully. Kvant had a duplicate port at its own aft end to support docking of both Progress and Soyuz spacecraft. On April 21, Progress 29 docked to the aft port of Kvant and was able to refuel Mir's propellant tanks through special plumbing lines routed through Kvant.

Soyuz TM-3 was launched on July 22 in order to pay a visit to Romanenko and Laveikin. Soyuz TM-3 flight engineer Alexandrov replaced Laveikin as Romanenko's companion of the long-duration team. Laveikin, Viktorenko, and Faris returned to Earth in the older Soyuz TM-2 spacecraft, leaving the fresher Soyuz TM-3 for the resident crew.

Cosmonauts Vladimir Titov, Musa Manarov, and Anatoli Levchenko were launched on December 21, 1987, in Soyuz TM-4. Their mission was to replace Romanenko and Alexandrov and then remain on board Mir for an entire year. When this team took over residence of Mir, it marked the first time one space station crew was completely exchanged for another without any lapse in time in between, thus demonstrating a permanent occupation of space.

Consequences

When the Soviet Union lost the Moon race, the direction of its space program was changed toward the development of Earth-orbiting space stations, a goal often publicized as a more important one than exploration of the Moon. Initial efforts in the Salyut program were geared toward upstaging the efforts of American astronauts aboard the Skylab space station.

Salyut 1 was launched on April 19, 1971, and was home to a trio of cosmonauts for twenty-four days, a record-setting endurance, in June of that year. The mission restored much of the diminished Soviet pride in its space program until the tragic loss of three cosmonauts during reentry in Soyuz 11, when a valve unexpectedly bled cabin atmosphere rapidly into space. Soyuz 11 landed safely, but its crew died. After a detailed investigation of the Soyuz 11/Salyut 1 accident, the space station program renewed its efforts to be successful prior to the launching of the United States' Skylab.

Salyuts 1 through 5 were not meant for repeated use and had no significant capability for repair and/or resupply. Salyut 6 was launched on September 29, 1977. Before it was deorbited destructively (in July, 1982), it was home to numerous cosmonaut teams for a total of 676 days. Lessons about physiological and psychological adaptation to long-duration exposure to life in an orbiting space station were learned and a confidence in operations grew. Progress and Soyuz resupply flights restocked consumables and propellants and delivered new scientific equipment, samples, and film. Salyut 7 operations (1982 to 1986) increased endurance to 237 days. When abandoned in orbit, Salyut 7 was replaced by Mir, a space station with six docking ports and a capability for permanent occupation.

David G. Fisher

President Ferdinand Marcos Flees Philippines

The People's Revolution of February, 1986, forced Ferdinand Marcos to flee the Philippines, ending more than a decade of dictatorship.

What: Political reform
When: February 25, 1986
Where: Manila, Philippines
Who:

FERDINAND MARCOS (1917-1989), president of the Philippines from 1965 to 1986

BENIGNO "NINOY" AQUINO, JR. (1932-1983), leader of Liberals in the Philippines

CORAZON AQUINO (1933-   ), his wife, who became president of the Philippines in 1986

JUAN PONCE ENRILE (1924-   ), secretary of national defense of the Philippines from 1970 to 1986

FIDEL RAMOS (1928-   ), deputy chief of staff of the Philippine Armed Forces from 1981 to 1986

Years of Oppression

In 1972, President Ferdinand E. Marcos declared martial law in the Philippines, claiming that communist and Muslim rebels had created a "national emergency." Actually, the rebels were not a real threat to the country, but Marcos's term of office was nearly over, and under the Philippine constitution he was not eligible to run for reelection. The national emergency was his way of hanging onto power.

Marcos then dissolved the Philippine congress and began to rule by decree. Political opponents, including Senator Benigno Aquino, were arrested. Newspapers and television and radio stations were shut down.

Through the next nine years, judges in the Philippines obeyed Marcos's wishes. The media were censored, and strikes were forbidden. La-bor leaders, lawyers, civil rights workers, and some religious leaders were harassed.

The armed forces had numbered 62,000 in 1972; by 1986 they had grown to 230,000. The Philippines had Asia's fastest-growing military and spent six hundred million dollars on defense each year. The military resources were concentrated in the capital, Manila, to support the rule of martial law.

Meanwhile, the Civilian Home Defense Forces (CHDF), bullies and small-time criminals armed and paid by the government, terrorized the people in the countryside. Between 1976 and 1986, the CHDF was involved in more than two thousand murders.

In 1965, the Philippines seemed likely to become an economic power in Asia, second only to Japan. But twenty years later the Philippine economy was in shambles. About two hundred members of the Marcos family and friends of Marcos had gained ownership and control of huge sections of the Phillipine economy; this came to be called "crony capitalism." Meanwhile, the country's foreign debt piled up to $26.5 billion by 1986.

In 1965, 28 percent of the Filipino people lived below the international poverty level; by 1986 that figure had risen to 60 percent. It was estimated in 1986 that 70 percent of Filipino schoolchildren suffered from malnutrition.

People Power

In the early 1980's, the crony-run corporations started to collapse, and Marcos's friends fled abroad with very large sums of money. Aquino, who had been sent into exile in 1980, returned to Manila on August 21, 1983, but was gunned down at the airport.

This was the beginning of the end for Marcos. The government opened an investigation of the

Aquino assassination, and the military officers accused of the crime were found innocent. But the Filipino people believed that Marcos's government had been responsible for the assassination.

As more and more voices in the Philippines and overseas called on Marcos to resign, he decided to hold presidential elections on February 7, 1986. Many groups—the Catholic Church, the labor movement, civil rights activists, students, business people, and members of the middle class—united to support Corazon Aquino, Benigno Aquino's widow, as a candidate for president.

Both Marcos and Aquino claimed victory in the election, in which Marcos supporters were shown to have cheated extensively. The congress supported Marcos, but on February 16 Aquino called her supporters to begin a protest campaign.

On February 22, Marcos began trying to get rid of military officers who were seeking reform, and Defense Minister Juan Ponce Enrile and army deputy chief of staff Fidel Ramos joined Aquino in demanding that Marcos resign. In the

next three days, as many as a million Aquino supporters moved into the streets to form a human barrier between armored vehicles and tanks and the opposition forces.

On February 25, 1986, faced with this "People's Revolution" and realizing that he had lost American support, Marcos and his family agreed to be airlifted out of the presidential palace into exile in Honolulu, Hawaii.

Consequences

The new president, Corazon Aquino, was considered upright and unable to be corrupted. She restored the independence of congress and the courts, and more than five hundred political prisoners were released. A commission on human rights was set up to investigate reports of torture under the Marcos regime. Censorship of the media was ended, and workers' right to strike was restored.

Hundreds of corporations that had been taken over by the state or by Marcos's cronies were returned to private ownership. Japan, the United States and international banks supported the Aquino government by making it easier for the Philippines to make payments on its foreign debt. The economy began growing again, though most of the country's poor people continued to suffer.

Though a number of military reformers had supported Aquino at first, many of them became unhappy with her, thinking that she was too reluctant to crush the communist and Muslim rebels. Soon after she took office her government released the rebel leaders from prison, offering them amnesty if they agreed to stop the armed struggle; but the Communist New People's Army continued to resist the government. At various times military groups tried to take power from Aquino as it became clear that the problems of the Philippines could not be dealt with quickly and easily.

A Filipino man strikes a painting of Ferdinand Marcos as rioters swarm the presidential palace in Manila after the resignation and flight of Marcos.

M. Leann Brown

U.S. and Libyan Forces Clash over Gulf of Sidra

> *The clash between American and Libyan forces in the Gulf of Sidra in March, 1986, was one of the two countries' many political and legal disagreements.*

What: International relations
When: March 24-25, 1986
Where: Gulf of Sidra (or Sirte) in the Mediterranean Sea
Who:
MUAMMAR AL-QADDAFI (1942-   ), head of state and commander in chief of the Armed Forces of Libya from 1969
RONALD REAGAN (1911-   ), president of the United States from 1981 to 1989

U.S.-Libyan Antagonism

On November 3, 1985, *The Washington Post* reported that President Ronald Reagan had allowed the Central Intelligence Agency (CIA) to carry out a secret operation to "destabilize" the rule of Colonel Muammar al-Qaddafi in Libya. On November 10, Qaddafi responded that if this report was true, he would "undermine America from inside." There were terrorist attacks in the airports of Rome, Italy, and Vienna, Austria, on December 27, and a radical Palestinian group was believed to be responsible. The United States accused Libya of helping this group.

On December 31, 1985, the Egyptian newspaper *al-Ahram* reported that about two thousand Soviet advisers had arrived in Libya in the previous few days to operate SAM-5 (surface-to-air) antiaircraft missile batteries. A week later, President Reagan ordered that all American economic ties with Libya be cut. He told Americans working in Libya that they should leave.

On January 15, Qaddafi declared that Libya would train, arm, and protect "suicide and terrorist missions" to help liberate Palestine. On March 4, the Libyan cabinet stated that hit squads should be formed to attack Israeli and American targets.

As early as 1981, the United States had accused Libya of plotting to assassinate President Reagan. In May of that year, the United States had closed its embassy in Tripoli, the Libyan capital; then the United States had stopped importing Libyan oil. Qaddafi vowed to keep opposing the United States.

The two countries also came into conflict over the Gulf of Sidra. In 1958, before Qaddafi's time, Libya had signed the Law of the Sea Convention, agreeing that each country would have sovereignty over ocean waters and the seabed up to twelve nautical miles from its coastline. Since 1973, however, Libya had been insisting that the entire Gulf of Sidra—which stretched well beyond twelve miles—was included in its territorial waters. The United States did not accept this claim, and neither did most other countries.

In August, 1981, U.S. fighters had shot down two Libyan jets that had tried to intercept American aircraft over the Gulf of Sidra. In December, 1985, Qaddafi drew a "line of death" across the gulf along 32° 30′ latitude north; he warned everyone not to cross it without permission.

Breaching the "Line of Death"

Late in January, 1986, the U.S. Sixth Fleet was sent north of the "line of death," supposedly to prepare for practice maneuvers. By March 20, the largest armada since World War II had been gathered offshore from Libya: three aircraft carriers (the *America, Coral Sea*, and *Saratoga*) with 240 planes on board, escorted by twenty-seven other U.S. ships.

On March 24, Libya fired SAM-5 missiles at U.S. fighter planes that crossed the fateful line. Missiles were fired again on March 25. In response, U.S. aircraft made attacks on missile and radar bases near the town of Sirte on the gulf

2203

coast. The United States reported that it had sunk two Libyan ships. Three U.S. ships sailed into the disputed waters on March 24. Libya's air force did not become involved in the conflict, and the American fleet left the area on March 27.

The next day, Colonel Qaddafi spoke at a "victory rally" in Tripoli; he claimed that three American aircraft had been shot down (though the United States denied this claim). Meanwhile, the commander of the U.S. Sixth Fleet thanked his crews for being "the spear and shield of American policy in a troubled and volatile region."

Publicly, the U.S. government insisted that the sea exercises had been intended to provide a challenge to Libya's claim to the entire Gulf of Sidra. Privately, though, U.S. officials admitted that they had wanted to prove that the United States was ready to punish the Libyan leader for backing terrorists.

On March 27, the foreign ministers of the Arab League announced that they were quite displeased with the American action. Libya demanded that the Arab League nations break their diplomatic relations with the United States and start an economic boycott of the United States, but the Arab foreign ministers were not willing to go this far. The United Nations Security Council discussed the Gulf of Sidra clash between March 26 and 31, but it then adjourned without passing a resolution.

Consequences

Many Americans were pleased that their government had sent planes and ships over Qaddafi's

"line of death," but close allies such as Great Britain and Israel were not enthusiastic. France, West Germany, and Japan agreed that Libya had no right to claim the entire Gulf of Sidra, yet they worried that the U.S. action would increase tensions in North Africa and the Middle East.

Greece, Spain, and Italy stated that they were displeased with the United States' open challenge of Libya, and the Soviet Union joined in denouncing the United States. (It is not known whether any Soviet advisers were killed or injured during the American attacks on the Libyans' coastal bases.) But the sharpest criticism came from Arab leaders, though some of them had their own scores to settle with Qaddafi.

On April 5, 1986, a bomb exploded in a discotheque in West Berlin, and two people (including a U.S. serviceman) were killed. This terrorist act was blamed on a Libyan plot. Ten days later, American planes dropped bombs on Libya. Clearly, the conflict between the two countries had not been resolved.

Peter B. Heller

Accident at Ukraine's Chernobyl Nuclear Plant Endangers Europe

> *An accident at a nuclear power plant at Pripyat, in what was then the Soviet Union, spread pollution over the nearby area, and clouds of radiative fallout drifted across Europe.*

What: Environment; Disasters
When: April 26, 1986
Where: Pripyat, Soviet Union (now in Ukraine)
Who:
VIKTOR BRYUKHANOV, director of the Chernobyl nuclear power plant
NIKOLAI FOMIN, chief engineer at the Chernobyl nuclear power plant
ANATOLY DYATLOV, deputy chief engineer at the Chernobyl nuclear power plant

The Accident

Construction of Chernobyl, planned to be one of the largest nuclear power plants in the Soviet Union, began at Pripyat in 1972. The first reactor was started up in 1977, and the next three began operating in 1978, 1981, and 1983. The fifth was scheduled to begin working late in 1986; meanwhile, its construction pit served as a dumping place for the debris from reactor number 4.

In 1986, the Chernobyl plant provided one-seventh of the Soviet Union's nuclear-generated electric power. Yet in March of that year, an article by Liubova Kovalevska in a Ukrainian magazine criticized the plant: It had been built sloppily, it was managed poorly, its workers were unhappy, and there were never enough construction materials for the building projects.

At 1:23 A.M. on April 26, 1986, because a night crew was doing an experiment on the number 4 reactor, the emergency water-cooling and shutdown systems were closed, along with the equipment for regulating power output. Then the technicians took almost all the controlling rods from the reactor's core but allowed it to keep running at 7 percent. It seems that the technicians were trying to see whether reactor number 4 could be kept going for a short time during a power failure.

The result, however, was that the balance of the fission reaction was upset; there was a sudden surge from 7 to 50 percent, and a section of the reactor near the top of the core quickly overheated. The overheated zirconium fuel rods reacted with the water in the cooling system to produce hydrogen, which leaked out.

This chain reaction in the core went out of control, setting off two explosions that blew off the reactor's heavy steel and concrete lid. A two-hundred-ton crane, used to place fuel in the core, broke loose and came crashing down on the reactor. Large amounts of radioactive materials were released into the atmosphere—six or seven tons of twenty-three different isotopes (from krypton 85m to plutonium 242), along with their alpha, beta, and gamma rays of decay.

If the prevailing winds had been moving southeast, the fallout would have been carried over Kiev, a large city only sixty miles away. Instead, the wind was blowing toward the northwest, and the atomic plume was carried great distances over the Ukraine, Belorussia (later Belarus), Poland, and Scandinavia. Traces of it even reached as far west as France and Italy.

Pollution Spreads

Close to the plant, the effects were quick and severe. Within twenty miles around Chernobyl, the soil and groundwater were badly contaminated by radioactivity. Beginning at 2:00 P.M. the day after the accident, 25,000 people were evacuated in a caravan of 1,216 large buses and 300 trucks that stretched for fifteen kilometers along the road between Kiev and Pripyat. Another

2205

9,000 people left in local buses and private cars. Then about 34,500 children were evacuated from Kiev. By May 2, another 116,000 persons and 86,000 head of cattle were also evacuated from a 30-kilometer zone around Chernobyl.

Meanwhile, graphite was burning in the plant, and firefighters were working within one hundred feet of the fire until helicopters arrived to douse it with bags of silicates, dolomite, and lead. Two control workers died immediately at the plant, and six firefighters died a little later. The situation was brought under control by May 1.

A commission led by Soviet premier Nikolai Ryzhkov and two other Communist Party officials, Yegor Ligachev and Vladimir Shcherbitsky, visited Pripyat on May 3 and then reported its findings to Mikhail Gorbachev, general secretary of the Soviet Communist Party. On May 14, Gorbachev made a twenty-five-minute speech on Soviet television; he announced that nine people had died; 299 others had been hospitalized, thirty-five of them with severe radiation sickness. (Later, more than thirty died, and another two hundred became seriously ill.)

Cleanup work continued at the Chernobyl plant. Military helicopters dumped more than four thousand tons of sand, clay, lead, and boron on the number 4 reactor, and workers filled a tunnel underneath the reactor with molten lead; under the lead tunnel they poured a concrete slab. Above the ground, they completely enclosed the number 4 reactor with steel girders and concrete walls and roof.

By mid-July, another sixty thousand children had been evacuated from the Gomel area in Belorussia, one hundred miles away. By September, plant officials were on trial, and twenty-seven Communist Party officials in the Pripyat region were expelled for "cowardice and alarmism."

In August, 1987, plant director Viktor Bryukhanov, chief engineer Nikolai Fomin, and deputy chief engineer Anatoly Dyatlov were given prison sentences of ten years each. Three others were sentenced to from two to five years in prison.

Consequences

Outside the Soviet Union, the first country to learn of the accident was Sweden. Within two

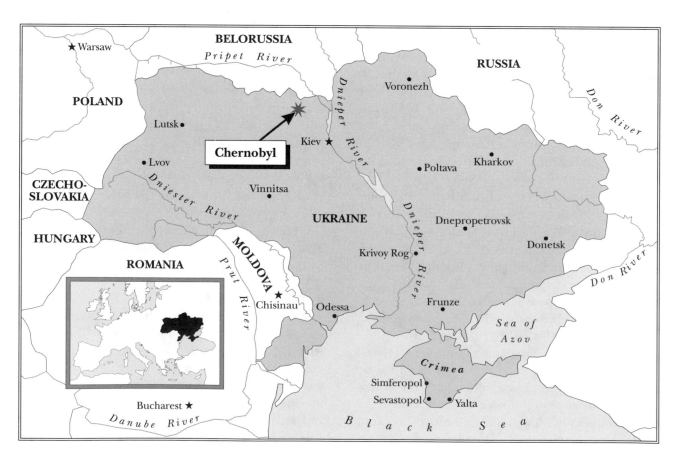

The remains of the number four reactor are visible in front of the chimney at the Chernobyl nuclear power plant.

AP/Wide World Photos

days of the disaster, some Swedish nuclear power plants recorded a 20 percent increase in radiation; the Soviets then admitted that there had been an accident at Chernobyl.

Radiation levels rose across Europe, and there was great fear and concern. Italy's largest vegetable market, in Milan, was ordered to dump all of its fresh vegetables. On April 29, the Polish government banned the sale of milk from grass-fed cows and restricted the sale of vegetables. All babies and young children were required to take doses of iodine to help prevent thyroid cancer. Austria advised parents to keep their children indoors and to take frequent showers.

Boris Yeltsin, head of the Communist Party in Moscow, gave an interview during a trip to Bonn,

West Germany, and stated that radiation around the Chernobyl plant had reached two hundred roentgens per hour—far above the level that is considered lethal. The alarming news brought a flood of petitions and demonstrations to ban nuclear energy in Europe.

In response, the Vienna Conference of August 25-29, 1986, was organized to tighten nuclear safety. Conference participants came up with a thirteen-point program: higher standards for nuclear plants, more frequent international inspections, setting up better early-warning systems, more advanced fire-fighting methods and equipment, and more research on the connection between radiation and health.

Donald E. Davis

2207

United Nations Drafts Plan for African Economic Recovery

> *After several years of severe famine in Africa, the United Nations made a plan to help African countries overcome the famine emergency and start back on the road to economic health.*

What: Economics
When: June 1, 1986
Where: New York and Africa
Who:

JAVIER PÉREZ DE CUELLAR (1920-   ), secretary general of the United Nations from 1982 to 1991

ABDOU DIOUF (1935-   ), president of Senegal from 1981, and chair of the Organization for African Unity

STEPHEN LEWIS (1937-   ), chair of the committee that planned the United Nations Special Session on African Recovery

Famine and Economic Crisis

A terrible drought struck Africa between 1983 and 1985, and the result was a severe famine. Famines happen in places that are very poor, and in the early 1980's twenty of the world's thirty poorest countries were in Africa. Many of these countries had gained their independence only fifteen or twenty years earlier, and some of their governments were unstable or corrupt. There were civil wars to determine who would control these governments.

As governments spent more on weapons and armies, they had less to spend on developing agriculture. Often the civil wars prevented harvests from being gathered and caused people to flee to neighboring countries. Yet these countries were equally poor and could not cope with even more poor people to feed.

Most African countries continued to depend economically on the European countries from which they had won independence. To buy oil, farm machinery, and cars—all the things a poor country could not produce at home—the African nations had to sell farm products such as peanuts, cocoa, and coffee. When drought struck, farm production fell. Even when harvests were good, a product such as coffee could not fetch enough money to pay for much oil or many cars.

Populations were growing, and few jobs were available in rural areas. Many Africans moved to cities, where they demanded cheap food. To avoid riots, governments cut food prices, but this hurt the farmers, who no longer had a reason to produce more food than they needed for themselves.

As food production decreased, African countries had to rely on imported food. Some governments chose socialist economic policies, putting tight controls on prices and supplies. This made the economic situation even worse, and drought brought total disaster.

In 1984, the United Nations' secretary general, Javier Pérez de Cuellar, visited camps for famine victims in Ethiopia. Returning from Africa in December of that year, he asked governments to help the Africans in their hour of desperate need. He also created the Office for Emergency Operation in Africa, which began its work early in 1985.

In July, 1985, African leaders gathered in a meeting of the Organization for African Unity and agreed on a program for economic recovery. They accepted responsibility for their countries' problems but also asked the United Nations to help. In December, the U.N. General Assembly voted to call a special session on African recovery in the late spring of 1986. Stephen Lewis of Canada was named chair of a committee that would make plans to help Africa.

2208

A Plan for Recovery

The special session was held from May 27 to June 1, 1986, and nearly one hundred speakers were involved in the debate. After six days of hard work, the General Assembly announced a five-year plan for recovery. It would cost $128 billion: $82 billion from the African nations themselves and $46 billion from foreign aid.

All the participating countries finally agreed with the basic principles of the plan. They knew that the economies of African countries needed to be reformed, but that richer nations would need to give financial help as well.

The discussions included a long debate about what Africa needed most—reform or aid. Those who argued that the most important need was reform pointed out that governments in Africa usually neglected peasants in rural areas; they said that African nations needed to recognize that women and children were an important part of agriculture. Also, these governments should do more to encourage trade and investment in African businesses.

Others argued that aid was Africa's most crucial need. They reminded their listeners that Africans could not invest if they had to pay back huge debts, or if they could not count on stable prices for the goods they traded to other countries.

In the end, the delegates agreed that African nations needed both reform and aid to address the problems of poverty, ignorance, disease, famine, and starvation. The Africans took responsibility for many of the problems they faced, and they agreed that reform was needed. They asked for aid to make those reforms possible.

The plan called for rich governments to help the poor African governments in several basic areas. First, African countries needed to begin feeding themselves. This meant that they needed to help farmers and peasants become more productive. Farmers needed to receive a fair price for food they produced—which meant that food prices could not be kept artificially low for the benefit of city dwellers. Also, better storage facilities were needed, since poor storage causes food to spoil.

The plan also asked rich countries to help Africa improve its environment and stop the spread of deserts. It called for programs to strengthen the human resources of Africa through education and literacy campaigns, for people who can read and write are more productive. Education of women and children, especially, helps to improve their capacity to work and earn. It also helps to reduce population growth.

Another concern was to help Africa develop its roads, hospitals, schools, power facilities, water resources, factories, and agricultural systems. If these resources and institutions were not strengthened, the poor countries would always suffer famine in times of drought. Finally, the plan asked African countries to encourage economic growth and reduce corruption.

Consequences

Progress in the continent's economic recovery has been slow. After the five-year recovery plan ended in 1990, there were still many poor countries in Africa, and some of them had grown even poorer. These countries still needed more foreign aid, better opportunities to earn money from trade, and cancellation of old debts.

Yet other countries did show signs of improvement, and some governments were beginning reforms. The five-year recovery plan did not succeed in helping all African countries to achieve their goals, but for many of them, hopes of a brighter future were still alive.

Robert F. Gorman

2209

Rehnquist Becomes Chief Justice of the United States

Controversy surrounded William H. Rehnquist's 1986 nomination, and more votes were cast against him than against any other successful Supreme Court nominee in the twentieth century. The Rehnquist Court established a reputation for conservative rulings on controversial issues.

What: Law
When: September 26, 1986
Where: Washington, D.C.
Who:

WILLIAM H. REHNQUIST (1924-   ), associate justice of the Supreme Court from 1971 to 1986; chief justice from 1986

RONALD REAGAN (1911-   ), president of the United States from 1981 to 1989

WARREN E. BURGER (1907-1995), chief justice of the Supreme Court from 1969 to 1986

Education and Experience

William H. Rehnquist was born on October 1, 1924, in Milwaukee, Wisconsin. He attended public elementary and secondary schools in suburban Milwaukee. Following his service in the U.S. Air Force during World War II, Rehnquist attended Stanford University, where he earned a bachelor of arts degree and a master of arts degree in political science in 1948. That same year, he was elected to the honorary fraternity Phi Beta Kappa. He earned a master of arts degree in government in 1950 from Harvard University and then returned to Stanford University, where he earned his law degree in 1952, graduating first in his class. One of his classmates was Sandra Day O'Connor, who later became an associate justice of the Supreme Court and joined Rehnquist on the bench. He served as a clerk to Supreme Court associate justice Robert H. Jackson from 1952 to 1953.

From 1953 to 1969, Rehnquist was engaged in the private practice of general law with emphasis on civil litigation in Phoenix, Arizona. As a practitioner, he became immersed in Republican state politics and was associated with the party's most conservative wing. He denounced the liberalism of the Supreme Court under the leadership of Chief Justice Earl Warren, and vigorously campaigned for Republican presidential candidate Barry Goldwater.

During the campaign, Rehnquist met and worked with Richard G. Kleindienst, President Richard M. Nixon's attorney general, who later appointed Rehnquist to head the Justice Department's Office of Legal Counsel. In January, 1969, he was appointed assistant attorney general in that office by President Nixon and became one of the Nixon administration's chief spokespeople on Capitol Hill, commenting on issues ranging from wiretapping to the rights of the accused. Rehnquist had the responsibility of reviewing the legality of all presidential executive orders and other constitutional law questions in the executive branch and frequently testified before congressional committees in support of the administration's policies.

Rehnquist on the Supreme Court

In 1971, Rehnquist was nominated by Nixon to the Supreme Court to replace retiring Associate Justice John Marshall Harlan. He was confirmed by the Senate by a 68-26 vote. After serving on the Court for fifteen years, Rehnquist was promoted to chief justice by President Ronald Reagan, the third sitting associate justice in history to be elevated to chief justice, the sixteenth U.S. chief justice.

In his Supreme Court career, Rehnquist became involved with presidential politics on four notable occasions: the Watergate presidential tapes, a case in which Rehnquist recused himself because Nixon had nominated him to the bench and Rehnquist had dealt with the tapes issue in his career in the Justice Department; the Paula Jones case in which the Arkansas state employee accused President Bill Clinton of sexual harassment; the President Clinton impeachment hearings conducted by Rehnquist; and the Florida election dispute during the 2000 election between presidential candidates George W. Bush and Al Gore.

Before becoming an associate justice, Rehnquist had been portrayed as an enthusiastic conservative whose ideology was often apparent on the bench. During his tenure on the high court, however, his ultraconservative position shifted, becoming somewhat modified, presumably in part because other justices, such as Antonin Scalia and Clarence Thomas, occupy the ultraconservative position.

Consequences

According to a Senate summary of the opposition to Rehnquist's 1986 nomination to be chief justice, his views on civil rights were questioned, and he was accused of harassing and attempting to disenfranchise African American voters in Phoenix during the 1950's and early 1960's. Rehnquist denied these allegations. Opposition to his elevation to the position of chief justice was evident when he was confirmed by the Senate by a 65-33 vote.

As chief justice, Rehnquist wrote the majority opinion for such notable cases as *Morrison v. Olson* (1988), in which the Court ruled that a special court can appoint special prosecutors to investigate crimes by high-ranking officials, and *Hustler Magazine v. Falwell* (1988), in which the Court ruled that public figures cannot be compensated for mental distress caused by a parody published in a national periodical. Rehnquist

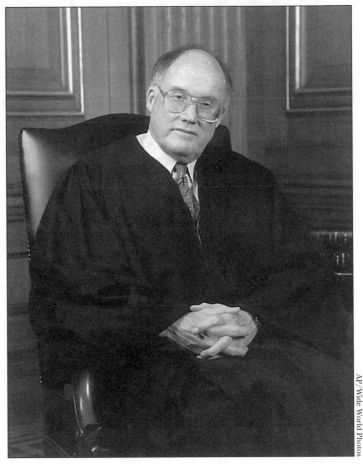

William H. Rehnquist.

dissented in *Texas v. Johnson* (1989), in which the Court ruled that the burning of an American flag in protest is protected by individual rights in the First Amendment to the U.S. Constitution. He also dissented in cases dealing with the right of a woman to choose to have an abortion (*Roe v. Wade*, 1973) and the legality of the death penalty (*Furman v. Georgia*, 1972).

Rehnquist is the author of several books, including *The Supreme Court* (2001), *All the Laws but One: Civil Liberties in Wartime* (1998), *Civil Liberty and the Civil War* (1997), and *Grand Inquest: The Historical Impeachments of Justice Samuel Chase and President Andrew Johnson* (1992).

Marcia J. Weiss

Rutan and Yeager Fly *Voyager* Nonstop Around World

An American team set a record by piloting an innovative aircraft around the world without landing or refueling.

What: Space and aviation
When: December 14-23, 1986
Where: Edwards Air Force Base, California
Who:
BURT RUTAN (1943-   ), the designer of the *Voyager* and other unconventional aircraft
DICK RUTAN (1938-   ), the brother of Burt Rutan, copilot of the *Voyager*
JEANA YEAGER (1952-   ), a copilot of the *Voyager*

Record Breaking

To fly around the world is not an easy task. In 1924, four U.S. Army open-cockpit biplanes attempted the first round-the-world air trip; after sixty-nine stopovers and 175 days, only two completed the journey. In the early 1930's, American aviator Wiley Post twice flew around the world in a Lockheed Vega; 1933, two years after his first trip, Post became the first person to accomplish the journey alone, completing it in seven days and eighteen hours. On an attempted round-the-world flight in 1937, Amelia Earhart vanished over the Pacific Ocean. In 1938, Howard Hughes, flying his Lockheed 14, reduced the world record to three days and nineteen hours.

In 1949, the first nonstop flight around the world was accomplished by a U.S. Air Force B-50, which was refueled in the air and required ninety-three hours for the trip. In 1962, a U.S. Air Force B-52 set the record for the longest nonrefueled flight, 20,180 kilometers. (The shortest distance recognized as "around the world" by the Fédération Aéronautique Internationale is 36,800 kilometers, which is equal to the circumference of the Tropic of Cancer or Capricorn. This distance is somewhat smaller than the circumference of the equator, which is 40,000 kilometers.)

The project of creating an aircraft to reach aviation's "last milestone," a nonrefueled global flight, began in 1980, when Burt Rutan sketched a possible design for such an aircraft on a restaurant napkin. Actual construction of the aircraft began in 1981. Upon completion, *Voyager* was the largest composite airplane ever built, with a wingspan of 33.8 meters.

Voyager was a "flying fuel tank." Altogether, seventeen fuel tanks were built into the plane's interior, leaving only a very small space—2.1 meters by 1.0 meters—for the pilots. The airplane was loaded with fuel until it weighed 4,497 kilograms, of which 3,180 kilograms was fuel—an incredible 3:1 ratio of fuel weight to aircraft weight.

Voyager was powered by two American-built Continental engines; the front engine was air cooled, while the rear engine was water cooled. The plane was built largely of high-technology materials such as carbon-fiber spars and Nomex honeycomb sandwich, which was used to cover the wings. Even the propellers were of a radical new design. Each propeller required sixteen different airfoil sections, which were carved from a computer-directed milling machine.

The use of such exotic lightweight construction materials, however, meant that the plane's frame was flexible, making it very difficult to fly in turbulent air. This problem was partially solved by the use of specially designed autopilots that could assist the pilots in turbulent condi-

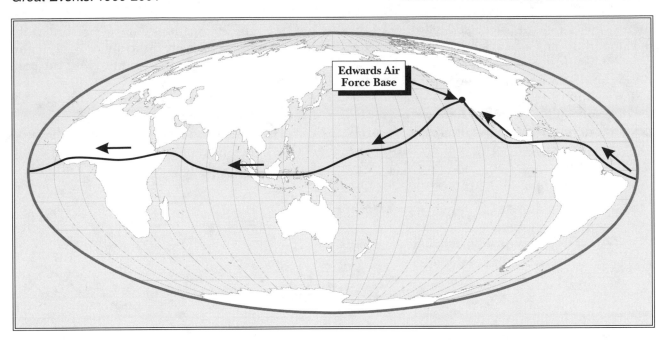

tions. After sixty-seven test flights, the *Voyager* was ready for its dramatic world flight.

Around the World in Nine Days

Voyager's flight began at 8:00 A.M. on Sunday, December 14, 1986, and almost ended in immediate disaster. Because of the weight of the fuel in the wings, the aircraft almost refused to become airborne. *Voyager* required all but 240 meters of a 4,600-meter runway to lift off, setting another world record of sorts for the longest takeoff roll. Both of the plane's wingtips were damaged during the takeoff, but the damage did not greatly affect its overall performance. The *Voyager* flew around the world from east to west. After flying southwest from California to the Indonesian archipelago, it continued almost due west, staying close to the equator until it crossed Africa and the Atlantic Ocean. After nearing the eastern tip of Brazil, it skirted the northern coast of South America, crossed Central America, and flew up the west coast of Mexico on its return to California.

Dick Rutan flew the plane for fifty-five of the first sixty hours, while Jeana Yeager monitored the plane's systems and recorded flight speeds, positions, winds, and fuel supplies. From time to time, Rutan and Yeager would switch positions, with Yeager doing the flying and Rutan resting; such movement was difficult in the limited living space. They took about 60 kilograms of provisions along, and they prepared hot meals by placing meal pouches around a heating duct.

On the sixth day, they noticed that the oil-warning light for the rear engine was flashing. By turning a hand crank in the wrong direction, Rutan had mistakenly taken a quart of oil out of the engine instead of adding it. The problem was diagnosed quickly, however, and the oil was replaced. On the seventh day, severe turbulence from a thunderstorm in the South Atlantic forced the *Voyager* into a ninety-degree bank, and both pilots experienced considerable discomfort. A short distance from home on day nine, off the west coast of Baja California, the rear engine quit. The front engine, which was kept as a backup, needed starting, but the aircraft lost more than 1,500 meters of altitude before this could be done.

Finally, on December 23, 1986, at 8:05 A.M., after nine days, three minutes, and forty-four seconds of elapsed time, the *Voyager* landed at Edwards Air Force Base in California before thousands of well-wishers. Less than 50 kilometers' worth of fuel remained in the tanks.

Consequences

The *Voyager* had traveled 40,212 kilometers during its record-setting flight. The trip was also notable for a number of additional accomplish-

2213

ments, including the successful piloting of the low-powered plane in a variety of extreme weather conditions and the pilots' ability to endure the phone booth-sized cabin for the flight's duration.

Public interest proved to be greater than the participants had expected. There were daily reports about the airplane and its crew in the press long before the actual flight. Initially, the project was funded by grass-roots donations from aviation fans. As the operation grew, so did the num-ber of engineering and technical personnel. In the latter stages of the project, major aircraft and avionics companies agreed to help with the construction of the *Voyager.* The Hartzell corporation, for example, provided the engineering for the new propeller shapes.

Perhaps the greatest achievement of the project was the dedication of a small group of people to the accomplishment of a very difficult task.

Michael L. Broyles

Muslims Riot Against Russians in Kazakhstan

> *Mikhail Gorbachev's policy of glasnost helped reveal that Communist Party chiefs had supported oppressive rulers in Kazakhstan and other Soviet republics.*

What: Civil strife
When: December 17-19, 1986
Where: Alma-Ata, Kazakhstan, Soviet Union
Who:
NIKITA S. KHRUSHCHEV (1894-1971), first secretary of the Soviet Communist Party from 1953 to 1964, and premier of the Soviet Union from 1958 to 1964
LEONID ILICH BREZHNEV (1906-1982), first secretary of Kazakhstan from 1954 to 1956, and first secretary of the Soviet Communist Party from 1964 to 1982
MIKHAIL GORBACHEV (1931-   ), first secretary of the Soviet Communist Party from 1985 to 1991
DINMUKHAMED KUNAYEV (1912-1993), first secretary of Kazakhstan from 1956 to 1962 and from 1964 to 1986
GENNADII V. KOLBIN (1927-   ), a Russian who was appointed as Kunayev's replacement in 1986

A Muslim Republic

After Tamerlane's great empire fell apart in the early 1400's, a group of Mongol Turks who called themselves the Kazakhs, or "free people," began to roam over the northwestern areas of Turkestan in Central Asia.

After the Crimean War (1854-1856), Russian peasants began moving into the fertile lands of Kazakhstan, confiscating nearly one hundred million acres. The native Kazakhs were pushed up into the hills or into the harsh desert regions of the south and southwest, where most of them died.

Kazakhstan joined the Russian Empire in 1893, and in 1916 Czar Nicholas II tried to draft Kazakh men into his army to fight in World War I. The Kazakhs rebelled for two reasons: They would have to leave at harvest time, and as Muslims they did not want to risk their lives to benefit a Russian "infidel." Their revolt lasted four months; thousands of Kazakhs were killed, and about one million fled to Xinjiang (Sinkiang), China.

After the Russian Revolution (1917-1918), the Kazakhs suffered in the new collective farms. The number of Kazakh households decreased by half—from 1.2 million to 565,000—and most of the native leaders were removed from their posts.

In the mid-1950's, Nikita S. Khrushchev decided to bring farm reforms to Kazakhstan, and he appointed Leonid Brezhnev first secretary of Kazakhstan to create these changes. Unlike the farmlands of the Ukraine, where much fertilizer was needed, the "virgin lands" of Kazakhstan could be sown immediately after plowing. They were so easy to farm that young people from the cities were brought in to work on the farms.

Within two years Brezhnev, aided by a Kazakh tribal chief named Dinmukhamed Kunayev, had turned the Kazakh grazing lands into productive wheat and cotton fields. These products were shipped out to benefit other areas of the Soviet Union, so that the Kazakhs now had to rely on imports to meet their own needs.

Corruption Is Exposed

When Brezhnev left Kazakhstan in 1956, Kunayev became first secretary. Khrushchev removed Kunayev from this post in 1962, but he was reappointed when Brezhnev became first secretary of the Soviet Communist Party two

years later. Kunayev increased his power by making alliances with Russian communists, tribal leaders, and others. These people were given government jobs, and few people questioned their actions.

Kunayev reigned like a tyrant in the Middle Ages. He encouraged the building of factories and was careless about the needs of farmers. A worse consequence of his policies was that so much water was drained from the Aral Sea that it practically dried up. The health and well-being of many people in the surrounding area suffered greatly.

When Mikhail Gorbachev became the leader of the Soviet Union in 1985, he began a new policy called *glasnost*, or openness. So that Soviet citizens could gain freedom of choice, corruption and unfair practices began to be exposed.

From the beginning, Gorbachev realized that Kunayev was a corrupt but very powerful leader. He knew that Kazakh officials had faked statis-

tics and changed government records to make Kunayev look good. Gorbachev would have to be very careful in trying to get rid of this despot.

In his political report in 1985, Gorbachev criticized Kunayev for the poor economy of Kazakhstan. By February, 1986, Gorbachev had dismissed two-thirds of the Party committee and about five hundred Party officials in Kazakhstan. When Kunayev requested funds for a major water project, the funds were denied; the Central Asians had to face the consequences of poor planning in their irrigation program.

On December 16, 1986, the Communist Party of Kazakhstan dismissed Kunayev and replaced him with Gennadii Kolbin, a Russian. This broke the tradition that the first secretary of Kazakhstan would be a Kazakh.

In protest, Kazakhs in the capital city, Alma-Ata, gathered in a demonstration, which turned into a riot when Soviet troops tried to break up the crowd. Strong feelings of nationalism filled the people of Alma-Ata. Demonstrators armed with wooden sticks and metal rods overturned public buses and beat Russian passengers. They also burned a food store, smashed store windows, looted, and tore up flower beds in front of the Party headquarters.

Then Soviet troops, supported by armored vehicles, attacked the crowd with billy clubs and water cannons. About 250 soldiers ended up in the hospital, and demonstrators continued to surround the Party building through the night. They renewed their attack on the building the next morning; the troops again fought back with water cannons and rubber bullets. Dozens of demonstrators were killed and hundreds injured.

By December 20, everything was back to normal. According to official Soviet reports, the rioters had been "hooligans." Actually, the rioters were fairly organized and had been supported by factory workers and some school organizations.

Gorbachev allowed news of the riots to be broadcast nationwide, serving notice to other corrupt leaders that their days, too, might well be numbered.

Consequences

With their rebellion, the people of Central Asia signaled to the world that they would no longer tolerate oppression. They became eager to gain national, religious, and cultural freedom, and to throw off the domination of the Soviet system.

The riots in Kazakhstan had a ripple effect throughout other Soviet republics, where other ethnic groups began seeking independence. The Soviet Union as a nation was dissolved at the end of 1991, and most of the republics decided to become independent states within a commonwealth.

Iraj Bashiri

Black Workers Strike in South Africa

Newly formed black trade unions were successful in challenging the apartheid system through massive strikes.

What: Civil rights and liberties; Labor
When: 1987
Where: South Africa
Who:
PIETER W. BOTHA (1916-), prime minister of South Africa from 1978 to 1984
NELSON MANDELA (1918-), leader of the African National Congress (ANC)
JOE SLOVO (1926-1995), general secretary of the South African Communist Party

The Unions Grow

The Republic of South Africa was an unusual nation. Although its economy was capitalist like those of other Western nations, it was different in one enormous way: Most of its working class, because of their race, did not enjoy human rights. South African society was divided into various "nations" by race; this system of segregation was called apartheid, meaning "apartness."

One of apartheid's most important goals was destroying black trade unions. This is not surprising, since under apartheid the only reason for a black to be in a "white" area is to work for the whites.

During the 1960's and early 1970's, the South African economy grew quickly, mostly because more and more blacks were hired as laborers. As black workers came to be needed in many businesses and factories, they gained more power. There was a shortage of white workers, and for the first time blacks were recruited to be skilled laborers. South Africa's parliament even considered giving black workers the right to strike.

White South African business leaders were forced to watch as their power was challenged. Most employers did not recognize black unions, yet these unions gained more and more members. Black workers' wages rose more quickly than white wages did. Some of the strikes were political; for example, some strikes protested the imprisonment of African National Congress leader Nelson Mandela. May Day celebrations sponsored by the South African Communist Party attracted huge crowds.

In late November of 1985, thirty-four of the most important black trade unions joined to form the Congress of South African Trade Unions (COSATU), which claimed 450,000 members. From the beginning, the new union federation was political. Some of its members wanted South Africa to come under a socialist system, and the federation supported the campaign for international disinvestment—withdrawing foreign investment in order to pressure the government to bring an end to apartheid.

By 1986, more than 1 million days of work had been lost because of economic strikes, and another 3.5 million strike days had been given to political protest. The next year, however, was a historic period for strikes in South Africa. On May Day, 1987, alone, 2.5 million working days were lost.

The Year of Strikes

More than ten thousand retail workers went on a ten-week work stoppage that began in February of 1987. There was also a forty-six-day strike against Mercedes-Benz, and postal workers were involved in four different disputes with management. This was the first time government workers had become involved in union strikes. A strike by twenty thousand railroad workers began on April 22 and lasted until June 5.

Perhaps the most important strike of 1987 was the strike by the National Union of Mineworkers (NUM). Gold mines brought almost half of South Africa's foreign earnings. During a strike in Au-

2218

gust, 1946, seventy-five thousand mine workers had been forced back to work at gunpoint; twelve had died, and more than a thousand had been injured. In the years that followed, the right-wing Mineworkers' Union, representing white workers, had formed an alliance with the owners. Black miners were given low pay and prevented from taking the better jobs in the mines.

The NUM, the first union for black mine workers, had been born in August, 1982. It had grown quickly, and some mines had even officially recognized this new union. Still, during the first legal strike by black miners in 1984, a number of strikers had been killed.

To win a national wage agreement, the NUM launched a strike on August 9, 1987. The strike lasted until August 30, and at its highest point it involved 340,000 workers. Sixty thousand workers were fired to punish them for their participation, and the NUM had to accept a pay agreement that was only a little better than the old one. Still, the mine workers succeeded in showing their strength.

Challenged by a railroad strike, the government of Pieter W. Botha claimed that the strikes were part of a Communist strategy, led by the Communist Party and its general secretary, Joe Slovo. In September, 1987, the Labour Relations Amendment attempted to keep workers from striking in sympathy with workers in other companies or industries. Yet the government and the white businesses could not crush the growing labor movement.

A "National Living Wage Campaign" sponsored by COSATU began in April, 1987. The campaign was political as well as economic; labor leaders called for the abolishment of apartheid. In response, the government invaded union headquarters and arrested many leaders.

Consequences

Most white South Africans were well-to-do, and South African companies had become accustomed to making huge profits. But for their prosperity they depended on black laborers who worked under dangerous conditions for very low pay. If blacks were treated better, white individuals and companies could not keep getting richer and richer.

So when hundreds of thousands of black workers went on strike in 1987, it was an attack at the heart of apartheid. As individuals, the black workers had no power, but when they banded together they were able to make a difference. They began to realize that there was hope of change in South Africa, while white employers feared that the old days of white control were forever gone.

The 1987 strikes in South Africa were more than the conflicts between laborers and owners that occur in many companies in other nations. The black workers were demanding not only better wages but also human dignity and freedom. They had found a way of protest between peaceful demonstrations and guerrilla warfare. Their new unity brought hope that they could keep working together to bring equality for South Africans of all races.

William A. Pelz

Anthropologists Find Earliest Evidence of Modern Humans

A team of scientists dated a modern-looking Homo sapiens *fossil at ninety-two thousand years, more than doubling the length of time that modern humans had been known to exist.*

What: Anthropology
When: 1987-1988
Where: Qafzeh cave, near Nazareth, Israel
Who:

HELÈNE VALLADAS, a French scientist at the Centre des Fables Radioactivités in France

BERNARD VANDERMEERSCH, a French physical anthropologist and archaeologist

DOROTHY ANNIE ELIZABETH GARROD (1892-1968), an English archaeologist and physical anthropologist

SIR ARTHUR KEITH (1866-1955), an eminent Scottish anthropologist

THEODORE DONEY MCCOWN (1908-1969), an American archaeologist and physical anthropologist

RENÉ VICTOR NEUVILLE (1899-1952), a French archaeologist

Theories in Question

The origin of modern human beings has been a persistent and vexing problem for prehistorians. Part of the difficulty is that there is widespread disagreement over the relationship between modern humans and their closest extinct relative, Neanderthal man. Neanderthals differ from modern and some archaic humans by the extreme robustness of their skeletons and by their heavy brow ridges, extremely large faces, and long and low skull caps. Neanderthals were once believed to have lived from approximately 100,000 to 50,000 years ago, while modern humans had been thought to have existed for 40,000 to 50,000 years. Although Neanderthals were originally as-

sumed to have been the ancestors of modern humans, prehistorians now agree that they were too localized, too extreme, and too recent to have been forerunners of modern people.

Archaic fossil humans are less robust and make better candidates as ancestors of modern humans. These have been found in sub-Saharan Africa and also in the Middle East, where they overlap in both location and time with Neanderthals. No specimens of this type have been found in Europe.

Arguments regarding the origin of modern humans center on the issue of one ancestral group as opposed to many ancestral groups. One view assumes that modern humans evolved from many local archaic types, including Neanderthal. A variation of this perspective holds that, while the ancestors of modern humans may have originated in one locality, they interbred with local peoples they met as they spread throughout the world. Both these views hold that this intermixture of genes explains the physical differences between modern populations.

Opposed to this view is the single-origin perspective, which maintains that earlier humans were replaced completely by physically and technologically more advanced members of a new group. The most favored homeland for this new and improved type is sub-Saharan Africa, where there are early examples of possible forerunners of modern humans, but where there is no known example of Neanderthals.

A Bottleneck

The region called "Levant," which includes Israel, has been of considerable interest to holders of both of the theories just described because it is the only "land bridge" between Africa and the rest of the world. Any population moving from

2220

one region to the other had to pass through the Levant. This fact became particularly important when the first Neanderthal found outside Europe was discovered in Galilee in 1925.

Pursuing this lead, the English archaeologist Dorothy Annie Elizabeth Garrod excavated a series of caves on Mount Carmel, now in Israel, between 1929 and 1934. She was assisted by a young American, Theodore Doney McCown. In two of these caves, Tabun and Skhul, McCown discovered human remains with stone tools that had characteristics associated with Neanderthal remains.

Back in England, McCown worked with the eminent Scottish anthropologist Sir Arthur Keith to analyze the bones. The fossils at Skhul, although archaic, resembled modern humans in most characteristics; those at Tabun resembled Neanderthals, although these skulls' features were less exaggerated than those of classic Neanderthals.

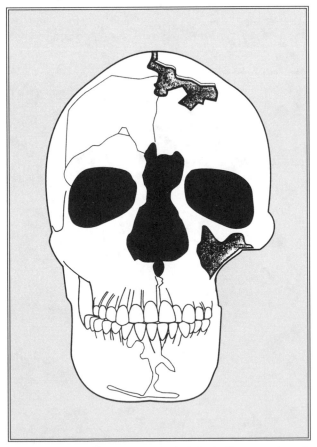

An artist's rendering of the Qafzeh 9 skull, approximately ninety-two thousand years old.

Overlapping with Garrod's excavations at Mount Carmel were those by René Victor Neuville from the French consulate at Jerusalem. Neuville excavated the Qafzeh cave between 1933 and 1935, finding the remains of five individuals. Unfortunately, World War II (1939-1945) intervened, followed by the Israeli-Arab conflict of 1947. Neuville died without analyzing his material. From 1965 to 1975, the French archaeologist and physical anthropologist Bernard Vandermeersch continued Neuville's excavations, finding the remains of eight individuals who resembled the non-Neanderthals from Skhul.

The meaning of these remains was interpreted variously. McCown believed that the Mount Carmel population was in the process of diverging into two groups from a more generalized ancestor and that neither were modern humans. Others thought that the fossils represented a cross between Neanderthals and modern humans. A few others thought that Neanderthals had been caught in the act of evolving into modern humans.

A major problem in making sense of the Levant fields lies in the inaccuracy of dates. Radiocarbon dating methods do not help because they are inaccurate for sites as old as Qafzeh, Tabun, or Skhul. Until recently, all that could be known was that humans of some sort had been in the Levant more than sixty thousand years ago and had lived there for an undetermined time.

Another method of dating, called "thermoluminescence," helped to clarify the dates. Thermoluminescent dating is used on objects such as pottery that were "fired," or heated during the time that they were used. When such objects are heated again in the laboratory, photons are released, producing thermoluminescence, or glow. The longer ago the object was fired, the more glow results. The greater the glow, the older the object. Thermoluminescence can be used to date much older material than can radiocarbon methods; unfortunately, it is not as accurate as radiocarbon.

The first objects to be dated by thermoluminescence in the Levant were burnt flints from the Neanderthal sites at Kebara. The dating was done by a French-Israeli team headed by Helène Valladas, with results being published in 1987.

2221

The Neanderthal site was dated at sixty thousand years, meaning that if the date is correct, Neanderthals were in the Middle East much later than had been thought.

In 1988, a team led by Valladas published a thermoluminescence date of ninety-two thousand years from Qafzeh. If this date is correct, then there were forerunners of modern humans living in the Levant twice as long ago as had been suspected. Furthermore, these individuals were there either before or at the same time as the Neanderthals.

Consequences

Since thermoluminescence gives only a rough estimate, confirmation by another form of dating is desirable. In the meantime, there have been two dominant reactions by scientists. Those subscribing to the single-origin, "out-of-Africa" model see the Qafzeh date as confirmation of this hypothesis.

Others, such as the American Milford Wolpoff, dispute this assessment. Wolpoff believes that Neanderthals contributed to the genetic makeup of modern Europeans. He points out that the late Neanderthals in Europe are more like modern Europeans in some respects than are the more modern-looking fossils from Skhul or Qafzeh.

The date from Qafzeh raises many questions, which will not be answered quickly. Perhaps the most perplexing is: If genetically modern humans appeared at least 92,000 years ago, why did it take more than 50,000 years for modern-appearing culture, such as art, to arrive?

Lucy Jayne Botscharow

Surgeon General Koop Urges AIDS Education

> *To the surprise of critics and supporters alike, Surgeon General C. Everett Koop urged television stations to broadcast commercials for condoms as part of the campaign to educate Americans about AIDS (acquired immune deficiency syndrome).*

What: Medicine; Social change
When: February 10, 1987
Where: Washington, D.C.
Who:

C. EVERETT KOOP (1916-   ), surgeon general of the United States from 1981 to 1989

RONALD REAGAN (1911-   ), president of the United States from 1981 to 1989

WILLIAM BENNETT (1943-   ), secretary of education from 1985 to 1988

Values in Conflict

When President Ronald Reagan nominated C. Everett Koop as surgeon general of the United States in 1981, it was seen as a victory for Reagan's conservative (right-wing) supporters. Koop had come to President Reagan's attention because of his strong stand against abortion.

At his confirmation hearings in the Senate, Koop was questioned about both his medical career and his attitude toward abortion. Liberal Democrats feared that Koop would use his position as surgeon general to advocate conservative attitudes toward abortion and homosexuality instead of providing objective information on health issues. Koop was confirmed only after he promised not to use the power of his office to oppose abortion.

As surgeon general, Koop was anything but a puppet for the president or the president's conservative supporters. For example, he spoke up much more forcefully against smoking (and therefore against the tobacco industry) than oth-

ers in the Reagan administration did. Yet it was his position on AIDS (acquired immune deficiency syndrome) that made Koop famous.

The controversy over AIDS was part of a long conflict over social values that went back to the late 1960's. During that time Americans became divided over civil rights, prayer in public schools, poverty, the Vietnam War, drug use, women's rights, abortion, and sexual liberation. Some Americans wanted to reject traditional values, which to them seemed corrupt and hypocritical. Others believed that the new values and lifestyles would cause political and social problems in the United States. Most Americans, however, were somewhere in the middle.

When Ronald Reagan was elected president in 1980, he was supported by many moderates—but also by many people who desperately wanted to return to the traditional values that had been rejected in the 1960's. Trying to satisfy both groups, the president found himself walking a political tightrope. Koop's appointment was intended to please Reagan's right-wing supporters.

Somewhere along the line, however, Koop seemed to change sides, especially regarding AIDS. Many conservatives saw the AIDS epidemic as a punishment for homosexuals, since in the United States the disease first spread quickly among male homosexuals. But Koop did not use the AIDS problem as a chance to preach about traditional values.

Koop Takes a Stand

On February 10, 1987, Koop urged television stations to broadcast condom advertisements as part of the nation's campaign to control AIDS. Speaking at a hearing in Congress, Koop said

2223

that condoms gave the best protection against AIDS for people who would not abstain from sex or have sex with only one faithful partner.

Koop was especially concerned that condom advertising reach African American and Hispanic people in the United States, because larger percentages of people in these groups suffer from AIDS. He also recommended that commercials contain clear instructions on how condoms should be used.

In the previous year, Koop had written a report on AIDS. After interviews with doctors, teachers, and AIDS patients, Koop recommended strongly that education be the main tool for fighting the spread of AIDS. This educational effort, he said, should include general information about human sexuality and should begin at the earliest age possible, maybe even in kindergarten.

Other members of the Reagan administration were unhappy with Koop's report and his

recommendations to Congress. Education secretary William Bennett was especially displeased; he said that Koop, who was a devout Christian, should make a proposal that fit better with his religious beliefs. Bennett and others in the administration believed that everyone at risk for AIDS should have to be tested for the syndrome, and that the names of those who had AIDS should be made public. Republican members of Congress joined in the criticism, saying that Koop was encouraging irresponsible sexual behavior.

Yet Koop's proposals also drew some enthusiastic support. Many liberal Democrats said that Koop was being realistic and compassionate toward AIDS victims. So Koop's career had taken a surprising turn: He had angered his conservative supporters, while the liberals who had opposed his nomination now made him a hero.

Koop's answer to his critics was simple. He said that, as surgeon general, he had to care for

C. Everett Koop (second from left) visits Coming Home Hospice, a facility housing terminal AIDS patients, in the Castro district of San Francisco in March, 1987.

the health needs of all Americans, not only those who shared his religious and social values. That was why he had tried to help even those whose sexual decisions he did not agree with.

Consequences

As the nation's surgeon general, Koop could not make his ideas into rules or laws; he could only inform people and encourage them to make healthy decisions. He could not make tele-vision stations use advertisements for condoms; he could not force people to use condoms.

Koop did help to educate the American public about AIDS and safe sex. After some hesitation, some television stations did run condom commercials, taking the risk of losing sponsors and angering viewers. Yet Koop's message did not bring all the changes he had hoped for. The AIDS epidemic continued to be a public health problem in the United States and abroad.

Ira Smolensky

Supernova 1987A Supports Theories of Star Formation

Observation of supernova explosion and measurement of neutrinos reaching Earth confirmed theoretical physics of star structure and evolution.

What: Astronomy
When: February 23, 1987
Where: Cerro Tololo InterAmerican Observatory, Chile
Who:

TYCHO BRAHE (1546-1601), a Danish astronomer who observed a supernova with the naked eye in 1572

JOHANNES KEPLER (1571-1630), a German astronomer who observed a supernova with the naked eye in 1604

IAN SHELTON, an astronomer who discovered a supernova with the naked eye on February 23, 1987

ALBERT JONES, an amateur skywatcher who observed a supernova at the same time as Shelton

A Rare Event

Supernova 1987A (SN1987A) appeared as a bright point of light in the Large Magellanic Cloud, one of a pair of small galaxies some one hundred thousand light-years from the Milky Way and situated for earthbound observers high in the southern sky. It was bright enough to be seen with the naked eye; an astronomer walking outside Cerro Tololo InterAmerican Observatory in Chile spotted the phenomenon almost by accident, the first supernova visible to the unaided eye in nearly four centuries. Ian Shelton of the University of Toronto southern station in Chile observed a supernova on February 23, 1987; Albert Jones, an amateur skywatcher, viewed the supernova with the naked eye almost simultaneously with Shelton. (The Danish astronomer Tycho Brahe was on hand to describe one in 1572, and the German astronomer Johan-

nes Kepler observed a supernova in 1604.)

Supernovas occur only at the end of the life cycle of very massive stars. When a massive star depletes its supply of hydrogen, it collapses, and its internal heat and pressure increase to the point where helium is converted to carbon. Elements with increasingly higher atomic numbers are formed as the collapse continues. Once the core is the element iron, the process cannot continue until more energy is added. At this point, the outer layers hit the core and bounce. The star explodes, sending a large portion of its mass into space. The remainder of the supernova collapses to become a neutron star or a black hole, depending on its mass.

At one point during the initial collapse, the ever-increasing gravitational force actually squeezes electrons into nuclei, forming an almost unimaginably dense neutron core. Neutrinos, nearly noninteractive particles with little or no mass, are driven away from the star by this process, carrying energy with them and thus hastening the star's gravitational collapse.

Theory-Confirming Observations

Inasmuch as nearly all theory in nuclear physics and astrophysics developed in the twentieth century, sighting of the first visual supernova in four hundred years offered an unprecedented opportunity to test ideas of star behavior. Within hours of the first sighting of Supernova 1987A, scientists all over the world turned their attention to it. The first problem was to identify which star was involved. Since supernovas are extremely hot in their early stages, they should radiate strongly in ultraviolet wavelengths. The International Ultraviolet Explorer satellite, already in orbit, confirmed quickly that the supernova had been discovered in an early stage but was cooling

2226

rapidly. Astronomers identified the exploding star as a supergiant of about 50 solar masses.

Supernovas release nearly all of their explosive energy in the form of neutrinos. The upsurge of neutrinos from SN1987A should have been detectable on Earth. Because of their physical characteristics, almost all neutrinos reaching Earth actually pass through the planet as if it did not exist. A few, however, collide with particles. These collisions may be detected in a number of large tanks constructed deep underground to protect them from cosmic radiation and filled with water or chlorine-rich liquids. Occasionally, neutrinos collide with atoms in the water and generate minute flashes of light or, in the case of chlorinated liquids, impact with a chlorine atom and create radioactive argon. Although these underground detectors originally were constructed merely to ascertain the existence of neutrinos in the universe, they were excellent facilities for detecting incoming neutrino bursts from SN1987A.

Neutrino detectors installed deep in an abandoned South Dakota gold mine, in a salt mine near the southern shore of Lake Eerie in Ohio, in Europe, and in Japan all detected an upsurge in these rare neutrino collisions, confirming crucial aspects of astrophysical theory. Measurements of the energy possessed by neutrinos detected in these underground tanks permitted astrophysicists to reconstruct in considerable detail events in the final stages of the collapse and explosion of the supergiant. SN1987A is estimated to have given off more energy when it exploded than an entire galaxy of average size emits in a year. Temperatures in the core at the time of explosion must have exceeded 10,000,000,000 degrees.

Consequences

As a phenomenon that confirmed a large and critically important series of astrophysical theories developed over the last century, SN1987A was an extremely significant and reassuring event. Physical models of the universe, which depended heavily upon these theories, could be used with much greater confidence.

SN1987A represented an unusual opportunity to determine whether the neutrino possesses mass. This question illustrates the many interconnections between theories of stellar evolution and cosmology. Scientists already realized that neutrinos were nearly without mass; an electron carries at least twenty-five thousand times the mass of a neutrino. Nevertheless, theory predicts that neutrinos are about one hundred million times as abundant as electrons in the universe, so that if they prove to have any mass at all, neutrinos become a significant portion of the overall mass of the universe. Determining this overall mass could help to settle the most basic debate in cosmology: Is the universe destined to expand forever, or does it contain sufficient mass for gravitational forces eventually to halt the expansion and cause contraction, presumably ending in another beginning similar to the big bang

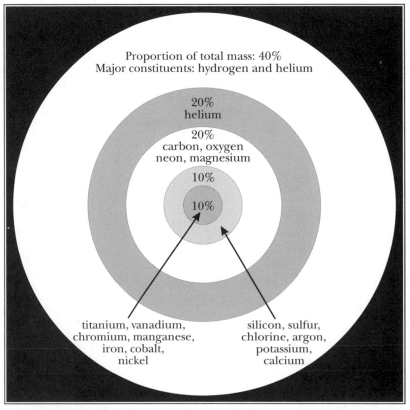

Proportion of total mass: 40%
Major constituents: hydrogen and helium

20%
helium

20%
carbon, oxygen
neon, magnesium

10%

10%

titanium, vanadium,
chromium, manganese,
iron, cobalt,
nickel

silicon, sulfur,
chlorine, argon,
potassium,
calcium

A supernova is born at the end of a massive star's life.

event thought by most cosmologists to have given birth to the present universe?

Unfortunately, neutrino measurement experiments did not yield conclusive results. Theoretically, if the neutrino does not possess mass, all neutrinos should travel at the speed of light and, according to relativity theory, arrive at a given observation point at the same time. If they possess even a minute mass, those with more energy should arrive slightly ahead of others. All neutrino detections from SN1987A occurred within a span of less than ten seconds. The different time of arrival could have been caused by differing energy levels for particles with mass, but also by processes of neutrino emission from the supernova that, argues the theory, probably take

about ten seconds to occur. Smaller differences also could have been caused by differing locations of the detectors. SN1987A added greatly to the store of data on neutrinos, but three years after sighting the explosion, physicists still had not solved the mystery of neutrino mass.

SN1987A quickly became one of the most intensively studied astronomical phenomena in history and continued to provide an additional wealth of information as later phases of the event unfolded. Through the end of the 1990's, it ranked as the closest Supernova ever observed through telescopes. It is likely to remain a focus of considerable research well into the twenty-first century.

Ronald W. Davis

Gallo and Montagnier Are Recognized as Codiscoverers of HIV

> *An American researcher and his French rival were declared codiscoverers of the virus that causes AIDS.*

What: Medicine
When: March 31, 1987
Where: Bethesda, Maryland, and Paris, France
Who:
ROBERT C. GALLO (1937-   ), the leading U.S. AIDS researcher
LUC MONTAGNIER (1932-   ), a French medical researcher
JONAS E. SALK (1914-1995), an American medical research scientist and Nobel laureate

A New Virus

Robert Gallo's initial medical research at the National Cancer Institute (NCI) in Bethesda, Maryland, focused on retroviruses, mystifying organisms that present mirror images of themselves genetically. As early as 1910, such retroviruses were identified as a cause of cancer in hens. Later, scientists established a link between retroviruses and cancer in other animals.

During the early 1970's, Gallo, spurred by generous governmental funding for medical research that could lead to a cure for cancer, searched for a link between retroviruses and cancer in humans. By 1975, he had isolated HL-23, a human retrovirus taken from the blood of a leukemia patient. Others in the scientific community questioned and disparaged his findings when they were unable to replicate his results; such critics suggested that his results were invalid, possibly because of contamination of Gallo's sample.

Late in 1978, researchers at the NCI discovered an atypical T-cell cancer in the lymph node of a patient and identified it as reverse transcriptase—seemingly a retrovirus—which they cultured and grew for the next two years. They tried to find the same retrovirus in other patients; two, from a large patient population, also tested positive for it. Gallo labeled this virus—human T-cell lymphoma virus—"HTLV." It seemed identical to the adult T-cell lymphoma virus, ATLV, isolated in Japan.

Controversy and Recognition

In 1981, the acquired immune deficiency syndrome (AIDS) was becoming a medical fact of life in the United States. James Curran of the Centers for Disease Control talked about it at the National Institutes of Health (NIH) in that year. Gallo, focusing on his own research, however, had little interest in what Curran presented. When Curran returned to the NIH in 1982 with the suggestion that AIDS attacks T-cells, Gallo became intrigued and involved, surmising that a link might exist between AIDS and HTLV, which also attacks T-cells.

Early in 1983, Luc Montagnier of the Pasteur Institute in Paris called Gallo to share information about a retrovirus identified in an AIDS patient there. Montagnier requested that Gallo send him some antibodies of HTLV for comparison with this patient's cells. This marked the beginning of what ultimately led these two scientists to the joint discovery of the human immunodeficiency virus (HIV).

The discovery of HIV—and the assignment of credit for it—was marked by considerable rancor on the parts of the two men most directly involved in the research. Gallo, obviously pursuing a Nobel Prize, was accused of plagiarizing important research findings related to the Pasteur Institute's lymphadenopathy-associated virus (LAV). The personal and professional relationship between Gallo and Montagnier cooled. Gallo publicly badgered Montagnier when he

2229

presented some of his findings at international scientific meetings.

Despite this contention, after considerable public wrangling and after revelations that cast suspicion on Gallo's scientific honesty, Jonas E. Salk, representing a group of Nobel Prize winners, became a shuttle diplomat between the United States and France, seeking some accord in this discordant affair. On March 31, 1987, U.S. president Ronald Reagan and French prime minister Jacques Chirac made a joint announcement that made public a French American agreement declaring Gallo and Montagnier codiscoverers of the human immunodeficiency virus. This meant that the royalties from all HIV blood tests would be shared equally by both scientists.

Consequences

Despite the acrimony involved in ascribing credit for the discovery of HIV, the achievement was monumental. It provided a vital link for dis-

covering the means of preventing, treating, and controlling one of the most threatening of human diseases. The isolation of HIV enabled subsequent researchers to focus upon means of dealing with the virus. Although AIDS has not yet been defeated medically, the identification of HIV provided the necessary initial step.

In *AIDS and Its Metaphors* (1989), the writer Susan Sontag noted that AIDS has become the most dreaded disease of its day, replacing cancer—many forms of which are now under medical control—as the seemingly incurable disease of the 1980's and 1990's. AIDS has a maximum incubation period of a decade or more. Those detected as HIV-positive face not only the almost certain possibility of developing full-blown AIDS but also many forms of discrimination in the workplace, in seeking health insurance, and in countless interpersonal relationships.

Without the discovery of HIV, nevertheless, the research path that may ultimately lead to immunization against the disease and cures for it would undoubtedly be blocked. For HIV-positives, the discovery has in many cases been a mixed blessing. Many equate their status with certain death when the seemingly inevitable onset of AIDS occurs. On the other hand, knowing their status enables HIV-positives to embark on courses of treatment that can delay the onset of AIDS.

As investigation into AIDS has progressed, the relationship between HIV and AIDS has been questioned. Studies, however, have concluded that HIV infection is an absolute concomitant for the development of AIDS. Some respected medical researchers consider such evidence specious, however, arguing that it is impossible to distinguish between HIV and other causes that may trigger AIDS. Such researchers seek absolute evidence, presently unavailable, for biochemical activity of HIV in AIDS.

Although quibbling over salient details continues, the impact of the Gallo/Montagnier discovery of HIV has been enormous. Their findings have been an important first step in understanding what causes AIDS and in moving toward the resolution of modern medicine's most perplexing dilemma.

R. Baird Shuman

AP/Wide World Photos

Robert C. Gallo.

Iraqi Missiles Strike U.S. Navy Frigate *Stark*

While on patrol in the Persian Gulf, a U.S. Navy frigate was struck and seriously damaged by missiles fired from an Iraqi warplane.

What: Military
When: May 17, 1987
Where: Persian Gulf
Who:
RONALD REAGAN (1911-   ), president of the United States from 1981 to 1989
GLENN BRINDEL (1943-   ), commander of the *Stark*
SADDAM HUSSEIN (1937-   ), president of Iraq from 1979

Conflict in the Gulf

In September, 1980, the surface calm of the Persian Gulf region was shattered when Iraq, led by President Saddam Hussein, invaded Iran. Iraqi forces quickly captured about ninety square miles of disputed territory in the oil-rich border province of Khuzistan. The invasion was the beginning of the Iran-Iraq War, which raged from 1980 to 1988. This war was very costly not only to the two warring nations but also to shipping in the gulf region.

The fighting spread into the Persian Gulf itself in 1984, when Iraq attacked oil tankers that were loading at Iran's Kharg Island. Iran struck back at tankers in Saudi Arabian and Kuwaiti ports. Yet Iraq was responsible for more than two-thirds of the attacks on civilian ships (mainly oil tankers) during the war.

Over the years of the war, Iran and Iraq fought almost to a standstill, with neither side in the conflict able to make major gains in territory. Both sides suffered many casualties—hundreds of thousands of people died or were wounded. The war also did great damage to the economy of both countries.

Fearing attacks on its oil tankers, the government of Kuwait appealed to the United States for protection. President Ronald Reagan had become concerned that Iran would attack Kuwaiti ships, and his administration began negotiating a plan to place these tankers under the U.S. flag so that they could be escorted and protected by American warships. Seven American ships were assigned to patrol the Persian Gulf as part of the U.S. Middle East Force.

An Accidental Attack?

On May 17, 1987, the U.S. Navy frigate *Stark* was struck in the Persian Gulf by missiles fired by an Iraqi warplane. Thirty-seven American sailors were killed.

The *Stark*, a guided missile frigate, weighed 3,600 tons and carried a crew of about 220. At the time of the attack, it was on patrol in international waters about eighty-five miles (135 kilometers) northeast of Bahrain Island. No other ships were nearby; in fact, the nearest commercial ship was about thirty-five miles (fifty-five kilometers) away.

Although Iraq had often attacked oil tankers at the northern end of the gulf, around the Iranian port at Kharg Island, it was rare for Iraqi aircraft to launch attacks in the southern gulf. Also, this was the first time that Iraq had attacked any U.S. ship.

According to Pentagon reports, an Iraqi Mirage F-1 fighter took off from an air base in Iraq and flew south along the coast of Saudi Arabia. After turning toward the east, the Mirage came down; at a distance of about ten miles (sixteen kilometers) from the ship, it fired at least one Exocet missile.

The *Stark*'s radar and an AWACS (Airborne Warning and Control System) radar plane both tracked the approach of the Iraqi jet. Following international custom, the *Stark* radioed the Iraqi jet twice, warning it to steer clear. But before turning away, the Mirage launched an Exocet

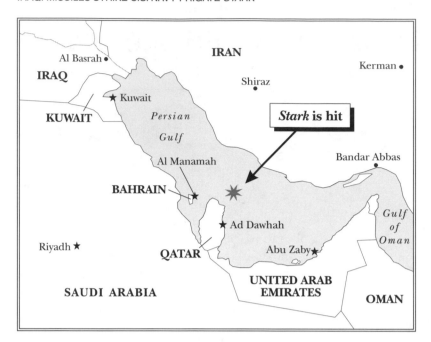

The *Stark*'s Phalanx gatling gun, which could fire 20-millimeter ammunition at the rate of three thousand rounds per minute, could probably have intercepted the Exocet. But Captain Brindel explained that in order to avoid accidentally shooting down friendly aircraft, his crew had left the Phalanx in the manual position rather than the radar-controlled automatic position. Brindel said that the first warning of an attack was given by a crew member on lookout, who spotted the Iraqi missile as it approached the ship.

Consequences

An investigation following the attack on the *Stark* raised questions about why the ship had not defended itself. It was not clear whether the electronic gear on board the ship had failed or whether the commanding officer or crew members had made mistakes.

President Reagan strongly protested the attack, and U.S. warships in the Persian Gulf area were put on a higher level of alert. Calling the attack an accident, Iraqi president Saddam Hussein made an apologetic statement.

Officials of the U.S. State Department informed a congressional subcommittee that the United States and Iraq would cooperate in investigating the incident. On May 22, 1987, at a naval base in Mayport, Florida, President Reagan and his wife led a memorial service for the thirty-seven sailors who had been killed on the *Stark*.

Francis Poole

missile. About ninety seconds passed between the time the missile was launched and the moment it struck the ship.

The Exocet hit the *Stark* below the bridge and landed in a crew sleeping compartment, where it exploded. Many of the victims of the explosion and the fire that followed had been asleep in their bunks. The *Stark*'s crew acted quickly to contain the flames and extinguish the fire; the ship was then towed to port in Bahrain.

On May 20, Captain Glenn Brindel described the attack. He admitted that the *Stark*'s sophisticated electronic systems had not detected the launch of the Exocet. He said that the ship had been surprised by the attack because in the past Iraqi warplanes had been considered friendly to American ships in the gulf area.

Hundreds Die in Riots at Mecca's Great Mosque

At the Great Mosque in Mecca, a clash between Saudi Arabian police and rioting pilgrims from Iran resulted in 402 deaths and hundreds of injuries—and tensions in the Persian Gulf region grew even worse.

What: International relations
When: July 31, 1987
Where: Mecca, Saudi Arabia
Who:
AYATOLLAH RUHOLLAH KHOMEINI (1902-1989), rahbar (religious leader) of Iran from 1979 to 1989
FAHD IBN ABDUL AZIZ (1923-   ), king and prime minister of Saudi Arabia from 1982

An Age-Old Dispute

The conflict between Sunni and Shiite Muslims dates back to the seventh century C.E. About 90 percent of Muslims around the world are Sunni, but Iran is mostly Shiite. Throughout history, Sunnis persecuted Shiites as heretics; the Sunnis tend to adapt their beliefs to changing times, while the Shiites are radicals who are willing to die to preserve their ancient faith.

After the Ayatollah Ruhollah Khomeini came to power in Iran in 1979, Iran tried to take its radical Islamic beliefs to Lebanon and to Persian Gulf countries where there were minority communities of Shiites. In the Iran-Iraq War of 1980-1988, most Arab nations backed Iraq rather than Iran, whose people are mostly Persian.

Some of the most holy places of Islam are in Saudi Arabia, a large nation dominated by moderate Sunni Muslims. Khomeini urged his followers to work together and seek "deliverance from the infidels." Iranian leaders called on Muslims to take Islamic holy sites from the Saudi government; they said that the Saudi monarchy (and other monarchies in the Persian Gulf region) needed to be overthrown.

Because Saudi Arabia feared that Iran would win its war with Iraq, the Saudis gave financial help to Iraq. This move made Iran even more angry. Concerned about its security, Saudi Arabia bought huge quantities of weapons from the United States to build up its armed forces. Again, Khomeini pointed to these purchases of weapons to show his followers that the Saudis were friendly to Western "infidels," or unbelievers.

In 1979, the Great Mosque in Mecca, Saudi Arabia, had been seized by about two hundred Muslim extremists. During a two-week siege, more than two hundred people were killed. The Saudi government was shaken by this event, for it showed that many Saudi Arabians were critical of the great power and wealth of the Saudi royal family. It does not seem that Iran was involved in this incident, but the Saudis were very concerned about the threat posed by Shiites from Iran.

Riots and Reprisals

Each year, Muslims from various countries stream into Mecca for the hajj, a special time of pilgrimage. During the 1980's, there were some reports of Iranian pilgrims trying to smuggle weapons and explosives into Saudi Arabia during the hajj, but the Saudis chose to deal with these cases quietly.

In 1987, Iranians made up the single largest group in the pilgrimage to Mecca: 155,000 out of a total of 2 million. On July 31, 1987, thousands of Iranian pilgrims broke Saudi Arabia's ban on political protests during the hajj. They blocked streets, held up banner-sized photographs of the Ayatollah Khomeini, burned effigies of United States president Ronald Reagan, and chanted, "Death to America! Death to the Soviet Union! Death to Israel!"

When Saudi police tried to break up the crowd with water cannons, the demonstrators threw rocks, turned over cars, and—according to police reports—stabbed bystanders and security officers with knives they had hidden in their clothing. According to the Saudis, the Iranian protesters stampeded in their retreat, so that hundreds, including many women and children, were killed.

Iranian officials claimed that the demonstrators were protesting peacefully until the police began using automatic weapons and suffocating gases against them. Saudi officials, however, denied that they had used firearms. By the end of the riot, 402 people had died: 275 Iranians, 85 Saudi police officers, and 42 pilgrims from other countries. Hospitals reported that 649 people had been injured.

Immediately after the riots, Iran claimed that the Saudis had provoked the fighting in response to a request from the United States. The ayatollah often made such accusations; his favorite name for the United States was "the Great Satan." On August 1, mobs in Iran's capital, Tehran, attacked the embassies of Saudi Arabia and Kuwait in revenge for the Iranian deaths. The French embassy was also stoned.

It was hard for outsiders to figure out exactly what had happened during the Mecca riot, since Saudi borders are closed off during the hajj and non-Muslims are not allowed in Mecca. When the first Iranian planeload of injured and dead pilgrims arrived in Tehran on August 5, the Iranian news agency reported that the survivors said they had been beaten and shot at by Saudi policemen. One man claimed that he had been hit

by fragments of explosive, soft-nosed bullets.

Lebanese pilgrims returning to Beirut on August 6 told reporters from *The Washington Post* that they had heard gunfire during the riot. Still, the United States and most Arab countries supported Saudi Arabia's version of the events and publicly criticized Iran for causing unrest.

Consequences

The riots in Mecca increased tensions in the Persian Gulf region, where the United States had sent warships to protect Kuwaiti oil tankers from expected Iranian attacks. The U.S. State Department accused Iran of "escalating tension, intimidation and destabilization" in the gulf region and promised that the United States would continue to escort Kuwaiti tankers under the American flag.

Saudi Arabia's King Fahd was said to have received messages of support from the leaders of Egypt, Jordan, Morocco, North Yemen, Sudan, Bahrain, Kuwait, Iraq, and other countries. Even

Syria, a country that had backed Iran in the Iran-Iraq War, expressed its sympathy to the Saudi king. Only the Shiite community in Lebanon stood up for Iran.

With most Arab nations united against him, Khomeini's position began to weaken. Within a year he was forced to accept a United Nations-sponsored cease-fire in the war with Iraq. The war had completely wrecked Iran's economy.

In 1988, the Saudis set up a quota system that limited the number of Iranian pilgrims attending the hajj, and the next year Iran refused to send any pilgrims at all. Yet this did not keep terrorism out of Mecca. In September, 1989, Saudi Arabia executed sixteen citizens of Kuwait for acts of terrorism, including two bombings in Mecca during hajj in July. The Saudi government said that Iranian diplomats in Kuwait had recruited and trained the terrorists and supplied their explosives.

JoAnn Balingit

Miller Discovers Oldest Known Embryo in Dinosaur Egg

A team led by Wade Miller discovered a fossilized dinosaur egg containing the oldest known embryo of any kind, about 150 million years old.

What: Earth science
When: September, 1987
Where: Cleveland-Lloyd Dinosaur Quarry, Emery County, Utah
Who:

WADE MILLER (1932-   ), an American geologist from Brigham Young University

KARL HIRSCH, an American paleontologist

KENNETH L. STADTMAN, an American geologist from Brigham Young University

JAMES H. MADSEN, JR. (1932-   ), an American paleontologist

Dinosaur Nests

The Cleveland-Lloyd Dinosaur Quarry in Emery County, Utah, is where more than twelve thousand different bones have been found to represent more than seventy individual dinosaurs, of at least twelve different kinds. The huge amounts of remains found in the area can be explained by the possibility that the area may once have been a marsh or a shallow lake in which the huge reptiles became stuck.

This quarry is where two geologists from Brigham Young University, Wade Miller and Kenneth L. Stadtman, along with colleagues Karl Hirsch and James H. Madsen, Jr., are credited with finding a fossilized egg from the 100-million-year gap in the fossil record in geologic time between the Lower Jurassic period and the upper Lower Cretaceous. The egg, from the Upper Jurassic period, contains an embryo that is approximately 150 million years old—the oldest known embryo yet found of any kind of living creature. The egg was discovered among the thousands of other dinosaur bones instead of in a nest.

Such nests were first found in great concentration on a Montana cattle ranch at the eastern boundary of the Two Medicine formation. Nests are small hollows, or concave depressions, believed to be the location in which the eggs were possibly laid, hatched, and nurtured for a year or more by the mother and/or father. Several of these hemisphere-like depressions, each of which contained the bones of baby dinosaurs, all at the same stage of development, were found.

An interesting speculation that has arisen from the idea of nests is that dinosaurs were warm-blooded, not cold-blooded like modern reptiles. Modern reptiles take a long time to grow; if the dinosaurs were cold-blooded, the growth process would keep them in the nest for as long as a year. In comparison, warm-blooded creatures, such as birds, grow very quickly and do so while in the nest. The evidence seemed to indicate that these baby dinosaurs were growing in the nest and at a fast pace.

Old Egg, New Dinosaur

The fact that the egg from the Utah site containing the embryo was not found in a nest, along with two other factors, suggests that the egg was retained in the oviduct, or tube, through which the egg passes from the ovary during birth. This is known as "oviductal retention." The theory here is that the mother for some reason was unable to lay her egg in the nest. The excellent condition in which the egg was found is perhaps a result of its being held together in the oviduct after fracturing until it was preserved by sediments.

Second, the distortion of the shell and its inverted curvature are both indications that, at some point during the breakage and deforma-

2236

tion, the shell was at least semipliable. The only place that can happen is in the oviduct.

Third, the theory of oviductal retention is suggested by the multilayered structure of the shell. In the case of the egg found by Miller, the primary, or original, shell is perfectly preserved and coated over by another layer called the "pathological" (or disease-related) layer. Such secondary layering may have occurred because of pathological conditions arising during the egg's development in the oviduct. The process would thus have started during retention. These characteristics are found in the eggs of modern and fossil reptiles and mammals alike and can be caused by stress, disease, or other environmental conditions.

Once the added layer has formed, the embryo may live for a short time and even continue to develop, but the exchange of vital gases can be hindered by problems in the development of pores connecting the first and second layers. The end result, when that happens, is death by suffocation for the embryo. This is what appears to have taken place in the egg found by Miller.

The egg cannot be identified either by the known species and genera of dinosaurs or by the other dinosaurs found in the dig. In addition to finding the oldest known embryo, the team of four paleontologists may have found a new dinosaur.

Wade Miller.

Consequences

The discovery of a dinosaur embryo in the fossil beds of the Cleveland-Lloyd Dinosaur Quarry added to modern knowledge in the field of paleontology. This oldest known and complete embryo, preserved perfectly for 150 million years, gives researchers a chance to learn more about the process of egg formation and dinosaur embryo development. In addition, the fact that the dinosaur cannot be classed with any of the other dinosaurs found at the same site is an indication that the egg might be of a new type of dinosaur.

The state of the embryo from Utah, which is most likely the result of oviductal retention, is a new addition to the study of egg retention and pathological eggshell formation. Because there has been no systematic study of such phenomena, these are processes that are still being questioned; however, because oviductal retention still occurs in the mammals and reptiles of the modern era, there is good information available on the possible cause of the abnormality.

The discovery of dinosaur eggs in their entirety, eggshells that are pathological, and remains of embryos are all extremely rare, especially from the time before the Cretaceous. The discovery in Utah has had a great impact on the world of paleontology, especially in view of the possibility of a new dinosaur in the discovered fossilized embryo.

Earl G. Hoover

2237

Palestinian Intifada Begins

Beginning in the Jebalya refugee camp and then in the rest of the Gaza Strip and the West Bank, Palestinians joined in the Intifada, or uprising, to resist the twenty-year Israeli occupation.

What: Civil strife
When: December, 1987
Where: The West Bank and the Gaza Strip
Who:
YITZHAK RABIN (1922-1995), defense minister of Israel from 1987 to 1989
YITZHAK SHAMIR (1915-   ), prime minister of Israel from 1983 to 1992
YASIR ARAFAT (1929-   ), leader of the Palestine Liberation Organization

Chafing Under Occupation

In 1967, Israel began occupying the West Bank and the Gaza Strip, lands that had previously been governed by Egypt. The residents, mostly Palestinian Arabs, found their lives changing as Israel started to bring the economy of these areas under its control. Roads and power systems were constructed to link the territories with Israel, and Palestinians were paid poorly to work as unskilled laborers. Soon most of the products imported to the territories came from Israel.

After the Likud Party came to power in Israel in 1977, the elected mayors of the West Bank and Gaza Strip were replaced with Israeli military officials. More and more land was taken by Israel, and Jewish settlements were built quickly. Though Israel was profiting from its control over the West Bank and Gaza Strip, it did not provide good education or health care for the Palestinian residents.

On August 4, 1985, Israeli defense minister Yitzhak Rabin announced his "iron fist" policy. The Palestinians became virtual prisoners, for to travel abroad they had to obtain permits. More and more Palestinians were put in jail, detained without charges or trial, or forced into exile. Fur-thermore, strict curfews were imposed, many schools were closed, and Palestinians' homes were demolished to make way for more Jewish settlers.

The Palestinians believed that Israel intended to drive them out. In 1982, Israel had invaded Lebanon and had forced the Palestine Liberation Organization (PLO) out of that country. As a result, Palestinians living in Lebanese refugee camps had been left without protection—and many of them had been massacred by a right-wing Lebanese militia.

On December 4, 1986, Israeli soldiers shot dead two Birzeit University students who happened to be from Gaza. On April 10, 1987, another Birzeit student from Gaza was killed. Then, on October 1, the Israeli army ambushed and killed seven Gaza men believed to be members of the Islamic Jihad (Holy War) movement. Five days later, four more Gaza men and an Israeli prison official were killed in a shootout. Several days later, a Jewish settler shot a Palestinian schoolgirl in the back.

The Uprising

Demonstrations against Israel became more frequent, and Palestinian youths pelted Israeli cars with stones. On December 7, 1987, a Jewish merchant was stabbed in the city of Gaza. The next day, an Israeli tank transporter slammed into a line of cars loaded with Palestinian laborers; four Palestinians from the Jebalya camp were crushed to death, and seven others were injured.

That night, ten thousand Jebalya residents gathered to protest. The burial procession the next morning became another demonstration, and the army fired into the crowd. Another Jebalya resident was killed—the first casualty of the Intifada.

Protests spread across the Gaza Strip, and by late December, Balata and other West Bank refugee camps had become involved. There were 288 demonstrations that month, two thousand arrests, and twenty-nine killed.

By mid-January, 1988, leaflets from the Unified National Leadership of the Uprising (UNLU) were being distributed. (Later it was revealed that the UNLU was a group of four activists representing four major groups within the PLO.) The leaflets, called *banayat*, gave instructions to the Palestinian people for confronting the military and forming action committees.

Intifada demonstrations were constant. Palestinian youths threw stones at Israeli soldiers, who responded with tear gas, rubber-coated or plastic-coated bullets, and sometimes live ammunition. Rabin began allowing soldiers to beat the demonstrators; some of these brutal beatings were televised across the world. In April, 1988, there were 416 demonstrations, and sixty-one Palestinians were killed.

Another Intifada method was strikes by Palestinian merchants. At first the Israelis tried to force the merchants to reopen their shops, but the Palestinians were persistent; from March, 1988, onward, the shops were open for only three hours each day.

The UNLU encouraged each neighborhood to set up committees to distribute food and other necessities. Youths cleared, plowed, and planted every available scrap of land so that the Palestinians could grow their own food. Eventually, the UNLU called for a boycott of all Israeli products that the Palestinians could produce themselves.

The Israelis closed Palestinian schools and universities, but the Palestinians formed education committees to school their children. Health-care committees taught first-aid procedures and cared for the wounded and the sick. The "popular committees" actually became a system of government for the Palestinians, helping them to become independent from Israel.

Meanwhile, Israel used greater force to try to control the Palestinians. During the first two years of the Intifada, more than 200,000 Israelis served in the West Bank and the Gaza Strip. Yet the Palestinians were determined not to give in.

Consequences

The Palestinians in the occupied territories had become tired of the Arab governments' attempts to solve their problems; the Intifada changed them from refugees in their own land to a society that was ready to become a nation. Though many of them were injured, jailed, or killed, the Intifada increased their pride, self-reliance, and unity.

Because of the Intifada and the demands of the UNLU, the PLO changed its political positions. The PLO decided to recognize Israel's right to exist and to stop using terrorism. Through the Palestine National Council, the PLO announced on November 15, 1988, that a separate Palestinian state was being established, and this solution gradually gained support around the world.

Israeli military leaders began to realize that they could not stop the Intifada, and more and more Israelis wanted the government to start negotiations with the PLO. The "two-state solution" of Israel and Palestine existing side by side began to seem reasonable.

Mahmood Ibrahim

Yasir Arafat.

United States and Soviet Union Sign Nuclear Forces Agreement

> *The Intermediate-Range Nuclear Forces (INF) Treaty of December, 1987, limited the development of new nuclear weapons but also called for eliminating one whole category of already-existing weapons.*

What: International relations
When: December 8, 1987
Where: Washington, D.C.
Who:

RONALD REAGAN (1911-   ), president of the United States from 1981 to 1989

MIKHAIL GORBACHEV (1931-   ), first secretary of the Soviet Communist Party from 1985 to 1991

GEORGE SHULTZ (1920-   ), U.S. secretary of state from 1982 to 1989

ANDREI GROMYKO (1909-1989), Soviet foreign minister from 1957 to 1985

EDUARD SHEVARDNADZE (1928-   ), Soviet foreign minister from 1985 to 1990

The INF Talks

In 1977, the Soviet Union placed intermediate-range nuclear missiles—which could travel from three hundred to thirty-four hundred miles—in its western lands and in Eastern Europe. Because these SS-20's and other missiles were aimed at Western European nations, the members of the North Atlantic Treaty Organization (NATO) reacted by asking the United States to develop similar weapons and place them in Western Europe, aimed at the Soviet Union.

Discussions within NATO between 1977 and 1979 brought the "two-track" decision in December, 1979. One "track," or strategy, would be to negotiate with the Soviets, to persuade them to withdraw their intermediate-range force (INF).

The other track was to place American INF systems in Europe if the Soviets did not agree to withdraw theirs.

There were negotiations between the United States and the Soviet Union between 1981 and 1983, but little progress was made. As a result, American missiles began moving into place in November, 1983. The ground-launched cruise missile (GLCM) was deployed in Great Britain, Italy, West Germany, the Netherlands, and Belgium. When the GLCM's began to arrive in Europe, the Soviets broke off negotiations with the United States.

More than a year later, the Soviet government reconsidered its decision, and the talks started up again in March, 1985. Mikhail Gorbachev became the new leader of the Soviet Union that same month. By that time, many cruise missiles had arrived in Europe.

For more than two years, each side offered different proposals. In the end, it seems that the Soviet Union compromised the most. Soviet negotiators wanted to withdraw some missiles eastward, farther into Soviet territory, or to keep some set up near the border with China. But the Americans wanted a "zero-zero" arrangement, with all U.S. and Soviet INF systems destroyed. Also, the Soviets insisted that British and French INF systems should be counted in the final agreement, but they finally gave in to the Americans, who refused to consider any weapons except those of the superpowers.

The Treaty

The two sides reached final agreement in fall, 1987, and both governments prepared for a summit meeting in Washington, D.C. On De-

2240

cember 8, 1987, President Ronald Reagan and First Secretary Gorbachev officially signed the treaty in the White House.

The Intermediate-Range Nuclear Forces Treaty was long, about 170 pages, and quite complicated. It stated that all U.S. and Soviet INF weapons (except for their nuclear warheads) would be destroyed, along with their ground-launcher support systems, over a period of three years. More than 850 U.S. INF devices and more than 1,750 Soviet devices would be destroyed. During the next ten years, groups of Soviet inspectors would travel to the United States, and vice versa, to make sure that all procedures were being followed. Under the treaty, no more weapons of this type could be produced or tested.

The missiles had to be taken apart and destroyed according to special procedures. Each side formed inspection teams, and the missiles began to be removed in early 1988, even though the treaty had not yet been ratified.

In late March, 1988, the Foreign Relations Committee and the Armed Services Committee recommended that the U.S. Senate approve the treaty. The Senate began its debate on the treaty on May 17 and passed it on May 27 by a vote of ninety-three to five. The Presidium of the Supreme Soviet followed the next day by voting in favor of the treaty.

Reagan flew to Moscow on May 29, and the two leaders exchanged the formal ratification papers on June 1. This final step made the treaty a legal obligation for both nations.

Consequences

Soviet media reported that the first Soviet INF missiles were destroyed in early August; the first American missiles were destroyed on September 1. Inspectors witnessed these events. At the end of 1988, the U.S. government reported that the Soviets were honoring their end of the bargain, and the United States continued to remove its missiles from Europe and destroy them.

Soviet leader Mikhail Gorbachev (seated left) and U.S. president Ronald Reagan sign the Intermediate-Range Nuclear Forces Treaty.

AP/Wide World Photos

2241

Not all American leaders were pleased at the results of the INF Treaty. Some worried that it created a gap between short-range weapons that would be used on the battlefield and long-range strategic missiles that could be sent all the way from the United States to the Soviet Union. Others were critical for another reason: Destroying the INF would take out only 5 percent of all the world's nuclear armaments, leaving huge numbers of dangerous weapons still in existence. The weapons of China, France, and Great Britain had not been touched.

Supporters of the treaty, however, argued that it was an important step toward better relations between the United States and the Soviet Union. The INF Treaty brought hope that the two superpowers could continue to reduce the numbers of nuclear weapons and cooperate on other projects as well.

Taylor Stults

Ethnic Riots Erupt in Armenia

With considerable bloodshed, the Soviet Republic of Armenia laid claim to an area within the Soviet Azerbaijani Republic.

What: Civil strife; Ethnic conflict
When: 1988
Where: Armenia and Azerbaijan, Soviet Union
Who:
KAREN DEMIRCHIAN (1932-   ), first secretary of the Armenian Communist Party from 1974 to 1988
MIKHAIL GORBACHEV (1931-   ), first secretary of the Soviet Communist Party from 1985 to 1991

Soviets Divide and Conquer

In 1988, the Armenian people were scattered across much of the Middle East and other parts of the world. But of 5.4 million Armenians around the world, about 4.7 million lived in the Armenian Republic, located in the Caucasus in the southwestern Soviet Union.

When the Soviets first took power over this area, the mostly Armenian Christian region of Nagorno-Karabakh and the mostly Muslim area of Nakhichevan were given to Armenia. But Turkey raised a protest, and the Soviets then gave Nakhichevan to neighboring Azerbaijan. (Most of Nakhichevan's population was Azerbaijani.) In 1923, Nakhichevan and Nagorno-Karabakh became "autonomous regions" within the Azerbaijan Republic.

The government of the Soviet Union wanted to get rid of religion, but at the same time it wanted to keep good relations with its Islamic neighbors. One of the Soviet leaders' main concerns was that the republics never unite against the central government.

Some experts think that Joseph Stalin, Soviet dictator from 1924 to 1953, wanted to draw the republics' boundaries in a way that would keep ethnic tensions stirred up. Then the republics would not band together against the Moscow government. At the same time, Stalin spread industry and agriculture among the republics so as to make them depend on one another— and so that no one republic could survive on its own.

After World War II, the Soviet government invited Armenians all over the world to come back to their Armenian homeland in the Soviet Union. At first, many did come; but some came to regret their decision.

The Armenians were allowed a certain amount of self-government, especially in the 1960's and beyond. The Armenian Apostolic church, which was not regulated by the Soviet government, became a symbol of Armenian independence. The Armenians especially longed for the return of Nagorno-Karabakh, which they called Artsakh. They sent many petitions to Moscow, but the Soviet government did not respond.

Rising Conflicts

When Mikhail Gorbachev became head of the Soviet government in 1985, he began to liberalize the nation with his policies of *glasnost* (openness), *perestroika* (restructuring), and democratization. Across the Soviet Union, various ethnic groups took the opportunity to stand up and demand fair treatment.

Among these groups were the 120,000 Armenians in Nagorno-Karabakh. Some of them joined in a demonstration in Stepanakert, the capital of Nagorno-Karabakh, on February 11, 1988; they demanded that Nagorno-Karabakh be reunited with Armenia. The demonstrations soon spread to Yerevan, the capital of Armenia, and then to Sumgait, an Azerbaijani city on the Caspian Sea.

2243

These early protests broke out at a time when Karen Demirchian, the leader of the Communist Party of Armenia, was being criticized by the Soviet government for not doing enough to end corruption in Armenia. Some historians have thought that the Armenian communist leaders encouraged the protests as a tool against the Soviet government. On February 26, the Armenian Communist Party asked Moscow to set up a commission to consider the problem of Nagorno-Karabakh.

Soon violence began erupting in various parts of Armenia; both Armenians and Azerbaijanis lost their lives or were wounded. The Soviet army stepped in several times to restore order, but both Armenians and Azerbaijanis were displeased with the army's actions.

In June, 1988, the Armenian Republic voted to annex Nagorno-Karabakh, but Azerbaijan and the central Soviet government did not accept this resolution. In July, the Soviet government stated that Nagorno-Karabakh would remain part of the Azerbaijan Republic.

In the second half of 1988, there were demonstrations in Yerevan almost every day. In early December, Moscow sent troops and tanks to the Azerbaijani city of Baku to control the protests, which had become more and more violent.

An earthquake in December, 1988, killed at least twenty-five thousand Armenians and left about five hundred thousand homeless. The Armenian people were deeply grieved by this tragedy—but they also were even more determined to annex Nagorno-Karabakh.

The All-Armenian National Movement (AAM) grew out of the nationalist movement in Nagorno-Karabakh, uniting about forty groups behind the goal of independence for Armenia. The AAM also worked to strengthen the Armenians' culture and religion.

Consequences

In early 1990, Armenian and Azerbaijani leaders decided to try to resolve their differences on neutral ground. They began meeting at Riga, in the Latvian Republic. These talks continued off and on for months, but so did the violence.

After conservative communists tried to stage a coup against Mikhail Gorbachev in August, 1991, both Armenia and Azerbaijan proclaimed their independence from the Soviet Union, along with several other Soviet republics. In December of that year the Soviet Union was officially dissolved, and leaders of the various republics began discussing a looser form of union—a confederation of independent states. Though Armenia and Azerbaijan now had their wished-for independence, a solution to the issue of Nagorno-Karabakh still seemed far off.

Peter B. Heller

Hunger Becomes Weapon in Sudan's Civil War

Trying to win a long civil war against the Sudan People's Liberation Army, the Sudan government began blocking international shipments of food intended for hungry people.

What: Civil war; Human rights
When: 1988
Where: Southern Sudan
Who:
SADIQ AL-MAHDI (1936-   ), prime minister of the Sudan from 1985 to 1989
CHARLES LA MUNIERE (1930-   ), coordinator of disaster relief for the United Nations

A Famine Approaches

As the dry season began in southern Sudan in October, 1987, the harvest was the poorest it had been in almost one hundred years, for there had been almost no rain during the growing season (May to October). Late that year, more than ten thousand people of the Dinka tribe—mostly women and children—traveled northward to try to find food.

The Dinka had close ties with a rebel movement known as the Sudan People's Liberation Army (SPLA), which had been winning many victories in the south over the army of the Republic of the Sudan. Unable to defeat the rebels, the Sudanese government decided to deny food aid to them during the famine.

International organizations such as the United Nations Children's Emergency Fund (UNICEF) and the Red Cross had come into the Sudan to help the famine victims. But the Dinka who took shelter in the towns of Meiram and Abyei found that the government would not let these organizations help them. The government even forced some relief organizations, including World Vision and Lutheran World Service, to leave the country. The Sudanese relief organizations were loyal to the government and did not give much assistance to the Dinka refugees.

However, the international organizations went ahead with their plans to bring food to the hungry Sudanese people. The United States Agency for International Development (U.S. AID) gave the Sudanese government's Relief and Rehabilitation Commission millions of dollars to help pay for transporting the food.

Meanwhile, the people of the southern Sudan were suffering. Food prices shot up in the winter of 1987-1988. As Dinka people fled north, many were attacked by the Murahileen, an Arab Sudanese militia that had received weapons from the government. Between the famine and the Murahileen raids, tens of thousands of refugees died. Other groups of young Dinka men were fleeing eastward, into Ethiopia, to join the SPLA. Many of them died during the long march.

The Food Crisis

Darius Bashir, acting governor of the Bahr al-Ghazal Province in southern Sudan, kept sending messages to Prime Minister Sadiq al-Mahdi to ask that food be sent to his province, especially to the capital, Wau. But the government regarded the people of the south as "rebels, bandits, and terrorists" and was in no hurry to help.

In February, 1988, though World Vision had been expelled from the Sudan, it managed to send a convoy carrying 167 tons of food to Wau. But this was the last food the people of Wau would receive for many months.

To the east, in Upper Nile Province, food supplies had almost vanished. In Malakal, the provincial capital, twenty-four hundred metric tons of cereals had been received in March, 1987. The army had taken most of this food, and shop owners hoarded the rest to sell at high prices. Most of the Nilotes people of this region had nothing to eat, and many of them moved north in search of food. By the end of February, 1988, more than thirty thousand hungry people had arrived in Kosti, where many of them accepted low-paying jobs so that they could survive.

Most of the food aid supplied by Western nations and relief organizations—about sixty thousand tons—was sent westward, into Kordofan and Darfur, rather than to the south where people were starving. Kordofan and Darfur were strongholds of Prime Minister al-Mahdi's Umma Party, a Muslim group that opposed the SPLA.

As the SPLA won more victories in Equatoria Province to the south, people swarmed into the cities of Yei and Juba in the hope of finding food and escaping the fighting. But they found little relief.

Other southern Sudanese began moving eastward, to refugee camps set up in Ethiopia by the SPLA. In April, 1988, more than 250,000 people made this trek; when they arrived, relief workers said many of them looked like walking skeletons. The United Nations High Commission for Refugees was providing most of the food for the camps, but there was not enough food and medicine for so many needy people, and many of them died there.

In late 1988, representatives from the Organization of African Unity visited one refugee camp in Ethiopia and reported that of the forty-three thousand refugees there, 60 percent were children under age twelve who had lost their parents to famine or civil war. By the end of the year, a British official estimated that more than 700,000 southern Sudanese had arrived in Ethiopia, and the same number had died trying to make the journey.

The international organizations had continued trying to get food to those who needed it most in the Sudan, but their efforts kept being blocked or delayed by the government and the trucking companies. Food had finally begun to reach the south in August, 1988, but most of it was taken by the army and the shop owners.

Consequences

Because the U.S. government and other Western governments did not want to start a conflict with the Sudanese government, they did not raise a storm of protest over how the southern Sudanese had been treated. But in late February, 1988, a letter of protest to Prime Minister al-Mahdi and to John Garang, leader of the SPLA, was signed by several members of Congress: Howard Wolpe, of the House Subcommittee on Africa; Senator Ted Kennedy; Thomas "Mickey" Leland, chair of the House Select Committee on Hunger; and Paul Simon, of the Senate Subcommittee on Africa.

That same month, the British press exploded with angry reports after the Sudanese government seized twenty Oxford Committee for Famine Relief (OXFAM) trucks to use in combat against the SPLA. Great Britain had just promised the Sudanese government a large sum of money to help fight hunger, so this move against a British relief organization made the British people especially angry. The trucks were immediately returned, but OXFAM and other aid agencies were prevented from carrying out their plans to help famine victims.

In response to the reports of many deaths, Charles La Muniere, U.N. Coordinator of Disaster Relief, traveled to Khartoum, capital of the Sudan. His efforts to pressure the government had some success. AID's Office of Foreign Disaster Assistance supplied one million dollars and sent a large shipment of grain to the Sudan; this food reached some needy areas before the dry season in October, November, and December.

According to one estimate, as many as 500,000 southern Sudanese died in 1988 (but other estimates are lower: 260,000-300,000 deaths). By December, 1988, food supplies were moving southward in the Sudan, but they were too little and too late.

Robert O. Collins

"Peace" Virus Attacks Mac Computers

Computing was changed forever when thousands of Apple Macintosh users found their computers infected by the first computer virus to receive major publicity.

What: Computer science; Business
When: 1988
Where: Worldwide
Who:
RICHARD WALTON BRANDOW (1930-   ),
    publisher of *MacMag* magazine
DREW DAVIDSON, a computer programmer

What Is a Computer Virus?

A biological virus is a tiny organism that causes illness and can multiply only inside living cells. A virus reproduces by tricking a cell in which it resides into making copies of it. The copies of the virus can then leave the cell and do likewise in other cells. A computer virus works in much the same way. It can damage computer programs, can reproduce only inside a computer, and can trick a computer into making copies of it.

Computers follow instructions that are contained in computer programs. Modern computers are capable of both writing programs and modifying existing programs. A computer virus is a small program that instructs the computer in which it is located to, among other things, modify some other program on that computer by adding in the instructions that make up the virus program. When that newly affected program is run, it will cause the computer to, among other things, modify another program. The process continues in this way, producing more and more copies of the virus. If the computer shares programs with other computers, the virus can affect those computers.

The "Peace" Virus

In late 1987, Richard Brandow unleashed the "Peace" virus (so called because it consisted of a message calling for world peace) on an unsuspecting world of Mac users. (The virus is also known as the *MacMag* virus, because it was commissioned by Brandow, the publisher of that magazine at the time; the Drew virus, because the programmer who wrote it was Drew Davidson; and the Aldus virus, because a demonstration disk distributed by the Aldus Corporation was infected with it.) The virus was set to go off on March 2, 1988. Until then it would reproduce.

"New Apple Products," the program that contained the Peace virus, was a HyperCard stack (a software package) that described a scanner that was being developed by Apple Computer,

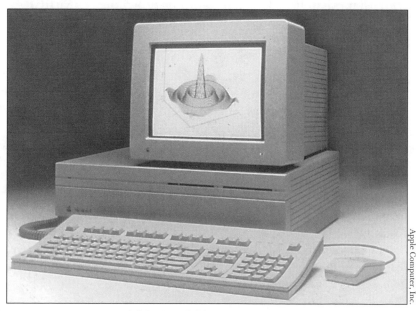

A Macintosh II series computer.

Apple Computer, Inc.

Incorporated. HyperCard stacks can be read with programs that are supplied free with every new Mac, so this stack was attractive to many Mac users. Such a program is called a "Trojan Horse"; it looks like a nice gift but contains a nasty surprise. Although Brandow readily admits his responsibility for the Peace virus (his name appeared on the screen along with the message calling for world peace), he denied that he uploaded it to any of the computerized bulletin-board systems that eventually permitted it to move into thousands of Macs. Someone, however, did do so, and few people worried about, or took steps to avoid, computer viruses at that time.

Computers that read "New Apple Products" were infected with the virus, which wrote itself into the system folder that controlled the computer. If a floppy disk that contained a system program was inserted into the computer, that disk would also be infected with the Peace virus. For several months, the virus continued to spread, unnoticed.

March 2, 1988, was the first anniversary of the introduction of the Mac II computer. When people whose computers had been infected turned on their machines, they saw the peace message displayed, after which the message and the virus that produced it disappeared. As quietly and mysteriously as it had arrived, it would vanish. Its last action was self-annihilation, erasing the program code that created it. Those who did not turn on their Macs that day would never see the message, or the virus, because it was designed to destroy itself the first time the computer was turned on after March 2. Some people, however, were not taken by surprise.

Although it was very small, even for programs of the time, the virus took up 1.7 kilobytes of memory. It was big enough to be noticed by some people, who were able to find it and decode it. They figured out what it would do and when it would do it, and then they spread the word. By March 2, many Mac users knew what to expect.

Had it occurred in isolation, the Peace virus might not have created a stir, but other virus programs were also creating havoc and making

headlines. The Scores virus was actually designed to destroy certain programs, and the nVIR virus spread much faster than the Peace virus did. The Peace virus, however, had been commissioned and written to make a point: Personal computers are vulnerable to virus attacks. The age of computer innocence was over.

Consequences

The most important thing about the Peace virus was not what it did, but what it could have done. Instead of displaying a harmless message about peace on the screen, it might have erased the hard disk, or even changed every thirtieth 7 in a spreadsheet program to a 3—a greater threat in many ways. It could have been very destructive. The ease with which this virus infected so many computers shocked the Mac user community—the largest community of personal computer users at that time.

Virus attacks were not new, but in the past, they had been restricted to mainframe computers. Understandably, news about them had been kept to a minimum, and only professional systems people were well informed about viruses. Personal computers were an inevitable target, and Brandow stated that the intention of the Peace virus was to dramatize the susceptibility of personal computers and make their users aware of the threat. The tactic succeeded remarkably well. The publicity surrounding the virus made virtually everyone aware of potential virus attacks. Dozens of programs were written to detect the presence of known viral strains, and other programs were developed that would alert a computer user if any program attempted to modify a system file.

Despite these safeguards, the problem of computer viruses continued to grow through the 1990's, aided by the rapid expansion of open networks, such as the Internet, which served as a kind of infection superhighway. In the mid-1990's, average computer users had about one chance in a thousand of having viruses enter their computers during a twelve-month period. By the year 2000, those odds had risen to one chance in ten.

Otto H. Muller

Recruit Scandal Rocks Japan's Government

A stock-trading scandal in 1988-1989 forced Japan's prime minister to resign and temporarily weakened the Liberal Democratic Party's hold on power in Japan.

What: National politics
When: 1988-1989
Where: Japan
Who:
YASUHIRO NAKASONE (1918-   ), prime minister of Japan from 1982 to 1987
NOBORU TAKESHITA (1924-   ), prime minister of Japan from 1987 to 1989
SOSUKE UNO (1922-1998), prime minister of Japan in 1989
TOSHIKI KAIFU (1931-   ), prime minister of Japan from 1989 to 1991
KIICHI MIYAZAWA (1919-   ), prime minister of Japan from 1991 to 1993
TAKAKO DOI (1929-   ), leader of Japan's Socialist Party

Japanese Money and Politics

After Japan surrendered to the Allies in 1945 at the end of World War II, it was occupied by American forces under the command of General Douglas MacArthur. MacArthur's goal was to bring a peaceful, democratic form of government to Japan.

Under MacArthur, Japan got a new constitution in 1947. Districts would elect members to the national parliament (called the Diet), and the political party whose members won the most seats in the Diet would normally choose the prime minister. The prime minister would form a cabinet by appointing members of the majority party to head the executive departments of the government.

After the end of the American occupation in 1952, only the Liberal Democratic Party (LDP) was able to gain enough votes in the Diet to choose the prime minister. The result has been a kind of one-party government. The main opposition parties—the socialists, the communists, and Komeito ("Clean Government")—have not been able to build up a large enough majority to take control over the government.

Over the years, the LDP has dominated politics in Japan, and the leaders of several LDP factions have taken turns being prime minister. Two questions are involved in most major elections in Japan: Can the LDP keep its majority, and if it can, which LDP faction will win the contest to name the next prime minister?

Elections in Japan are very expensive. Candidates are usually expected to reward their supporters with gifts. Some of the gifts for special supporters are costly, but even people who attend campaign rallies expect to get a souvenir of some sort—perhaps a necktie, a towel, or a pen. Not surprisingly, candidates are always running short of money.

With so much money flowing in elections, it is easy for companies and special-interest groups to influence candidates by offering contributions. The biggest contributions go to the biggest candidates—those who have the most supporters. These candidates are usually competing for leadership of the party and the position of prime minister.

Campaign contributions, even very large ones, are legal if they are reported publicly. But some contributors like their gifts to be secret, so that no one will know how they are trying to influence the election.

Scandal Leads to Downfall

In the summer of 1988, Japanese newspapers revealed that Recruit, a real estate company whose stock price had risen quickly when it offered shares to the public for sale in 1986, earlier had offered cheap shares to many leading members of the government. When the price rose, the early buyers had been able to sell their shares at huge profits. Some of those who had made money on Recruit stock in this way were close associates of Prime Minister Noboru Takeshita and former prime minister Yasuhiro Nakasone.

At first the stock trades seemed to have been legal, yet many people were angry at unfair profiteering by their officials. In the Diet, opposition politicians demanded an investigation.

The scandal soon got worse. In October, Finance Minister Kiichi Miyazawa admitted that he had received ten thousand shares of Recruit stock. In a related case, one of Nakasone's business supporters was convicted of bribery.

The investigation led to indictments for dozens of officials and politicians, including more associates of Nakasone, Takeshita, and other LDP leaders. Takeshita claimed at first that he himself had not received Recruit money; in April, 1989, however, he was forced to admit that he had received more than one million dollars in contributions from the company. He said that he had reported these contributions, as the law required, but the records were never found. Takeshita's popularity rating fell to 3.9 percent in the polls—a record low for any prime minister. The opposition demanded that he resign.

Then investigators found a contribution of fifty million yen ($380,000) from Recruit to Takeshita, made in 1987 through one of his political assistants. The assistant committed suicide when the report became public, and Takeshita announced his resignation on April 25.

So many LDP leaders were involved in the Recruit scandal that it became difficult to find a new prime minister. Takeshita's successor, Foreign Minister Uno Sosuke, served less than two months before he was forced to resign because of a sex scandal.

Uno's scandal prompted Japanese women to vote against LDP candidates in various local elections during the summer of 1989. The LDP lost twenty of its sixty-three seats in the Tokyo Metropolitan Assembly election, while more than two-thirds of its candidates for election to the upper house of the Diet were defeated. The Japan Socialist Party benefited from the LDP's losses, and the socialists' leader, a woman named Takako Doi, became an important political figure in Japan.

Consequence

After the resignations of Takeshita and Uno, the LDP looked desperately for a "Mr. Clean" candidate for prime minister. They chose a relatively unknown leader named Kaifu Toshiki. Kaifu, too, had received Recruit money, but he had reported it and so had stayed within the law.

In office, Kaifu was popular, but he had difficulty in governing because his party had lost so much power. During his two years in office, the old factions of the LDP continued to compete behind the scenes. In October, 1991, Kaifu was replaced as party leader and prime minister by former finance minister Kiichi Miyazawa.

Donald N. Clark

Pakistani Bank Is Charged with Money Laundering

After major international investigations of mismanagement and illegal transactions, the Luxembourg-based Bank of Credit and Commerce International was forced to stop its operations.

What: International relations; Economics
When: 1988-1991
Where: United States, Great Britain, and Pakistan
Who:
AGHA HASSAN ABEDI (1923-1995), the Pakistani financier who founded BCCI
ROBERT M. GATES (1932-   ), deputy director of the Central Intelligence Agency from 1986 to 1991
CLARK CLIFFORD (1906-1998), longtime presidential adviser and chair of First American Bankshares
ROBERT ALTMAN (1944-   ), president of First American Bankshares
ROBERT MORGENTHAU (1919-   ), Manhattan district attorney, who brought charges against Clifford, Altman, Abedi, and others in 1992

Banking in an International Economy

The Bank of Credit and Commerce International (BCCI) was established in 1972 by Agha Hassan Abedi, a financier from the kingdom of Mahmudabad (which was made part of the new country of Pakistan in 1948). Abedi had founded his first bank in 1959, with old friends from Pakistan as his financial advisers. These friends were not professional bankers.

Like many other South Asian business executives of his generation, Abedi relied on borrowers' honesty and integrity, not their credit record or collateral, when he made decisions about lending money. After founding BCCI, Abedi managed his rapidly growing network of international finance in a similar way. He placed confidence in handshakes and promises rather than the kinds of financial analyses used by American and European bankers.

Under Abedi's leadership BCCI grew quickly; millions of Pakistanis deposited their savings in the bank, along with oil-rich Middle Eastern countries. BCCI became one of the largest banks in the world, with offices in seventy-two countries. By 1988, however, bad loans and Abedi's declining health were problems for the bank. New, highly trained managers recognized that BCCI needed new sources of cash. Some of them approved the creation of shaky subsidiary companies; some made high-risk investments in currency and bond markets. Usually these decisions brought further losses for BCCI.

BCCI's managers also turned to many kinds of illegal transactions—illegal trade in currencies and laundering the money of drug traffickers. (BCCI accepted the drug traffickers' money and passed it through several accounts so that it would be hard to trace the money—and hard to prove that it was earned illegally.) The bank also managed secret accounts for intelligence agencies, including the U.S. Central Intelligence Agency (CIA). BCCI managers ran accounts for terrorist organizations, including such infamous organizations as the Abu Nidal group.

By the end of the 1980's, however, these illegal or high-risk investments were not enough to keep BCCI from losing hundreds of millions of dollars each year. The British government's central bank, the Bank of England, led a worldwide investigation of BCCI's operations. The investigation was secret, and many governments took part. Robert Gates, deputy director of the CIA, had begun investigating BCCI by at least 1986; the CIA found that BCCI was linked with international criminal organizations and had laundered money for narcotics dealers in Central America.

2251

Death of a Giant

By early 1991, agents of the Bank of England had found that BCCI was involved in money laundering and illegal investments. Still, a secret audit of the bank's debt and loan records was made before a decision was made to shut down BCCI. Investigators learned that during 1990-1991 alone, BCCI had lent almost two billion dollars to borrowers who could not repay their debts. Many of these borrowers were large Pakistani corporations run by friends of Abedi.

If BCCI collapsed, the international banking industry would be in serious trouble—and so would countries such as Pakistan and the United States, where BCCI controlled important institutions. Together, government officials from Great Britain, the United States, and other countries coordinated the plan to end BCCI operations.

At the last minute, there were offers to give BCCI emergency assistance so that it would not close down. It is said that on July 1, 1991, the wealthy royal family of Abu Dhabi in the United Arab Emirates offered cash and loan guarantees worth nearly five billion dollars in order to save the bank. Before this deal could be concluded, though, BCCI's operations around the world were suspended, and its international accounts were frozen.

On July 5, banking regulators from several European, Caribbean, and North American nations ordered the surprise shutdown of BCCI. The only country where BCCI was allowed to continue functioning as a local bank was Pakistan, because there more than four million small investors and depositors had placed their savings in BCCI. Closing BCCI in Pakistan might set off a general financial panic that could have damaged the entire economy of the country.

Consequences

After the bank was closed, American officials publicly claimed that BCCI was one of the most complex and secretive criminal organizations in the world. The U.S. government was especially concerned about BCCI's secret purchases of American banks in California, Georgia, Florida, and Washington, D.C. In making these purchases, BCCI's owners appeared to have committed many civil and criminal offenses, including conspiracy, fraud, and racketeering.

It was clear that more investigations were needed. In the United States, several institutions and agencies were involved in BCCI probes: the CIA, Congress, and state and federal prosecutors. They wanted to find out exactly what crimes had been committed, what mistakes had been made, and who was guilty.

By buying American banks, BCCI's directors had hoped to avoid strict U.S. laws governing foreign ownership of banks and how such banks should do business. In 1985 BCCI had illegally gained control of First American Bankshares, an institution based in Washington, D.C.; it had used the legal services of Clark Clifford, an influential American lawyer who had been an adviser to several U.S. presidents. In the investigations that followed the bank's closure, the Senate looked into Clifford's dealings, and he and protégé Robert Altman were indicted in August, 1992, on charges of fraud and bribe-taking. Abedi himself was also indicted in the United States, along with a number of others who had made decisions about BCCI's activities.

In the meanwhile, regulators in many countries seized BCCI's possessions and began offering them for sale, so that they could pay back at least some of the creditors who had loaned BCCI money in the late 1980's. Most Pakistani depositors—both those living in Pakistan and those residing in other countries—were temporarily protected against the loss of their savings, but they worried about the long-term safety of their money.

Perhaps more significant, several Saudi Arabians with ties to government and to the royal family were indicted; for example, in August, 1992, the head of Saudi Arabian intelligence, Sheik Kamal Adham, pleaded guilty to conspiring to help BCCI purchase First American Bankshares.

These developments suggested a web of relationships—both legal and illegal—among Arab states, Israel, and the United States under the Reagan and Bush administrations. It was suspected that these relationships were at the root of arms and money transactions in the Middle East, including the sale of arms to Iraq. Unraveling the knotted story of BCCI's operations was a long, complicated task involving many governments and investigative agencies.

Laura M. Calkins

2252

Canada and United States Sign Free Trade Agreement

> *The United States and Canada signed a free trade agreement making it easier for each country to invest in the other, and getting rid of almost all barriers to the flow of goods and services between them.*

What: Economics; International relations
When: January 2, 1988
Where: Palm Springs, California, and Ottawa, Canada
Who:
RONALD REAGAN (1911-), president of the United States from 1981 to 1989
JAMES A. BAKER III (1930-), U.S. secretary of the treasury from 1985 to 1988
BRIAN MULRONEY (1939-), prime minister of Canada from 1984 to 1993
MICHAEL WILSON (1937-), Canadian minister of finance from 1984 to 1991

The Road to Free Trade

For more than a century, the United States and Canada had worked to liberalize trade between the two of them. The road to an agreement has often been rocky, but the basic trend in U.S.-Canadian trade has been cooperation and loosening restrictions.

Even before Canada became an independent state, the United States and Great Britain recognized the importance of trade between the United States and Canada, and they made efforts to make it secure. The Reciprocity Treaty of 1854 was a limited free-trade agreement that did away with tariffs on many "primary products" (unprocessed goods such as farm products) and some manufactured goods. Trade flourished under this treaty, though on both sides of the border there were some complaints about treaty violations.

In 1866, however, the treaty was renounced by the United States. The reasons were mostly politi-cal, related to the Civil War. After Canada gained independence in 1867, its government tried to negotiate a new trade agreement with the United States but did not succeed.

Canada took a step back from free trade in 1879, when the conservative government of Sir John A. MacDonald set up barriers to trade with its National Policy. Over the years, these protectionist rules came to be cherished by business leaders in Canada. In 1911, the United States tried to start talks about lowering trade barriers, but the Canadian government was not interested.

In 1930, the U.S. Congress passed the Smoot-Hawley Tariff Act, which created tariffs higher than any that had existed in U.S. history. Protectionism was now popular in both the United States and Canada, and the spiral was not broken until 1935, when the two countries agreed to consider each other as a "most favored nation" for trading. This agreement led to further efforts to expand and improve trade.

At the end of World War II, the United States and Canada joined more than twenty other countries in the General Agreement on Tariffs and Trade (GATT). The GATT, signed in 1947, was a shared effort to free up international trade and commerce. Besides taking part in these international efforts, the United States and Canada continued working on private agreements between themselves. The 1965 Canada-U.S. Automotive Products Trade Agreement was one result of these efforts.

Opening the Markets

In the late 1970's and early 1980's, the energy crisis and recession led some Americans to favor protectionism, but by the mid-1980's free trade was once again the goal. Pierre Trudeau, who was

After signing the free trade agreement, U.S. president Ronald Reagan talks to Canadian prime minister Brian Mulroney. The president was in Palm Springs, California, for the New Year holiday.

prime minister of Canada from 1968 to 1979 and again from 1980 to 1984, appointed the Mac-Donald Commission to investigate Canada's economic problems. The commission presented its findings in 1985 and recommended that Canada negotiate a free-trade agreement with the United States. Brian Mulroney, who had become prime minister in 1984, was happy to accept this recommendation.

In March, 1985, Mulroney informed the U.S. government that he was interested in negotiating; President Ronald Reagan was pleased, for he was a firm believer in the virtues of free trade. The discussions between the two governments concluded on January 2, 1988, when Reagan and Mulroney signed the Canada-U.S. Free Trade Agreement (FTA) creating the world's largest free-trade area. The FTA became effective on January 1, 1989, after it was approved by the U.S. Congress and the Canadian legislature.

The specific agreements of the FTA were to begin operating over a period of ten years. It eliminated all tariffs on trade in goods and ser-

vices and reduced many other barriers to trade. The FTA also made it easier for investments to be made across the U.S.-Canada border. Included in the FTA is provision for resolving trade disputes and carrying out the various parts of the agreement.

Canada is the United States' largest trading partner, and the two countries' trade with each other is the largest of any two nations in the world. Because of the huge amount of trade and the benefits of free trade to consumers and manufacturers, both governments argued that the FTA was a "win-win" situation. Consumers would benefit by being able to buy high-quality goods and services at lower cost, since competition and the lack of tariffs would bring prices down. Manufacturers would be able to sell their products to a larger market, and productivity would improve as manufacturing operations grew.

Yet there were some people who criticized the FTA. In the United States, opponents argued that the agreement would favor Canadian companies by allowing them to sell to the much larger American market (Canada has a population of about 26 million, while the United States has about 245 million). The gains for American companies would be smaller.

At the same time, many Canadians feared that their country would be unable to keep its economic and political sovereignty in a free-trade arrangement with the United States. Edward Broadbent, leader of the Canadian New Democratic Party, claimed that within twenty-five years the FTA would turn Canada into the United States' fifty-first state.

Consequences

Economists have stated that one of the causes of the Great Depression of the 1930's was high protectionist "walls" countries had built around themselves to keep out foreign products. So the FTA is an important step away from the dangers of protectionism and toward free trade.

As inefficient businesses in both the United States and Canada faced new competition, the FTA was expected to cause some business failures and lost jobs. Yet many experts believed that the long-term benefits would more than make up for these losses. The FTA would bring increased exports for both countries, a rise in income for consumers, and improved efficiency because of competition.

Vidya Nadkarni

U.S. Cruiser *Vincennes* Shoots Down Iranian Airliner

> *The USS* Vincennes, *a missile cruiser in the Persian Gulf, shot down an Iranian airliner carrying 290 passengers and crew, leaving no survivors; the result was even stronger anti-American feelings in Iran.*

What: Military; International relations

When: July 3, 1988

Where: Persian Gulf

Who:

WILL ROGERS III (1939-　　), commanding officer of the *Vincennes*

RONALD REAGAN (1911-　　), president of the United States from 1981 to 1989

ALI KHAMENEI (1939-　　), president of Iran from 1981 to 1989

AYATOLLAH RUHOLLAH KHOMEINI (1902-1989), rahbar (religious leader) of Iran from 1979 to 1989

Tensions in the Gulf

Relations between the United States and Iran had gone quickly downhill after the overthrow and expulsion of Shah Mohammad Reza Pahlavi in 1979. Because the shah had been a staunch ally of the United States, the Iranian revolutionary movement, which was led by fundamentalist Muslims, was strongly anti-American. Fanning the flames of hatred for the "Great Satan" America, the revolutionary leaders in Iran encouraged an attack on the American embassy in Tehran, and staff members of the embassy were held as hostages for more than a year. Enmity between the two countries continued even after 1981, when the hostages were released.

Chances for restoring a normal relationship with Iran ended when the conflict between Iran and its neighbor Iraq broke into a full-scale war. That war was stepped up in 1984. In spite of the secret Iran-Contra deal, by which U.S. weapons were sold to Iran, the United States sided with Iraq in the war and did not renew diplomatic relations with Iran.

During the war, Iran threatened often to make sure that oil could not be shipped from the Middle East to the United States and its allies. Some oil tankers in the Persian Gulf were bombed by Iran and Iraq, and the United States decided to send warships to protect tankers belonging to Kuwait and other countries that were not involved in the Iran-Iraq War.

Protecting these tankers was difficult, because the Persian Gulf is rather narrow and ships have little time to fight off attacks from the air. This became especially clear in 1987, when an Iraqi pilot mistook the U.S. frigate *Stark* for an Iranian vessel and fired two missiles at it; the ship was heavily damaged, and thirty-seven crew members were killed. After the *Stark* disaster, the U.S. Pentagon made new "hair trigger" rules for doing battle in the gulf. Navy officers were to respond immediately to any sign of trouble.

A Tragic Mistake

On July 3, 1988, the USS *Vincennes*, a missile cruiser patrolling the Persian Gulf near the Strait of Hormuz, launched two electronically guided missiles against what Captain Will Rogers believed was a hostile airplane about to attack his ship. Within seven minutes, the plane was destroyed, long before it could do any harm or even be seen by the naked eye.

In the previous hour, the *Vincennes* had been doing battle with Iranian speedboats off the Persian Gulf's southern shore. Rogers believed that the enemy boats were about to get help from fighter planes and that he needed to react quickly.

Unfortunately, what the radar and transponder equipment on the *Vincennes* had identified as

2256

an F-14 Tomcat fighter turned out to be Flight 655, a much larger Iran Air Airbus. The civilian aircraft had left Iran from Bandar Abbas on a regularly scheduled flight to Dubai in the United Arab Emirates, on the other side of the Persian Gulf. Its crew and passengers, 290 persons, were all killed when the cruiser's missiles blew the plane apart.

The tragic accident resulted from human error combined with sophisticated technology. In the open sea, detecting and tracking a "bogey" (unidentified aircraft) can be done across hundreds of miles, but in the narrow Strait of Hormuz, the Iranian airliner was already at a dangerous distance from the *Vincennes* when it was first detected. After radio warnings were ignored by the plane's pilot, there was no time to check the aircraft's flight pattern or identity. For Captain Rogers there seemed to be no choice. He had to order the deadly missile launch.

It is hard to say exactly who was to blame for the tragedy. The *Vincennes* sent a total of seven warning messages to the airbus, and the U.S. frigate *Sides* sent five more. The plane's pilot did not respond, even though his radio was in perfect working order—the airport at Bandar Abbas had been monitoring his signals. Still, the U.S. Navy seemed partly to blame, because it had failed to track the movements of civilian aircraft in the area. The *Vincennes* had no on-board information about regularly scheduled flights.

Under great psychological pressure and with limited information, Captain Rogers had chosen to attack the unseen airplane rather than risk losing his ship. It was an unfortunate decision, but an understandable one.

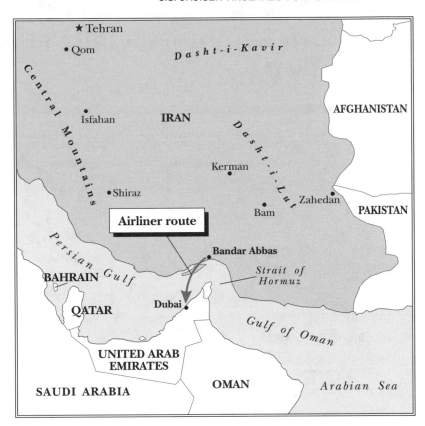

Consequences

While most of the world accepted the American claim that the destruction of the airliner was unavoidable, several Iranian leaders called it a criminal act. Mourners rushed through Tehran shouting, "Death to America! Death to Reagan!"

In spite of the angry accusations of many Iranians, there were some moderate leaders in Iran who believed that their government should respond with restraint to the tragedy. Interested in restoring relations with other countries and ending Iran's isolation from the world, these leaders were secretly trying to restore diplomatic ties to several countries, including the United States. Unfortunately, the downing of Flight 655 almost destroyed any chance of their success.

John W. Fiero

2257

Ayatollah Khomeini Calls Halt to Iran-Iraq War

> *After eight years of bloody fighting that cost a million lives and gained nothing for either side, Iranian religious leader Ayatollah Ruhollah Khomeini called an end to the war with Iraq.*

What: War
When: July 20, 1988
Where: Tehran, Iran
Who:

AYATOLLAH RUHOLLAH KHOMEINI (1902-1989), rahbar (religious leader) of Iran from 1979 to 1989

SADDAM HUSSEIN (1937-   ), president of Iraq from 1979

JAVIER PÉREZ DE CUELLAR (1920-   ), secretary general of the United Nations from 1982 to 1991

Eight Bloody Years

The Iran-Iraq War broke out in 1980, sparked by a long dispute over the Shatt al-Arab waterway. Shatt al-Arab ran along part of the Iran-Iraq border and was formed by the coming together of the Tigris and Euphrates Rivers; it provided Iraq's only access to the Persian Gulf. Launching an attack on Iran, Iraqi president Saddam Hussein believed that this conflict would be as short as the 1967 Arab-Israeli War. He also expected to bring the downfall of Iran's new religious leader, Ayatollah Ruhollah Khomeini, whom he despised.

The war's effect, however, was opposite to what Hussein had hoped: Khomeini's power was strengthened as Iranians united in a frenzy of determination to die for their country if necessary. After a few military successes, the Iraqi forces were forced out of Iran. What followed was a long, grueling war that used the old-fashioned methods of World War I—including "human wave" attacks, trench fighting, and the illegal use of chemical weapons.

During 1987 and the first half of 1988, it became more and more clear that Iran could not win the war. Seven years of fighting had not brought many positive results, and hundreds of thousands of lives had been lost. Late in 1986, Iran had made one final, massive human-wave assault against Basra, an important port that was the second largest city in Iraq. Like many smaller attacks Iran had made against Basra, this one failed, and Iran's losses were great.

During the following months, Iraq won a series of victories. In six weeks during early 1988, Iraq launched 180 missiles against Iranian cities—most of the missiles were aimed at the capital, Tehran. These attacks caused much damage and discouraged the Iranian people.

From April to August, 1988, Iraq recaptured territory it had lost to Iran, and nearly all Iranian forces were pushed out of Iraq. In June and July, Iraq gained the upper hand in the war by helping an Iranian dissident movement, the People's Mujahedeen, take some cities from Iran. Iraq then followed with chemical weapons attacks against Kurds within its own borders. (The Kurds, an ethnic group spread throughout Iran, Iraq, and Turkey, wished to form their own state. Their separatist movement was supported by Iran.)

No one had expected Iraq to win the war. Iraq had better technology and more weapons, but Iran had three times the population and four times the geographic size of Iraq.

Accepting a Cease-Fire

On July 3, 1988, the United States missile cruiser *Vincennes* accidentally shot down a civilian Iranian airliner, killing all 290 crew members and passengers. In Iran, people began to fear that the United States would soon join the war on the side of Iraq. American ships continued to patrol the Persian Gulf, and the Iranian military was running low on weapons and enthusiasm.

Iran's primary leader, an Islamic clergyman named Ayatollah Ruhollah Khomeini, knew that matters could not go on this way.

On July 20, Khomeini reluctantly announced an end to the war with Iraq. He said, "Taking this decision was more deadly than taking poison. I submitted myself to God's will and drank this drink for his satisfaction."

The war-weary Iranian leader had officially accepted the United Nations' Resolution 598, which called for a cease-fire, pulling troops back inside the recognized borders of each country, and exchanging prisoners of war. Hussein had accepted this resolution a year earlier, when the war was not going well for his country.

Although both countries had now accepted the resolution, the fighting continued. Encouraged by Iran's military and political weakness, Hussein ordered attacks on southern Iran. Iraqi soldiers, backed by tanks, warplanes, and chemical weapons, retreated only after capturing many Iranian soldiers for use as "bargaining chips" in the upcoming peace talks. Iraq wanted to prove its military strength and then back off, allowing the Iranians to pause and think through their position.

The U.N. continued to help in the peace process, sending a team to Iran and then Iraq to investigate how many prisoners of war each side had and what condition they were in. When the official cease-fire began on Saturday, August 20, 350 U.N. observers from twenty-four nations were present to supervise the 740-mile border between Iran and Iraq.

Five days later, the UN-sponsored peace talks, moderated by U.N. secretary general Javier Pérez de Cuellar, began in Geneva, Switzerland. At the talks, Iraq demanded control of the Shatt al-Arab waterway, which had been clogged by debris accumulated during the war.

Consequences

One million people died in the Iran-Iraq War, 1.7 million were wounded, and 1.5 million were forced to flee as refugees. Iran and Iraq both had large oil reserves and potential for great in-ternational power, but they had wasted $400 billion on the war; they also lost major cities and the chance to build up their economies.

The effects of the war stretched beyond the two combatants. Iran's search for weapons led to an entanglement with the administration of U.S. president Ronald Reagan. In the secret Iran-Contra affair, Iran was able to buy American missiles in exchange for the release of hostages held in Lebanon; some of the money from the weapons sale was used to help right-wing rebel fighters in Nicaragua.

The Iran-Iraq war also ended Khomeini's attempts to spread his conservative Muslim revolution abroad. During the war, Iraqi president Hussein had developed fighting strategies that he would later use against another neighboring country, Kuwait, which had been his ally during the war with Iran. In the Gulf War of 1991, however, Iraqi forces were expelled from Kuwait by a coalition force led by the United States.

Frank Wu

Ayatollah Khomeini in 1979.

Jordan Abandons Its West Bank

Jordan's King Hussein gave up all claims to the territories west of the Jordan River, held by Israel; the king declared that those lands would now belong to the Palestine Liberation Organization.

What: International relations
When: July 31, 1988
Where: Amman, Jordan
Who:
Hussein ibn Talal (1935-   ), king of Jordan from 1952
Yasir Arafat (1929-   ), chairman of the Palestine Liberation Organization
Yitzhak Shamir (1915-   ), prime minister of Israel from 1983 to 1992
Shimon Peres (1923-   ), leader of Israel's Labor Party

Conflicting Claims

On May 14, 1948, Great Britain gave up its Palestine Mandate and withdrew from Palestine. Jordan quickly joined Syria and Egypt in a war to prevent a Jewish state from being formed, but in the struggle that followed, Israel won its independence. During the war, however, King Abdullah of Jordan had gained control of a strip of territory that almost stretched to the Mediterranean Sea. He also took the entire walled city of Old Jerusalem, east of the modern Israeli part of the city.

The Jordanians claimed this conquered territory for themselves, though only two countries—Great Britain and Pakistan—ever recognized their rights over what came to be called Jordan's West Bank. During the next nineteen years, Arab inhabitants of the West Bank were named citizens of Jordan. All Jewish residents were expelled to Israel, for under Jordanian law Jews cannot be permanent residents of Jordan.

Jordan lost all of this territory in the Six-Day Arab-Israeli War of 1967. Israel began to rule the West Bank, which it renamed Judea and Samaria; yet judges appointed by Jordan's government still used Jordanian law to govern the territory. Only Old Jerusalem was actually made part of Israel.

Meanwhile, the Palestine Liberation Organization (PLO) had been founded in 1964. The PLO wanted to destroy Israel and create a secular Arab state in all Palestine; its preferred method was terrorism. Using Jordan as a base for their terrorist attacks, the PLO fighters became an embarrassment to Jordan in 1970 when they hijacked a series of airplanes. The Jordanian army drove the PLO out of the kingdom, and about three thousand PLO soldiers died in what became known as "Black September."

Syria might have taken this opportunity to invade Jordan, but the United States and Israel made it clear that they would come to Jordan's defense. Now Jordan had to thank Israel for its survival, and though the two nations were supposedly still at war with each other, they began to cooperate.

Many tourists entered Israel from Jordan, while farm products in unmarked crates moved from Samaria into Jordan. Israeli customs officials avoided stamping passports so foreign Arabs would not have the embarrassment of admitting that they had visited Israel. Jordan stayed out of the Yom Kippur War of 1973 and Israel's 1982 war against the PLO in Lebanon.

Giving in to the PLO

In December, 1987, Palestinians in Israel's occupied territories, including the West Bank, began what they called the intifada—an uprising to protest Israeli rule. The "open-bridges" policy between Israel and Jordan continued, but pressure from other Arab states made it necessary for Jordan to express support for the intifada in the West Bank.

The Palestinians' cause gained new sympathy around the world as newspapers and television showed images of young Arabs confronting heavily armed Israeli soldiers. In June, 1988, a conference of Arab nations, including Jordan, agreed that the Palestinians had the right to their own independent state, and that the PLO was the only acceptable representative of the Palestinian people. At the same time, the PLO showed a new willingness to accept the existence of Israel.

On July 31, 1988, King Hussein gave up all claims to the West Bank and announced that the territory now belonged to the PLO. He promised that the Palestinians there could count on his protection for the time being. Soon, however, the Jordanian government took actions that caused panic among the people of the West Bank. The king canceled a five-year development plan for the West Bank, and it was announced that Arabs living in the area would have their Jordanian citizenship reevaluated. King Hussein had already dissolved Jordan's parliament, and he stated that the West Bank would not be allowed to send representatives to the lower house in the coming election. Then, on August 4, the government of Jordan laid off twenty thousand government employees who lived in the West Bank.

It was perfectly clear that King Hussein had no intention of handing over his lost inheritance to the PLO and its leader, Yasir Arafat. The king hated and feared Arafat, considering him a dangerous revolutionary. King Hussein was certain that Israel would not agree to deal with the PLO, and the United States—which gave large amounts of aid to both Israel and Jordan—firmly opposed the creation of a new Palestinian state. So when the king gave up his claim to the West Bank, he seemed to be supporting the PLO but was actually counting on Israel's continued opposition to the PLO.

After King Hussein's announcement of July 31, Israeli nationalists began to demand that Judea and Samaria be immediately made part of Israel. Wisely, Prime Minister Yitzhak Shamir, however, avoided making any hasty decisions, while Foreign Minister Shimon Peres agreed with Jordan that Israel's borders would have to be changed someday. As King Hussein had expected, both right- and left-wing Israelis still refused to cooperate with the PLO.

Consequences

King Hussein's renunciation of the West Bank was part of a series of events that changed the balance of power in the world. The Palestinian Arabs and the PLO had won new popularity through the intifada, but they made the mistake of supporting Iraq during the Gulf War of 1991. At the same time, as the Soviet Union fell apart, the United States became less interested in aiding Israel.

New Middle East peace talks began in late 1991, and for the first time some Palestinian Arabs seemed willing to consider giving up the idea of an independent Palestinian state. Meanwhile, the majority of Israelis seemed ready to negotiate new boundaries, as long as Jordan took part in the agreement.

Arnold Blumberg

2261

Congress Formally Apologizes to Wartime Japanese Internees

> *In 1942, about 110,000 Americans of Japanese ancestry were forced to move into internment camps; in 1988, Congress apologized and paid reparations to the surviving internees.*

What: Civil rights and liberties
When: August 10, 1988
Where: Washington, D.C.
Who:

WAYNE COLLINS (1912-　　), an attorney for the American Civil Liberties Union (ACLU)

FRED KOREMATSU (1919-　　), a Japanese American internee

The Internments

When Japan attacked Pearl Harbor on December 7, 1941, the United States was forced to enter World War II. Within the United States, there was a great outburst of prejudice and suspicion against Japanese Americans. Were they loyal to their native land? Were some of them spies?

The "issei"—Japanese immigrants—were not allowed to become U.S. citizens, though their children born in the United States, known as the "nisei," were considered native-born citizens. Most Japanese Americans lived in the West Coast area, in the states of California, Washington, and Oregon.

In February, 1942, President Franklin D. Roosevelt issued Executive Order 9066, which gave the U.S. Army authority to protect "strategic" and "sensitive" areas—even to remove civilians from those areas, if necessary. The next month, Congress passed Public Law 503, which established the War Relocation Authority. This agency was to set up camps in which Japanese Americans would be held.

Ten camps were built: at Manzanar and Tule Lake, California; Poston and Gila, Arizona; Minidoba, Idaho; Heart Mountain, Wyoming; Granada, Colorado; Topaz, Utah; and Rohwer and Jerome, Arkansas. Eventually about 110,000 Japanese Americans were moved into the camps, where they were crowded into tiny spaces with almost no furniture. The camps were fenced with barbed wire and guarded by soldiers.

Before going to the camps, the internees had to sell most of their belongings, including land and home, at unfairly low prices. Upon arriving in a camp, a family would be given the task of stuffing bed-ticking sacks with straw to make their mattresses. Food was provided in a mess hall, but it was all army rations.

In 1940-1941, about five thousand nisei had been drafted into the U.S. Armed Forces. After Pearl Harbor, many of these men were discharged, and in 1942, the United States stopped drafting Japanese Americans. Yet many nisei were eager to fight for their country, so the Army agreed to accept volunteers.

Two all-nisei units were formed in 1943: the 442d Regimental Combat Team and the 100th Infantry Battalion. Both units served in Europe. The 442d Regimental Combat Team eventually received more awards than any other combat unit in U.S. history; its forty-five hundred members won more than eighteen thousand decorations for bravery, including one Congressional Medal of Honor and fifty-two Distinguished Service Crosses. Many of them lost their lives in winning the awards, so that the ribbons and medals were brought to their parents—who were still living behind barbed wire in the camps.

During the entire war, no Japanese American was ever shown to have committed an act of sabotage against the United States.

Protest and Apology

Immediately after World War II, Wayne Collins, a fiery lawyer who had dedicated himself to

2262

protecting civil liberties, began to challenge the government's action in interning Japanese Americans. He worked hard on the case of Fred Korematsu, a nisei who argued that the internments were illegal since they were used only against Japanese Americans. Collins and Korematsu took their case to the Supreme Court and lost—but they had helped keep the issue alive.

Gordon Hirabayashi and Minoru Yasui had been arrested during the war for breaking a curfew that applied only to persons of Japanese ancestry. Like Korematsu, they appealed to the Supreme Court and lost.

In 1975, President Gerald Ford issued a proclamation stating that the internments had been wrong. In 1980, Congress set up the Commission on Wartime Relocation and Internment of Civilians; its task was to investigate the internments and to interview survivors of the camps.

In 1983, Fred Korematsu had the satisfaction of finally winning his case against the government. The same year, the congressional commission published a report, *Personal Justice Denied*, stating that the laws against Japanese Americans had been racist and unjust. A class-action suit had been filed in court, asking more than four billion dollars in reparations for Japanese Americans.

In 1988, Congress began to debate whether reparations should be paid to surviving internees. Some conservatives, such as Senator Jesse Helms (a Republican from North Carolina), argued that no payments should be made until the Japanese government paid reparations for the attack on Pearl Harbor. Yet others pointed out that the people who had been put into camps were not Japanese citizens; they were mostly Japanese Americans who had been mistreated by their own government.

Representative Bill Frenzel, a Republican from Minnesota, said, "It is time for an apology." The Reagan administration agreed; President Ronald Reagan himself said, "No payment can make up for those lost years. What is most important in this bill has less to do with property than with honor. For here we admit wrong."

On August 10, 1988, the reparations bill became law. A tax-free award of twenty thousand dollars was offered to each survivor of the internment camps.

Consequences

Japanese Americans, especially surviving internees, believed that it was necessary to receive the U.S. government's apology. One said, "We were among the few who knew what we were fighting for during World War II. We were fighting for our rights but also for our parents and our children. Our rights were not just threatened, they were not even recognized. The nation needs to recognize our loyalty."

Perhaps the most fitting response was that of Fred Korematsu: "No money can compensate but an apology is appropriate. The greatness of a country lies in how it treats the weak. An apology is one way of making sure this never happens to anyone ever again."

Iraq Uses Poison Gas Against Kurds

> *In August, 1988, Iraqi aircraft dropped lethal gases on Kurdish settlements in the northern part of Iraq; thousands of Kurds were killed, and sixty thousand fled into Turkey.*

What: Human rights; Military conflict; Ethnic conflict

When: Late August, 1988

Where: Kurdistan (northern Iraq and southeastern Turkey)

Who:

SADDAM HUSSEIN (1937-   ), president of the Republic of Iraq from 1979

MASOUD BARZANI, a leader of the Democratic Party of Kurdistan

SERBEST LEZGIN, a leader of Kurdish guerrilla forces, the *pesh merga*

RONALD REAGAN (1911-   ), president of the United States from 1981 to 1989

GEORGE HERBERT WALKER BUSH (1924-   ), vice president of the United States from 1981 to 1989

A Beleaguered Minority

The Kurds, a Muslim people, live in the mountainous regions of northern Iraq, portions of southeastern Turkey, parts of the southwestern Soviet Union, and along Iraq's borders with Iran and Syria. Beginning in the early twentieth century, they tried to gain independence and form their own nation, but the European nations that signed the Treaty of Sèvres in 1920 did not abide by their promise to give the Kurds independence. Instead, the former Ottoman Empire was divided among five nations—Iraq, Iran, Turkey, Syria, and the Soviet Union—with some Kurds in each of these nations. In Iraq, for example, the Kurds made up almost a fifth of the population.

These nations were not sympathetic to the Kurds' culture. The Kurdish language was not used in schools attended by Kurdish children, and Kurdish books and folk music were often banned.

Iraq was particularly hostile to the Kurds, and

launched attacks on them in the 1960's. In March, 1988, just before the end of the Iran-Iraq War of 1980-1988, Iranian forces swarmed across the border and occupied Halabja, a mostly Kurdish city in Iraq. In response, Iraq dropped lethal gases on Halabja; most of the four thousand people who died were not Iranians but Kurds. Since the Kurds were considered troublemakers, the Iraqi government was not upset about these deaths.

The Poison-Gas Controversy

Iraq had already been using lethal gases, primarily cyanide, mustard gas, and nerve gas, against Iranian soldiers during the war. The Geneva Protocol of 1925 forbade chemical warfare, but in dropping lethal gases within its own borders, Iraq was not breaking international law.

Nations known to have chemical weapons in 1988 were the United States, the Soviet Union, France, and Iraq. Leaders of the Central Intelligence Agency (CIA) believed that Burma, Cuba, Egypt, Ethiopia, Iran, Israel, Libya, North Korea, South Africa, South Korea, Syria, Taiwan, Thailand, and Vietnam had these weapons as well. For years the United States and several European countries had been trying to broaden the Geneva Accord, so that manufacturing and storing lethal gas would be prohibited.

Production of poison gas in the United States had been stopped in 1969, when Richard M. Nixon was president. During the 1980's, however, President Ronald Reagan and his advisers were troubled by reports that the Soviet Union was developing lethal gases. Bills were introduced to allow the United States to start producing lethal gases once again, and Vice President George Bush twice cast tie-breaking votes in the Senate in favor of these bills.

On the morning of August 25, 1988, Iraqi warplanes dropped poison gas on the Kurdish town of Mesi in northern Iraq, close to the Turkish border, killing more than nine hundred people.

Some people who escaped across the border into Turkey said that the gas had smelled like rotting onions. They described how it burned their skin, eyes, and lungs, and how many people had collapsed, never to rise again. The survivors had oozing sores; their skin was scorched and badly discolored, and their hair fell out.

At about the same time, Iraqi planes dropped bombs armed with lethal gases on several other villages. In Butia, some people survived by drinking large quantities of milk, which absorbed some of the poison; others ran away. More than sixty thousand Kurds poured over the border into Turkey. Refugee camps were set up there.

Yet Turkey, which depended on Iraq for oil and for other trade, insisted that the United Nations did not need to investigate the gassing. Turkish officials said that 40 doctors and 205 other health officials had already investigated the situation and had found no evidence of chemical warfare in northern Iraq.

The Iraqis, who denied that they were gassing the Kurds, were trying to punish those who had sided with Iran during the Iran-Iraq War. Iraqi president Saddam Hussein wanted to wipe out the Kurdish guerrillas, or *pesh merga* (which means "those who face death"), who were fighting for an independent Kurdistan.

Masoud Barzani, a leader in the Democratic Party of Kurdistan, and Jellal Talibani, a leader in the Patriotic Union of Kurdistan, officially accused Hussein's government of practicing genocide against the Kurds. Western leaders believed the Kurds. U.S. secretary of state George Shultz called for immediate action, and the U.S. Senate quickly voted to cut back trade with Iraq.

The Iraqi government offered amnesty to the Kurds who had fled from Iraq; the refugees had to decide by October, 1988, whether to return to their homeland or to risk losing their citizenship. Realizing that Hussein's government considered them traitors, the Kurds did not trust the offer of amnesty. The leader of the *pesh merga*, Serbest Lezgin, said that he would return with his men only under orders from the nationalist leaders of Kurdistan.

Consequences

If the Kurds from Iraq, Iran, Turkey, Syria, and the Soviet Union had succeeded in forming their own nation, they would have created a country of approximately twenty million people—a country considerably larger than the population of Iraq, and more than twice the population of Syria. This new nation would have been more powerful than Iran, which had been much weakened by the Iran-Iraq War. The Soviet Union had its own problems with separatist movements among ethnic groups. None of the five nations with large Kurdish populations was ready to support the Kurds' goal of independence. Yet it was clear that Iraq was trying to exterminate the Kurds. Independence seemed their only hope for survival.

Most other nations did not rush to criticize Iraq, even though the Iraqis had used chemical weapons and were practicing genocide. Iraq owed France several billion dollars for weapons, so French leaders decided not to condemn Iraq openly. The British held back because they did not want to give the impression that they were siding with Iran against Iraq. It seemed that there were no simple solutions to the Kurds' problems and that no one was willing to protect them against Iraqi attacks.

Gorbachev Becomes President of Supreme Soviet

> *Communist Party chief Mikhail Gorbachev, who had called for the Soviet government and economy to be reformed, was chosen president of the highest legislative body in the Soviet Union.*

What: National politics
When: October 1, 1988
Where: Moscow, Soviet Union
Who:

MIKHAIL GORBACHEV (1931-   ), first secretary of the Soviet Communist Party from 1985 to 1991

ANDREI GROMYKO (1909-1989), Soviet foreign minister from 1957 to 1985, and president of the Supreme Soviet from July 2 to September 30, 1988

YEGOR LIGACHEV (1920-   ), a Conservative leader

BORIS YELTSIN (1931-   ), a member of the Communist Party Politburo, and first secretary of the Moscow City Party Committee from 1986 to 1988

Plans for Reform

In 1985 Mikhail Gorbachev became first secretary of the Soviet Union's Communist Party, and he immediately began trying to bring economic progress. *Perestroika* was the name he gave to his campaign to restructure industry and farming, giving managers some freedom from central planning. In spite of his reforms, however, production did not rise.

Gorbachev decided that political changes were necessary, because he believed that old-time party bureaucrats were standing in the way of reform. He thought that the national president needed to have greater authority, and that the bureaucrats would lose their power if the Soviet people were given more opportunities to express their wishes through voting.

In May, 1988, as Soviet troops began to withdraw from Afghanistan, Gorbachev and the Central Committee of the Communist Party called for changes in the political system. Gorbachev wanted limited terms for state and party officials, greater powers for a legislature that would be elected by the people, and reduced powers for the party in managing the economy.

The Russian Orthodox church celebrated its one thousandth birthday that spring. Also, the Soviet Supreme Court examined old records and decided that Grigory Zinoviev, Alexei Rykov, Nikolai Bukharin, and Lev Kamenev, who had been "purged" by Joseph Stalin (dictator of the Soviet Union from 1924 to 1953), did not deserve the punishment they had received.

A remarkable conference of five thousand party officials was held in Moscow from June 28 to July 2, 1988. There Gorbachev explained his proposals for political change so that the legislative bodies (soviets) would be given increased authority. He wished to be nominated as executive president of the Supreme Soviet. The president of this body would supervise the country's internal and foreign affairs and would also be chair of the powerful Defense Council.

Gorbachev also proposed that a new legislative body, called the Congress of People's Deputies, be elected in part through the popular vote. Television cameras recorded the proceedings as Gorbachev's ideas were discussed. His proposals were approved by the delegates on July 1.

Winning the Presidency

Andrei Gromyko, who had been Soviet foreign minister for many years, was then serving as chairman of the Presidium (president) of the Supreme Soviet. Gromyko resigned his post on September 30, and Gorbachev was unanimously elected to take his place the next day. (In seeking this office, Gorbachev had followed the examples of earlier Soviet leaders such as Leonid Brezhnev, Yuri Andropov, and Konstantin Chernenko.)

There were other important shifts in government positions around the same time. Anatoly Dobrynin, a former ambassador to the United States, lost his office as secretary of the Central Committee of the party. Vladimir Kryuchkov was appointed to head the Committee for State Security (KGB), and Yegor Ligachev, a powerful figure in the conservative wing of the party, was moved from the commission on ideology to chair the commission on agriculture—a difficult post.

On September 30, new appointments to the Politburo allowed Gorbachev to gain a clear majority of votes for his state proposals. The Politburo was quickly called into an unscheduled session—so quickly that Ligachev, who had been on vacation, had little time to organize opposition to Gorbachev's reforms. An opponent from the other side of the party, Boris Yeltsin, had argued that reforms needed to move more quickly, but

he had been removed from the Politburo earlier in the year.

The rest of Gorbachev's June proposals were approved on December 1, 1988, when the Supreme Soviet passed the proposed changes to the constitution, along with a new electoral law. The country began preparing for March elections to the new Congress of People's Deputies. The 2,250 delegates in this congress would be chosen every five years. They would meet each year and would elect their own chairperson; they would also elect the 450 members of the Supreme Soviet.

The new president would have broader powers than the former president of the Supreme Soviet. The president and other officials would be limited to two five-year terms of office. The new Supreme Soviet would deal with day-to-day issues of lawmaking, and it would supervise a smaller Council of Ministers.

After these changes were approved, President Gorbachev traveled to the United Nations center in New York to announce that the Soviet Union

Politburo members vote to remove Andrei Gromyko as Soviet president. They then named Mikhail Gorbachev (bottom right) as the new head of state.

2267

was reducing its military forces. He had to hurry home, though, to deal with the terrible earthquake that struck Armenia, which was then a Soviet republic. More than twenty-five thousand people died in this earthquake. Poor highways made it hard for government and private relief agencies to bring in supplies to help the survivors.

Consequences

President Gorbachev had won the changes he wanted, but his support was weakened by continuing problems in the Soviet Union. The government decided that bread prices could no longer be kept artificially low, but when they were allowed to rise the Soviet people became upset. Though prices rose, there were still shortages of food. Ethnic groups and leaders of some of the republics were clamoring for independence, and Gorbachev did not know how to respond to their demands.

The Soviet nation became sharply divided. Yeltsin, who had been elected president of the Russian Republic, urged the nation to adopt more aggressive reforms, while traditional party officials warned that reform would bring ruin.

Trying to steer a middle course, Gorbachev made both sides angry. Conservatives tried to force him out of office in August, 1991. Although he managed to survive this threat, the Soviet Union began to dissolve in the fall of 1991 as various republics chose to separate themselves from the union. Gorbachev lost his position as president, but remained an important figure in Russian politics.

John D. Windhausen

Congress Passes Comprehensive Drug Bill

A bill to fight drug abuse increased punishments for drug offenders and set aside more money for treatment programs.

What: Law; Social change
When: October 22, 1988
Where: Washington, D.C.
Who:
WILLIAM J. BENNETT (1943-   ), head of the Office of National Drug Control Policy from 1989 to 1991

Drug Plague Hits America

In 1988, the United States faced a serious drug problem. It was estimated that there were 5 to 6 million regular cocaine users and 1.5 million heroin or cocaine addicts. Use of cocaine and a cheap related drug, crack, had more than doubled since 1982. Deaths resulting from cocaine use rose from 470 in 1984 to 1,582 in 1988.

Also, a drug called "crank" had become a new menace. Known sometimes as "poor man's cocaine," it is made with ether and other chemicals. In 1988, authorities seized 667 crank laboratories, eight times the number closed down in 1981.

Trade in crack and crank was a part of life in the ghettos of many American cities, and drug-related murders had dramatically increased in Washington, D.C., Houston, New York, and Los Angeles. In 1987, 55 percent of the murders in Los Angeles were thought to be drug-related. Well-organized drug gangs had spread to small-town and rural America, bringing an increase in violent crime and challenging the skills and resources of local police. In 1988 it was estimated that people convicted of drug-related crimes made up a third of all federal prisoners.

Members of Congress decided to address the issue, but there was much debate in the House and Senate over what an antidrug bill should contain. A bill written in the House included the rule that all operators of transportation machinery and all nuclear-power workers would be required to be tested for drug use. This provision showed that many Americans were concerned about the dangers of using drugs in the workplace. In 1987, for example, an Amtrak worker who was partly responsible for a train accident was found to have used marijuana. In the final day of negotiations over the antidrug bill, however, the mandatory-testing rule was dropped.

Another issue that was debated was whether evidence that had been obtained illegally should be permitted in federal court. The House version of the bill, which was much tougher than the Senate version, stated that in some cases this evidence should be allowed. But eventually members of the House and Senate agreed that allowing this kind of evidence to be presented in court would be a threat to civil liberties.

Taking a Stand

An antidrug bill was passed in October, 1988, with support from both Democrats and Republicans in the House and Senate. The bill allowed more serious penalties for people who used or possessed illegal drugs such as heroin, cocaine, and marijuana—even in small amounts. For example, possessing less than five grams of cocaine would now be considered a felony.

Persons who possessed or used illegal drugs could lose student loans or mortgage guarantees; they could be forced out of public housing; pilots could lose their licenses. They could also be fined up to fifteen thousand dollars. These penalties were not mandatory; judges were given leeway to decide which penalties to assign. Also, the person convicted of drug possession or use would not lose welfare benefits, Social Security, Medicaid, or Medicare.

The bill made serious drug-related crimes subject to the death penalty. Murders committed

2269

or organized by drug traffickers and murders of police officers during drug-related crimes could now be punished by death.

The antidrug bill required U.S. chemical manufacturers to keep a record of sales of certain chemicals that react together to form illegal drugs, and of sales of chemicals that are used in processing raw materials into drugs. Many drug producers in South America and Asia had been buying these chemicals, and under the bill U.S. companies were required to report suspicious sales. The goal of this part of the bill was to cut down on the amount of illegal drugs produced.

Another important part of the antidrug bill was the appointment of a national director of drug policy. This person would serve on the cabinet and would oversee antidrug law enforcement, drug rehabilitation, and education to prevent drug abuse. Lawmakers believed that this position was needed because in the past, government agencies such as the Drug Enforcement Administration, the Customs Service, and the State Department had not cooperated very well together in the fight against drugs.

The bill showed a concern to change the focus of attack on drugs. Fifty-five percent of the new antidrug budget was to be used for treatment and education, with the goal of immediately treating any addict who wanted help. Before the bill, only 30 percent of the government's budget to fight drugs had been spent on prevention and treatment; the other 70 percent had been used for law enforcement and trying to stop drugs from being smuggled over the border. Lawmakers believed that the border patrols were fairly useless, because huge amounts of drugs were still flowing into the United States from South America.

Many people were pleased that Congress had passed this bill; it was considered an important step in the fight against illegal drugs. Critics pointed out, though, that only a small fraction of the $2.8 billion assigned to new antidrug programs was available right away. The rest had to come from budget cuts or tax increases within the following year.

Critics of the bill also claimed that the threat of the death penalty would hardly stop the top drug dealers, who were already working under the greater risk of being murdered by drug-dealing rivals. They could make so much money in their trade that they were willing to take serious risks.

Consequences

In his inaugural address in January, 1989, the new U.S. president, George Bush, made it clear that fighting drug abuse would be a top priority during his time in office. He called for a "war on drugs" and promised, "This scourge will stop."

Bush appointed William J. Bennett as the head of the new Office of National Drug Control Policy. In September, 1989, Bennett, who became known as the "drug czar," presented a plan that had the goal of reducing drug use in the United States by 50 percent over the following ten years. He pointed out that although casual drug use was decreasing, drug addiction was still increasing. Bennett predicted that the war on drugs would be long and expensive, and he asked for tougher law enforcement against those who sold drugs.

Bryan Aubrey

Bhutto Is First Woman Elected Leader of Muslim Country

> *For the first time since the seventh century, when Muhammad founded Islam, a woman was elected as the head of government in a country dominated by Muslims.*

What: National politics; Civil rights and liberties
When: November, 1988
Where: Pakistan
Who:
ZULFIKAR ALI BHUTTO (1928-1979), founder of the Pakistan People's Party, and president of Pakistan from 1972 to 1977
BENAZIR BHUTTO (1953-   ), his daughter, prime minister of Pakistan from 1988 to 1990 and 1993 to 1997
MOHAMMAD ZIA UL-HAQ (1924-1988), president of Pakistan from 1977 to 1988
GHULAM ISHAQ KHAN (1915-   ), president of Pakistan from 1988

Conflict of Values

Pakistan is a very new country, created in 1947 when the British Indian Empire was partitioned to make a homeland for the Muslims of the area. From the beginning, the Islamic religion was a strong influence on politics and government in Pakistan. But there was another powerful force: the Western traditions left behind by the British.

These two forces were not always in harmony. Islam is a spiritual commitment that is supposed to rule all aspects of Muslims' lives. To fundamentalist Muslims, it makes no sense to separate religion and the law of the state. They act according to a sense of religious duty rather than a sense of civic duty. For example, women can expect to be protected by men, but this is considered a man's obligation rather than a woman's right.

The British political tradition, on the other hand, is based upon democracy and the idea that the government should not favor one religion over another. In this tradition all people are considered more or less equal; the law should be neutral, and all citizens have a right to equal justice.

The attempt to combine these two traditions was very frustrating for Pakistan. Martial law was imposed in 1958 and again in 1969, both times after constitutional government had failed. In 1971, the eastern portion of the country seceded to become Bangladesh, and martial law was imposed for the third time.

A new government and a new constitution emerged out of this bitter experience, with the promise of a more open, democratic political system. Soon, however, Prime Minister Zulfikar Ali Bhutto and his Pakistan People's Party (PPP) began to be strongly opposed by conservative groups—wealthy landowners and the Ulema, or Islamic religious leaders. Bhutto tried to please these groups by prohibiting alcoholic beverages and closing nightclubs.

The Struggle Continues

Bhutto's attempts were not enough. In 1977, his government was dismissed by military leaders, who said that the country was unstable and needed to be under martial law. General Zia ul-Haq, commander of the army, promised that elections would be held soon and that civilian rule would return. But ten years later, this promise had still not been kept. The wealthy landowners had thrived, but others had not. Working-class and middle-class people became quite unhappy with military rule.

Zia intended to bring every part of Pakistani society into conformity with Islam. The PPP, which remained very popular, was an obstacle to this goal. Zia had Bhutto executed, but many Pa-

2271

kistanis continued to honor the memory of their fallen leader.

After Bhutto's death, his widow, Nusrat Bhutto, became the leader of the PPP; but she was in poor health and was not skilled in politics. So it was that the real leader came to be Bhutto's daughter, Benazir. (Because Bhutto's sons had decided to become involved in violent action against the government, they could not be directly involved in political affairs.)

During the decade of martial law, the PPP and other opposition groups were harassed and persecuted by Zia's government. Many opposition leaders, including Benazir Bhutto, spent time in jail or under house arrest.

In the mid-1980's, parties that were opposed to Zia banded together to form the Movement for the Restoration of Democracy (MRD). Led by Bhutto and the PPP, this coalition included very different groups, from conservative religious organizations to socialists. They joined in public demonstrations, but their efforts did not succeed in removing Zia from power.

By 1988, the situation seemed to be stuck. MRD members were becoming frustrated, but so was General Zia; his program of "Islamization" was going nowhere. Then, suddenly, an accident changed the whole situation: On August 17, 1988, a plane crash took the lives of Zia, the American ambassador, and several American and Pakistani generals.

The president of the senate, Ghulam Ishaq Khan, had to take over. He quickly appointed Benazir Bhutto, as the leader of the largest political group, to the post of acting head of government. Elections held in November confirmed the power of the PPP and Bhutto. She had become the first woman head of state chosen by popular election in a Muslim country.

Consequences

From the beginning, Bhutto's time in office was stormy. The MRD coalition fell apart, and many who had formerly allied themselves with the PPP began opposing it. Provincial governments in Punjab, Baluchistan, and North-West Frontier Province refused to follow the central gov-

ernment's orders, and conservative Muslims were very unhappy that a woman was running the country.

Bhutto also had personal problems: She was embarrassed by the behavior of her husband, who seemed indifferent to public opinion. Her government floundered, unable to gain a sense of direction.

Disorder and violence continued in various parts of Pakistan, and Bhutto opponents believed that the country was out of control. In the end, Bhutto's government was brought down by the same man who had placed her in power: Ishaq Khan. As president, he announced that Bhutto's government had shown that it was incapable of governing. By his order, she was formally removed from office on August 6, 1990.

Ishaq Khan then asked opposition parties to form a temporary government until new elections could be held. In the elections of October

Benazir Bhutto (right), with her husband, Asif Zardari, in Islamabad.

24, 1990, the PPP suffered a serious defeat. The new government, under Nawaz Sharif, announced that Sharia (Islamic law) would be the highest law of the land. With political power in the hands of their enemies, Bhutto and the PPP found themselves facing accusations of corruption and misuse of office.

In October, 1993, Bhutto staged a comeback, when the PPP won 86 of 207 seats in the national elections. After forming a tenuous coalition with several small parties, she became prime minister once again. Her position was strengthened by the election of a PPP candidate for president. However, Bhutto's popularity continued to suffer because of the prominent roles in government she assigned to her relatives, including her husband. At the same time, she was perceived as slow to act on the country's mounting problems. In new elections in early 1997, the PPP lost to the party of Nawaz Sharif, who again succeeded Bhutto as prime minister.

Louis D. Hayes

George Bush Is Elected U.S. President

With room to spare, George Bush won the 1988 presidential election, continuing the pattern of "divided government"—a Republican president and a Democratic Congress.

What: National politics
When: November 8, 1988
Where: United States
Who:

GEORGE HERBERT WALKER BUSH (1924-   ), vice president of the United States from 1981 to 1989, and Republican presidential candidate in 1988

DAN QUAYLE (1947-   ), U.S. senator from Indiana, and Bush's running mate

MICHAEL DUKAKIS (1933-   ), governor of Massachusetts from 1975 to 1979, and from 1983 to 1991, and Democratic presidential candidate in 1988

LLOYD BENTSEN (1921-   ), U.S. senator from Texas, and Dukakis's running mate

Bush Wins the Image War

George Bush had been vice president under Ronald Reagan during the 1980's, but in the presidential campaign of 1988 he had to convince skeptical Republicans that he would be a strong candidate. The Iowa caucus on February 8, 1988, was a blow to Bush, as he finished third. In the primaries that followed, however, he did better. On March 8, known as "Super Tuesday," Bush won all the sixteen primaries held in various states. This triumph guaranteed him the Republican nomination.

Meanwhile, a number of Democrats were competing for their party's nomination. Massachusetts governor Michael Dukakis gradually pulled into the lead, collecting enough delegates to be sure of the nomination before the Democratic National Convention in late July. Dukakis selected Texas senator Lloyd Bentsen as his vice-presidential running mate. The Democratic convention closed on an optimistic note, for public opinion polls showed Dukakis well ahead of Bush.

Trying to gain public support at the Republican National Convention in early August, Bush got off to a shaky start when he named Dan Quayle, a fairly young senator from Indiana, as his running mate. Quayle soon ran into trouble when reporters asked whether he had joined the National Guard in the 1960's to avoid combat in the Vietnam War. By sticking with Quayle, however, Bush helped to overcome his weak image. The rest of the convention went well, with Republicans unified behind their candidate.

Still behind in the polls, the Bush campaign began trying to put negative labels onto Dukakis. The Democratic candidate was pictured as a bad governor who did not care about the problem of violent crime and who had little experience in national politics. In speeches, Bush questioned Dukakis's patriotism and made him seem an extremist.

Instead of responding to these accusations, Dukakis pointed out that Bush had no experience in heading a state or national government. He also criticized the Republicans for the Iran-Contra scandal and for having no real plans to make health care more affordable for Americans.

When polls showed that Bush had moved ahead, the Dukakis campaign began criticizing Quayle and finally answered Bush's charges. But it was too late; Bush's negative campaign had worked. In fact, the polls showed that most Americans did not believe Bush's campaign promises (for example, that he would not raise taxes), but they still preferred him to Dukakis. In televised debates, Bentsen handled himself much better than Quayle, but Bush was able to hold his own with Dukakis.

2274

A Split Government

On November 8, 1988, Bush won the popular vote by about 54 percent to 46 percent; he won 426 of 538 electoral votes. Voter turnout was low, just around 50 percent. In both houses of Congress, the Democrats kept their majority.

Dukakis's chances of winning had probably been slim to begin with, because the Democratic Party had come to be full of divisions. As a Northeasterner, Dukakis had tried to balance his ticket by choosing Bentsen, a Southerner, as his running mate, but many Democrats considered Bentsen too conservative. Also, many African Americans were frustrated that Jesse Jackson, who had finished second to Dukakis in the primaries, had not been invited to be Dukakis's running mate.

In any case, the Democrats were no longer the majority party they had been from 1932 to the early 1970's. Republicans had made some gains, but the biggest change was that more and more voters were calling themselves independents.

Republican candidates faced the challenge of working as a minority party. Yet there was more unity within the Republican Party, and when Americans began turning away from many of the "radical" political ideas of the 1960's, many Republican candidates had benefited. Bush had followed Reagan's approach in campaigning: He appealed to conservative "right-wingers" on such issues as abortion and gun control, but he also invited the support of moderates by stating his commitment to a "kinder and gentler" America.

Congress, however, was another matter. There divisions in the party did not hurt Democrats, for they could focus their campaign on the people of one state and region rather than trying to please many different groups. So Republicans found it very hard to gain a majority in Congress.

When Ronald Reagan, called "the Great Communicator," was elected in 1980, Republicans hoped that their time had finally come. In fact, the Senate became Republican in 1980 for the first time since 1953. Yet Reagan was unable to keep this trend going; by 1986, instead of making progress in the House of Representatives, Republicans had lost their majority in the Senate.

The 1988 election continued this pattern of "divided government," with a Republican president and a Democratic Congress. This meant

George Bush.

that the executive and legislative branches would be in constant conflict over the budget (especially the budget deficit), social welfare programs, and other important issues.

Consequences

During his campaign, Bush had promised to be both the "education" and "environmental" president; he also promised, "No new taxes!" As president, however, he did not achieve any quick successes in improving schools or cleaning up the environment, and he backed off the "no-taxes" promise just before the 1990 elections for Congress.

Like Reagan, Bush seemed always to be fighting Congress about national issues, with neither side winning. He did manage to appoint two conservatives to the Supreme Court: David Souter and Clarence Thomas. In November, 1992, Bush lost in his bid for reelection and was succeeded as president by Arkansas's Democratic governor, Bill Clinton, the following January. Although Bush's presidency ended in 1993, the men he appointed to the Court were still making decisions on the bench into the twenty-first century.

Ira Smolensky

2275

Namibia Is Liberated from South African Control

On December 22, 1988, the Namibian people gained a plan for independence in 1990 that would end more than one hundred years of colonial domination and paving the way for majority rule. Namibia would be the last African nation to shed colonial rule.

What: Political independence
When: December 22, 1988
Where: United Nations, New York City
Who:

SAM NUJOMA (1929-   ), president of the South West Africa People's Organization (SWAPO), and president of Namibia from 1991

JOSÉ EDUARDO DOS SANTOS (1942-   ), president of Angola from 1979

FIDEL CASTRO (1926 or 1927-   ), premier of Cuba from 1959

FREDERIK W. DE KLERK (1936-   ), president of South Africa from 1989 to 1994

CHESTER CROCKER (1941-   ), U.S. undersecretary of state for African Affairs from 1981 to 1989

A Mandate Territory

The area now known as Namibia was ruled by Germany from 1884 until shortly after the outbreak of World War I. Until the Germans' conquest, various ethnic groups—the Ovambo, the Nama, the Herero, the Damara, the San, the Kavango, and the Tswana—lived in the area. The Germans had to fight and kill many Nama and Herero in order to establish their claim to what was then called South West Africa.

In 1915, during World War I, South Africa invaded, taking South West Africa from the Germans. During the next seventy-three years, the political and economic rights of the black Namibian people were denied, and South Africa used force against Namibians who rebelled.

For a short time, other nations accepted South Africa's right to govern South West Africa under a mandate from the League of Nations. Later, however, the United Nations told South Africa that it must make the territory a trust and prepare to give it independence. South Africa refused and began instead to introduce apartheid—a strict system of racial segregation—into South West Africa.

Most African trust territories gained independence in the early 1960's, and many of their governments argued that apartheid in South West Africa violated the principles of the U.N. Charter. By the Native Nations Act, South Africa created homelands for each ethnic group in South West Africa, so that the Africans would be divided up and could not unite into a single group. On October 27, 1966, the U.N. General Assembly voted to revoke South Africa's mandate because it had not governed the territory properly.

The next year, the U.N. appointed an eleven-person Council for South West Africa to govern the territory, which it renamed Namibia. On June 21, 1971, the International Court of Justice declared that South Africa's presence in Namibia was illegal and that it must withdraw immediately.

Freedom at Last

Within Namibia, church organizations organized petitions, and workers began a general strike. The South West Africa People's Organization (SWAPO) helped organize the resistance, even though Sam Nujoma and its other leaders had been living in exile for many years. The South African government responded by taking hundreds of people into custody and torturing them. It is not known how many Namibians were executed.

Over the years, South Africa used many different methods to punish Namibian rebels: assassination, mass murder, imprisonment in crowded, filthy jails, injections of depressant drugs, and electric shocks. Fighter planes were sent to drop bombs on the Nama. In the schools, black Namibian children were taught that South Africa was good and SWAPO was bad.

A paramilitary unit called the Koevoet (also known as Takki Squads or Etango) was South Africa's most feared weapon. The three thousand members of this fighting force were trained in brutality, and they did not hesitate to beat, torture, rob, rape, and murder.

Meanwhile, a number of transnational corporations were allowed to extract Namibia's mineral riches, even after the United Nations ruled that it was illegal for these corporations to be there. South Africa did not regulate how the corporations treated their black workers; most of

the laborers were underpaid and did not have good benefits.

Though South Africa defied the U.N. for many years, it finally began to show some willingness to back down. There were negotiations about the affairs of several southern African nations in the United Nations, and a regional agreement that included a plan for Namibian independence was signed on December 22, 1988. The main negotiators were representatives of José Eduardo dos Santos, president of Angola, and Frederik Willem de Klerk, president of South Africa. Chester Crocker, U.S. Undersecretary of State for African Affairs, helped work out many of the details of the agreement. Because Cuba had stationed troops in Angola for many years, Cuban premier Fidel Castro was also involved in the talks.

According to the agreement, if Angola sent the Cuban troops home and met certain other conditions, then the Namibians could have independence elections. The conditions were met, and a U.N. team supervised a series of elections in Namibia. SWAPO was the winning party in the

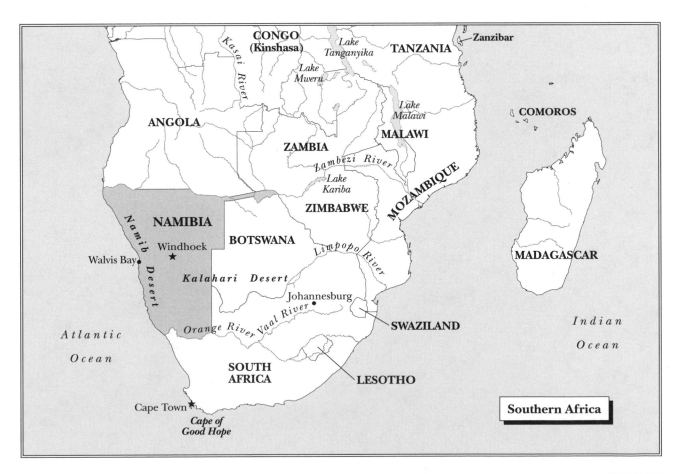

elections, and Namibia officially became an independent nation on March 21, 1990.

Consequences

After the first elections in Namibia, representatives from the competing political parties gathered to write a new constitution. After the constitution was presented to the people and some changes were made, it was approved. It is one of the world's most democratic constitutions.

The Namibian constitution guarantees the human and political rights of all of its citizens "regardless of race, colour, ethnic origin, sex, religion, creed or social or economic status." Apartheid laws and any other kinds of racial discrimination were made illegal, and women have equal rights with men.

In 1991, the Namibian government was working on writing a labor code that would respect the rights of workers as well as owners. The police squads were done away with, and a civilian police force and a new military were trained. The legal system guaranteed due process and a speedy trial. Newspapers were free to criticize the government.

Sam Nujoma was elected Namibia's first president. He remained in power through the 1990's. In the national elections of November-December, 1999, both he and his SWAPO allies in the National Assembly were reelected by wide margins. The president's term of office was set at five years, with a limit of two terms. The constitution spread political power among the president, the national assembly, the prime minister, the cabinet, the supreme court, and regional officials.

Eve N. Sandberg

Vietnamese Troops Withdraw from Cambodia

After ten years of occupation, Vietnamese armed forces pulled out of Cambodia in 1989.

What: Political independence
When: 1989
Where: Cambodia and Vietnam
Who:
POL POT (1928-1998), secretary general of the Kampuchean Communist Party from 1962, and prime minister of Democratic Kampuchea from 1976 to 1979
HUN SEN (1951-   ), prime minister of the People's Republic of Kampuchea from 1985
HENG SAMRIN (1934-   ), president of the People's Republic of Kampuchea from 1979
NORODOM SIHANOUK (1922-   ), the Cambodian monarch
SON SANN (1911-   ), leader of the Khmer People's National Liberation Front

Vietnam Invades Cambodia

When the Vietnamese communists finally won the battle to reunify Vietnam in 1975, Communist groups in neighboring Laos and Cambodia gained important victories in their own civil wars. Each of these three nations now had a Communist government, but they became rivals because their ideas and alliances were different.

The Cambodian communists (known as the Khmer Rouge) were backed by the People's Republic of China. The main Khmer Rouge leader was Pol Pot, whose original name was Saloth Sar. During the years of Khmer Rouge rule over Cambodia—1975 to 1979—Pol Pot's government tried to change the country into a self-sufficient communist state where all citizens would be "equal," Democratic Kampuchea.

The Khmer Rouge's policies were brutal. City dwellers were forced to leave their homes and belongings and move to the countryside. Teachers, government workers, and lawyers were considered dangerous, and many of them were tortured. Tuol Sleng, which had once been the largest primary and secondary school in the capital, Phnom Penh, became a place of execution. It has been estimated that perhaps a million Cambodians died from starvation and repression during the Khmer Rouge years.

Meanwhile, there were clashes between Cambodian and Vietnamese troops along the border. Vietnam, with the support of the Soviet Union, invaded Cambodia in late December, 1978. Soon Phnom Penh was captured, the Khmer Rouge government was thrown out, and a new pro-Vietnamese government was established: the People's Republic of Kampuchea.

The Khmer Rouge leaders had been driven out, but most of them survived the invasion. They set up bases near the border with Thailand and began recruiting fellow Cambodians to help in their struggle against the Vietnam-backed government. Their rebellion continued throughout the ten years of Vietnamese occupation, with help from China and the United States.

The Vietnamese Withdraw

The Treaty of Solidarity, signed on February 18, 1979, by representatives of Vietnam and the new Cambodian government, made it legal for Vietnam to keep part of its army posted in Cambodia. The leaders of the People's Republic of Kampuchea, Heng Samrin and Hun Sen, continued to cooperate with Vietnam. Because the new government had been imposed by the Vietnamese, the United Nations refused to recognize it; Cambodia was represented in the U.N. by Khmer Rouge delegates.

During the years of Vietnam's occupation, Vietnam claimed that its soldiers were helping with disaster relief and construction projects in Cambodia. Yet the Cambodian population was growing quickly, prices rose, and the Khmer Rouge kept fighting. Eighty percent of Cambodians were farmers, and they remained poor.

In the late 1980's, there were important political changes in the Soviet Union. Vietnamese leaders realized that they could not count on the Soviets to keep sending financial help—and also that the long rivalry between the Soviet Union and China was winding down. Perhaps it was time to pull out of Cambodia.

On April 5, 1989, Vietnam announced it would no longer insist that certain conditions be met before its troops left Cambodia. All of its military forces would leave by the end of September, 1989, even if Hun Sen's government had not reached a political settlement with the Khmer Rouge by that time.

World leaders were interested in trying to help the Cambodians reach a peaceful settlement of their differences. A major international conference was called in Paris in August, 1989, but there was no agreement on one very important issue: whether the Khmer Rouge should be allowed to be part of a new coalition government in Cambodia. Two other opposition groups were demanding participation as well: the Khmer People's National Liberation Front (KPNLF), led by Son Sann, and the forces led by Prince Norodom Sihanouk and his son, Prince Norodom Ranariddh.

Soon after the Paris conference disbanded, it became apparent that the opposition forces were preparing to fight harder than ever against the Phnom Penh government. The Khmer Rouge were receiving weapons from China, while the United States and the Association of South East Asian Nations (ASEAN) gave extra help to the KPNLF and the Sihanoukists.

Beginning in mid-September, as Vietnamese troops left western Cambodia, Khmer Rouge forces began attacking. It seemed the Khmer Rouge strategy was to capture as much territory as possible and in that way to win participation in the coalition government. The Khmer Rouge began forcing thousands of refugees living in camps near the Thai border to move to recently captured territory farther inside western Cambodia. There were reports that the other two opposition forces were doing the same.

In early September, Heng Samrin announced that the Phnom Penh government would not punish Cambodians who had supported opposition groups. But the government was using surprise raids in the city to round up young men and force them into the armed forces. A steady stream of army deserters, many between the ages of fifteen and nineteen, fled to Thailand.

Consequences

It has been estimated that about 23,500 Vietnamese soldiers were killed during the occupation of Cambodia, and that another 55,000 were seriously wounded. Those who returned to Vietnam were congratulated by government officials, but because of Vietnam's financial hardships, many were dismissed from service.

At the time of the withdrawal, the Khmer Rouge had about thirty thousand active soldiers, and its leaders were eager to become involved in the government of Cambodia. So the main result of the Vietnamese withdrawal was that the Cambodian civil war became more intense. In the northwestern part of the country, where most of the fighting took place, civilians suffered greatly. Many lost arms or legs when land mines exploded in rice fields; at least 10 percent of them had to leave their homes. In early 1990 there were 250,000 Cambodian refugees in United Nations camps near the Thai border; by the spring of 1991, the number had risen to 330,000.

In late 1991, Cambodia began forming a new coalition government that would prepare the country for free elections. Prince Sihanouk returned to Phnom Penh to work out an alliance with Hun Sen, Norodom Ranariddh, the former prime minister Son Sann, and Pol Pot. United Nations troops kept the peace while negotiations were hammered out. Yet by mid-1992, Pol Pot remained in hiding near the Thai border, refusing to disarm his forces until every Vietnamese soldier was out of the country, and Hun Sen was still in place as prime minister. The ancient legacy of strife between Vietnam and Cambodia made the likelihood of a lasting peace appear small.

Soviet Troops Leave Afghanistan

After a decade of bloody fighting, the Soviet Union ended its occupation of Afghanistan.

What: Political aggression
When: 1989
Where: Afghanistan
Who:
MOHAMMAD DAUD (1909-1978), president
of the Republic of Afghanistan from
1973 to 1978
NUR MOHAMMAD TARAKI (1917-1979), first
secretary of the People's Democratic
Party of Afghanistan (PDPA) from
1978 to 1979
HAFIZULLAH AMIN (1929-1979), the PDPA
leader who ruled Afghanistan from
September to December, 1979
BABRAK KARMAL (1929-   ), leader of
the PDPA from 1979 to 1986
NAJIBULLAH AHMEDZAI (1947-   ), leader
of the PDPA from 1986
MIKHAIL GORBACHEV (1931-   ), first
secretary of the Soviet Union from
1985 to 1991

The Soviets Invade

Afghanistan, a nation about the size of Texas, has been called "the highway of conquest," because invaders have swept through the region since the time of the ancient Persians and Greeks. Modern Afghanistan is a land of various ethnic and cultural groups; most of its people see themselves as members of their clan, family, or village, rather than as Afghans.

Afghanistan was also a poor nation. Even by the late 1970's there was no railroad system, and there were few paved roads. Life expectancy was around forty years, and the average Afghan earned only $168 a year. For most rural Afghans, the national government in Kabul seemed to have nothing to do with them.

For decades Afghanistan was ruled by a weak king who was friendly with the Soviet Union. In July, 1973, Prince Mohammad Daud led leftist military officers to overthrow the king and set up a republic. There was no real democracy, but Daud did promise reforms and modernization.

Within a few years, Afghanistan was tired of Daud's rule. The Muslims were unhappy when he opposed the Muslim Brotherhood, while he alienated workers by breaking up strikes. There were food shortages and growing unemployment. By the late 1970's, hundreds of thousands of Afghans had had to go abroad to look for work.

The People's Democratic Party of Afghanistan (PDPA) began recruiting new members among army and air force officers, and they prepared to overthrow Daud and create a communist government. They succeeded in April, 1978; the new leader of the country was Nur Mohammad Taraki, first secretary of the party.

The PDPA tried to bring land reform, schools, women's rights, and modernization, but the conservative country people wanted no part in the changes. These people were deeply religious, while Taraki's government was seen as atheistic.

Troubled by his government's failures, Taraki began to blame Hafizullah Amin, who led a dissident group within the PDPA. But Taraki's plan to arrest Amin backfired; Amin escaped and then returned to the People's Palace with his supporters. Taraki was captured, and on October 9 the new government announced that he had died from a long illness. The "illness" was probably a death sentence imposed by Amin.

The leaders of the Soviet Union were not at all pleased with Amin as Afghanistan's new ruler. He was blamed for making a mess of the reform program, and he had severely embarrassed the Soviet Union by murdering Taraki only days after Taraki's visit to Moscow. Also, Amin was making open statements about how Afghanistan needed to be independent.

Civil war was already raging in Afghanistan when, on the evening of December 27, 1979, a special Soviet assault force attacked the palace in Kabul and executed Amin. Four days later, Babrak Karmal arrived from the Soviet Union and became the new PDPA first secretary and president of Afghanistan.

The Pullout

Karmal brought noncommunists into his cabinet and softened many of the reform programs. Amnesty was offered to refugees, and the land-reform laws were changed. Also, Karmal tried to portray himself as a good Muslim. Official speeches would begin with an Islamic prayer; a new flag with the old Muslim colors—black, red, and green—replaced the former red flag.

Nevertheless, Karmal did not win the people's support. They did not trust him, because he seemed too dependent on the Soviet Union. Soviet troops remained in Afghanistan to protect Karmal's government; the war continued, with fierce fighting in the countryside.

The rebels, known as *mujahideen*, considered themselves to be fighting a holy war, and they were determined to resist the Soviets to the bitter

end. They were scattered throughout the rugged countryside, so the Soviet army could not simply plan a major battle that would wipe them out. Their tradition of "blood for blood" meant that every time the Soviets killed an Afghan, the survivors vowed to avenge their dead.

The rebels succeeded in winning other countries over to their cause. Rebels often slipped across the border into Pakistan or Iran, where the Soviets could not pursue them. The United States sent huge shipments of weapons and cash.

By the time Mikhail Gorbachev came to power in the Soviet Union, the war in Afghanistan was a major problem for the Soviet government. It seemed impossible to defeat the rebels, and the war was unpopular among the Soviet people. Gorbachev decided that there must be a settlement that would allow the Soviets to pull out.

In May, 1986, Karmal was replaced with Najibullah Ahmedzai, who was asked to make a political compromise with the *mujahideen.* On May 15, 1988, the Soviets announced that they intended to begin removing their troops. By the time the last Soviet soldier left in 1989, the

Soviet troops leave Afghanistan on the Salang Highway.

Hulton Archive

Soviets had suffered 13,310 dead and 35,478 wounded. The number of dead Afghans could only be guessed at.

Consequences

Unfortunately, the fighting and dying did not end when the Soviets left. The civil war between the PDPA government and the *mujahideen* continued to rage. As the Soviets left, the rebels boasted that the days of Najibullah and the PDPA were numbered; Western experts believed they were right. Yet soon the Kabul government made an unexpected show of strength.

Without the Soviets as a common enemy, the rebels began to fight among themselves over their disagreements. Furthermore, not all Afghans opposed the PDPA's reforms. Many people in the cities decided to stand by the government.

For many of the rebels, education was the same as atheism. Some teachers suspected of teaching women to read were executed in public. It was not just communism that was the enemy of the *mujahideen*; they were also opposed to what people in the West would call progress.

So it was that many urban Afghans—even those who opposed socialism—believed they had little choice but to support the government. These people did not want to see a conservative Islamic regime that would stop all modernization.

William A. Pelz

Kashmir Separatists Demand End to Indian Rule

> *The Kashmir separatist movement insisted that the Kashmiri people should have the right to make their own political allegiances.*

What: Civil rights and liberties
When: 1989
Where: Kashmir, India
Who:
SHEIKH MOHAMMAD ABDULLAH (1905-1982), a leader of Kashmir
JAWAHARLAL NEHRU (1889-1964), prime minister of India from 1947 to 1964
RAJIV GANDHI (1944-1991), Nehru's grandson, prime minister of India from 1984 to 1989
FAROOQ ABDULLAH (1937-   ), chief minister and prime minister of Jammu and Kashmir
JAGANMOHAN, governor of Jammu and Kashmir

A Disputed Land

A beautiful region in the north of the Indian subcontinent, Kashmir was ruled by Hindus until the fourteenth century, when Muslims took over. The Sikhs took Kashmir from the Muslims in 1819, but the British won control in 1846. Later the British sold Kashmir to the Hindu maharajah, making it a "princely state" under the protectorate of British India.

When India became independent on August 15, 1947, it was divided into two dominions, based on religions. The nation of India had a secular constitution, but the majority of its people were Hindu; Pakistan was an Islamic republic. The princely states, such as Kashmir, were given a choice: They could become part of India, become part of Pakistan, or stay independent.

The Kashmiri people were not convinced that independence was a possibility. Though most of them were Muslims, they accepted the leadership of Sheikh Mohammad Abdullah, who wanted Kashmir to become part of India with guarantees of self-government.

Yet there was some resistance to Sheikh Abdullah, mostly in the town of Poonch, in the southwestern corner of Kashmir. The maharajah's rule had been especially heavy-handed in that town, so that the people of Poonch were strongly hostile to India and preferred union with Pakistan to becoming part of India. Soon they were rising up in an armed rebellion, which lasted from July to October, 1947.

The Poonch rebels were supported by the Pathans, a tribal group from northwestern Pakistan. Inspired to help their kinfolk across the border, the Pathans invaded Kashmir from the west. The maharajah, Hari Singh, asked India for protection on October 26.

With this action, Kashmir legally became a part of India so that India would have the right to fight the Pathan invasion. The leaders of India were convinced that the invasion was backed by Pakistan's governor general, Mohammed Ali Jinnah.

On October 27, 1947, India sent a military force to Kashmir. A bloody conflict broke out: the Pathans on one side, supported by a growing number of Pakistani soldiers, and on the other side, the well-equipped forces of the Indian army.

The Separatist Movement

On December 31, 1947, Indian prime minister Jawaharlal Nehru complained to the United Nations about Pakistan's aggression. Pakistan responded on January 15, 1948, by questioning the maharajah's rulership; the Kashmiri people had not been given the opportunity to choose their leader.

The United Nations decided that Kashmir should have a plebiscite—a chance to vote. But hostilities continued between India and Pakistan until July 27, 1949, when the two countries signed the Delhi Agreement, agreeing to a cease-fire and dividing Kashmir into two parts. The northwestern third was claimed by Pakistan and named Azad Kashmir, or "free Kashmir." The southeastern two-thirds, called Jammu and Kashmir, was made part of India for the time being.

Over time, the leaders of India decided that the law did not really require a plebiscite in Jammu and Kashmir. Meanwhile, Pakistan would not let go of Azad Kashmir; two more wars were fought over the issue, in 1965 and 1971.

Though the maharajah agreed to make Jammu and Kashmir part of India, the people of Kashmir were not so easily satisfied. They believed that the Indian government treated them unfairly. During the elections of 1987, for example, some Kashmiris charged that Rajiv Gandhi's Congress Party worked with some members of the National Conference, a local political party, to rig the vote. Though the National Conference had been fighting for Kashmiri self-determination, the person elected as chief minister, Farooq Abdullah, did not have a strong commitment to independence.

In disgust, Kashmiri Muslims began demonstrating against Indian oppression, and on August 9, 1989, India sent its army into the state. Cooperating with Governor Jaganmohan, the army kept international journalists and representatives from human-rights groups out of the state. The Indian army then arrested, wounded, and killed many peaceful protesters; *The New York Times* reported that as of June 2, 1990, at least six hundred had been killed.

The army's worst attack was against a group of mourners who were carrying the body of their religious leader. Dozens of mourners were killed. This attack served only to fuel the anger of Kashmiri nationalists, who were more certain than ever that India meant to squash their rights.

Consequences

The separatist movement soon posed problems for the tourist industry, which had been the center of Jammu and Kashmir's economy. Long curfews, one of them sixteen days long, were enforced by the Indian military. Also, military and police groups flooded the state: Indian army troops, the paramilitary Central Reserve Police Force, the Border Security Force, the Indo-Tibetan Border Patrol, and National Security Guards. The region was no longer a holiday playground for tourists.

Second, and more important, the goodwill that had existed between Kashmiri Muslims and Hindus was damaged. After extremists among the separatists killed some Hindu government officials, thousands of Hindus became frightened and fled from their homes to camps that were hastily built by the Indian government. Some Muslims tried to defend their Hindu neighbors, but a cloud of mistrust had already formed between the two groups.

A third effect of the fighting was the danger of nuclear war between India and Pakistan. The United States sent a special envoy, Robert M. Gates, who worked with government leaders to reduce this danger. Still, the Kashmiri people did not gain the freedom they hoped for, and their land continued to be a place of tension and violence.

Hungary Adopts Multiparty System

As part of the democracy movement that swept Eastern Europe, Hungary began allowing other political parties to compete with the Communist Party.

What: Political reform
When: 1989
Where: Budapest, Hungary
Who:
JÁNOS KÁDÁR (1912-1989), first secretary of the Hungarian Communist Party from 1956 to 1988
KAROLY GROSZ (1930-   ), first secretary of the Hungarian Communist Party from 1988 to 1989
JOZSEF ANTALL (1932-   ), prime minister of Hungary from 1990
ARPAD GONCZ (1922-   ), president of Hungary from 1990

Steps Toward Change

In the years after Soviet leader Mikhail Gorbachev began his policies of *glasnost* (openness) and *perestroika* (restructuring), a flood of political and economic changes swept over not only the Soviet Union but also Eastern Europe. Hungary was one of the first Eastern European nations to start liberalizing.

The Hungarian Workers Socialist Party, also known as the Hungarian Communist Party, was already one of the more liberal Communist organizations in Eastern Europe. Yet the Hungarian people were eager to see rapid changes in their system, and in 1988 the Party began trying to satisfy them.

János Kádár, who had been first secretary of the Party since 1956, was removed in May, 1988, and replaced by Karoly Grosz. A year later, Kádár also lost the title president of Hungary and was removed from the Central Committee. Though Kádár had allowed some political and economic reforms over the years, Hungarians had come to think of him as the man who had betrayed the 1956 Hungarian uprising.

In the mid-1980's, Hungary had begun allowing noncommunists to participate in politics; by 1985, 10 percent of the seats in the Hungarian parliament were held by independents. A multiparty system began to develop in 1988, but most of the parties were actually formed in 1989.

The first organized opposition group was the Hungarian Democratic Forum, founded in September, 1988. Not all the opposition parties were new; the Smallholders and the Social Democrats, for example, had existed before World War II and were now revived. Between late 1988 and 1990, fifty-two parties were born. The Hungarian people were very eager to express their own political views.

At first, it seemed that the Hungarian Communist Party would be one of the few communist parties in Eastern Europe to survive. Party leaders moved boldly but cautiously in their reforms, remembering how the Soviets had invaded and crushed the reforms of 1956. When at last they were sure that the Soviet Union would not step in this time, it may have been too late for the party.

Gaining Strength

The opposition groups grew quickly. New laws were passed, and plans were made for elections in 1990. At first some people feared that the Communists would not accept a defeat at the polls. But they were encouraged in June, 1989, when the Communist government began a series of talks with the opposition parties.

The same month, the party reorganized its leadership so that power was shared among four men: Karoly Grosz, Imre Pozsgay, Mikos Nemeth, and Rezso Nyers. The new leaders were quite

2287

liberal. In October, 1989, the party renamed itself the Hungarian Socialist Party, and soon afterward the country's name was changed from "People's Republic of Hungary" to "Republic of Hungary."

The two opposition parties that gained the most support were the Hungarian (Magyar) Democratic Forum, or MDF, and the more liberal Alliance of Free Democrats, or SDS. Jozsef Antall, leader of the MDF, and Arpad Goncz, founder of the SDS, became important political leaders.

Parliamentary elections were held in March and April, 1990. The clear winner was the MDF, in coalition with the Smallholders and the Christian Democrats. Together these parties won 60 percent of the seats in the new parliament. The SDS and its allies became the democratic opposition. Antall became prime minister, and in July, 1990, Goncz was elected president—a

position without much real power.

Although both the MDF and the SDS were reform parties, they had different approaches. The SDS wanted Hungary to switch quickly to a free-market economy, while the MDF moved more cautiously, fearing that rapid changes would cause problems. About 40 percent of Hungary's people were living at or below the poverty level, and the MDF did not want to risk bringing even more people into poverty.

Consequences

Hungary's new multiparty system and the other reforms of 1989 helped the nation to become completely free of the Soviet Union at last. Yet many Hungarians were disappointed that the change to a more democratic system and a free-market economy was not quick or easy. The Hungarian government began taking steps to cooperate with Western Europe, especially Austria and

One of Hungary's reforms included dismantling the barbed wire fence that ran along its border with Austria.

Germany, but a communist economy cannot be taken apart all at once—and Hungary was struggling with a large foreign debt. Nevertheless, it, along with Poland and the Czech Republic, was accepted as a member of the North Atlantic Treaty Organization in 1999.

As in other countries of Eastern Europe, freedom in Hungary brought benefits but also risks. People began to express extreme nationalist views and prejudices against other races and ethnic groups, such as Jews. Then there was the problem of the ethnic Hungarians who had been living in Romania and suffering discrimination there. The new government of Hungary spoke up in support of these people but did not actually offer them the free right to emigrate to Hungary. Clearly, democracy and capitalism would not be a "miracle cure" for Hungary's problems.

Thomas Tandy Lewis

Three Colombian Presidential Candidates Are Assassinated

When three presidential candidates were assassinated, it became clear to the world that Colombia was hard-pressed to deal with the violence of drug lords, paramilitary groups, leftist guerrillas, and common criminals.

What: Assassination
When: 1989-1990
Where: Colombia
Who:

LUIS CARLOS GALÁN (1943-1989), a
   Liberal presidential candidate
BERNARDO JARAMILLO OSSA (1952-1990), a
   Patriotic Union presidential candidate
CARLOS PIZARRO (1952-1990),
   presidential candidate of the M-19
   Democratic Alliance
VIRGILIO BARCO (1921-1997), president of
   Colombia from 1986 to 1990
CÉSAR GAVIRIA (1947-   ), president of
   Colombia from 1990
PABLO ESCOBAR (1949-   ), a leader of
   the Medellín drug organization
GONZALO RODRÍGUEZ GACHA (1947-1989),
   a Medellín drug lord who was killed by
   security forces

A Bloody Land

Violence has been part of life in Colombia almost since the country was formed. During the nineteenth century there were fifty wars of various sorts; the War of the Thousand Days (1899-1902) took more than 100,000 lives.

The 1940's and 1950's included a period of bloodshed that came to be called simply *la violencia*—"the violence." It began with a split in the Liberal Party which allowed the minority Conservative Party to win the 1946 presidential election. When the Conservatives tried to replace Liberal officeholders throughout the country, the Liberals resisted—especially after their popular leader, Jorge Eliécer Gaitán, was assassi-

nated in 1948. Gaitán's death was followed by three days of burning, looting, and death in the capital city, Bogotá.

The fighting spread throughout the countryside and lasted for years. Guerrilla groups were formed, along with bands of Colombians who made their living by crime and bloodshed. In 1953, the military, led by General Gustavo Rojas Pinilla, took over the government to try to restore order.

Finally, the two major parties decided to patch up their differences so that they could take the government back. In the National Front pact, they agreed that Liberals and Conservatives would take turns choosing a president for four-year terms between 1958 and 1974; other political offices would be divided equally between the two parties. To make this possible, they had to overthrow Rojas Pinilla, which they did in 1957.

With the National Front agreement, the Liberals and Conservatives succeeded in bringing an end to the worst of *la violencia*. Presidency-by-turns was carried out as planned.

Yet political groups that were not included in the power sharing were not happy. The guerrilla groups did not disappear; instead, they grew. By the 1980's Colombia had six major leftist guerrilla groups and many other kinds of guerrillas as well. These bands ruled large areas of the countryside. Colombia's geography includes three mountain ranges and some regions of thick tropical rain forest, so there were many places for the guerrillas to hide from the government forces.

Other violent groups thrived in Colombia. Death squads tortured and killed labor leaders, social activists, homosexuals, prostitutes, and others. Criminal groups disguised themselves as political organizations to steal, kidnap, and kill.

Drugs and Politics

In the late 1970's, trade in illegal drugs began to flourish in Colombia. Drugs, especially cocaine, brought huge amounts of money into the country—and more foreign exchange than Colombia's famous mountain-grown coffee. Some of this money was used to protect drug-processing factories in areas controlled by guerrillas. Some of it was used to bribe police and judges.

The "drug lords"—the men who dominated the drug trade—competed hard against one another. Especially in the city of Medellín, they tried hard to please the public, hoping that their "business" would be made legal. The names of the drug lords—Pablo Escobar, Gonzalo Rodríguez Gacha, the Ochoa family, Carlos Lehder, and Gilberto and Miguel Rodríguez Orejuela— became familiar to Colombians.

Some American leaders believed that stopping the drug trade in Colombia would help solve the problem of drug abuse in the United States. With American help, the Colombian government began cracking down on drugs in the mid-1980's. Military teams raided drug-processing facilities and the drug lords' luxurious country estates.

An extradition treaty stated that Colombian drug lords could be sent to be imprisoned and tried in the United States, where they were less likely to bribe their way out of jail. In response, the leading drug traffickers named themselves "The Extraditable Ones" and began a campaign of terror; they said they would not stop until the extradition treaty was canceled.

With bombs and machine guns, the drug lords' security forces succeeded in assassinating many police officers, judges, and other government officials, especially in Medellín. Justice Minister Rodrigo Lara Bonilla was killed in 1984, and Attorney General Carlos Hoyos was slain in 1988.

Most terrifying to the Colombian people, however, was the machine-gun slaying on August 18, 1989, of a leading candidate for president, the Liberal Party's Luis Carlos Galán. Galán had spoken in favor of extradition, so the drug lords wanted to make an example of him.

Two other murders of presidential candidates appeared to be the work of the Medellín drug lords as well. Bernardo Jaramillo, from the Patriotic Union, was assassinated on March 22, 1990, at the Bogotá airport. On April 26, 1990, the candidate of the M-19 Democratic Alliance, Carlos Pizarro, was killed on board an Avianca airplane.

Consequences

The drug lords' attempts to intimidate the government worked against them. After Galán was assassinated, President Virgilio Barco refused to negotiate with the drug lords; instead, he tried to make peace with the guerrilla groups, cooperated with the United States, and ordered the police and military to work harder than ever to capture the drug traffickers. By 1990, the government's forces had killed Rodríguez Gacha and forced Pablo Escobar to go underground.

The next president of Colombia, César Gaviria, wanted to bring an end to the violence. He promised that drug lords who turned themselves in would not be extradited to the United States, but would be tried in Colombian courts. A constituent congress was elected to change Colombia's constitution; Gaviria hoped that this would help solve some of the country's problems.

Maurice P. Brungardt

Luis Carlos Galán.

Kenya Cracks Down on Dissent

During the 1980s, Kenya's government became increasingly repressive, and in 1989 the administration of Daniel arap Moi was especially hard on dissenters.

What: Civil strife; Human rights
When: 1989-1991
Where: Kenya
Who:

JOMO KENYATTA (1894-1978), president of Kenya from 1964-1978

DANIEL ARAP MOI (1924-   ), president of Kenya from 1978

GITOBU IMANYARA, editor of Nairobi Law Monthly

JOSIAH M. KARIUKI (1929-1975), an opposition member of Parliament

NGUGI WA THIONG'O (1938-   ), a novelist and scholar

OGINGA ODINGA (1911-1994), an opposition political leader

GIBSON KURIA (1947-   ), a lawyer and human rights advocate

Silencing Dissent

Kenya, which had long been a British colony, began its independence as a democracy in 1963. During the rule of Kenya's first president, Jomo Kenyatta, there was competition in elections for parliament, but eventually the government came to be dominated by an elite group from the Kikuyu tribe, including many members of Kenyatta's family. One of his nephews was the foreign minister, another was high commissioner to London, and his daughter was mayor of Nairobi, the capital. Kenyatta's political party was called KANU, for Kenya African National Union. He made it the only legal party.

The strong sense of tribal identity common throughout the colonial days continued through Kenyatta's presidency. The Kikuyu people, who made up about 21 percent of Kenya's population, were in control; the second largest group, the Luo, had little say in the country's affairs.

Several of Kenyatta's chief rivals were eliminated during his presidency. Tom Mboya, a Luo who served as minister of economic planning and development, was gunned down at midday on a Nairobi street in 1969. When people gathered for public protest, the government insisted that opposition organizations were illegal. Parliament investigated the murder and found that there had been a police cover-up.

In 1975, Josiah Kariuki, a Kikuyu member of parliament, was kidnapped from a Nairobi hotel and assassinated by the secret police. His crime was trying to form a coalition to challenge the government.

Jomo Kenyatta died in 1978, and Daniel arap Moi succeeded him as president. Moi was from the Kalenjin tribe, which made up only 11 percent of Kenya's population. Many people wondered whether Moi would be able to effectively govern a country where Kikuyus had been dominant for so long.

Moi quickly proved, however, that he would be as hard on political opponents as Kenyatta had been. Some air force officers tried to overthrow him in 1982; in response, he dissolved the whole air force. More than six hundred members of the force were convicted of mutiny, and twelve were executed.

The Crackdown

For the 1988 elections, Moi's government set up what it called an "African" system for voting. The Ministry of National Guidance announced that secret ballots were "un-African and against the will of God"; instead of marking ballots secretly, then, the voters were to line up behind photographs of the candidates they chose.

Not surprisingly, only about one-third of the eligible voters chose to vote in the election. When one KANU leader, Kenneth Matiba, complained that the elections were not run properly, he was expelled from the party, and police went to his home and beat his wife and daughter.

That same year, President Moi decided to change the constitution so that judges and courts would no longer be independent branches of government. Judges now served only at the pleasure of the president. Moi also extended the time a person could be held in jail without charges being filed.

Yet some Kenyans continued to speak out against the government. One leader, Oginga Odinga, announced that he was forming a National Democratic Party; however, this group seemed to be dominated by Luos, and many people from other tribes avoided it. Ngugi wa Thiong'o, a Kenyan novelist who had fled the country and was living in London, founded the United Movement for Democracy in Kenya. Other opposition groups included the Union of Nationalists to Liberate Kenya, the Kenya Patriotic Front, and the Mhetili Nationalist Movement.

A journal for lawyers and judges, the *Nairobi Law Monthly*, has been closed down several times since it was founded in 1987. Its editor, Gitobu Imanyara, was jailed because he called for multiparty elections. To protest, the International Bar Association canceled a meeting it had scheduled for the summer of 1990 in Nairobi.

From time to time, college students in Nairobi organized demonstrations. In July, 1990, twenty or more students were killed at one protest; many others were wounded, and hundreds were arrested.

A lawyer named Gibson Kuria, who had won the Robert F. Kennedy Human Rights Award in 1988, was jailed because he had often represented political dissidents in court. In mid-1990 he gained his freedom and took refuge in the United States embassy; some time later he fled to the United States.

Some church leaders added their voices to the outcry of dissent, and they, too, paid for their courage. Alexander Muge, Anglican Bishop of Eldoret, received a death threat from Moi's minister of labor in August, 1990; three days later, he died in a mysterious auto accident.

Consequences

Moi's repressive actions did not go unnoticed in other countries. In October, 1990, an editorial in *The New York Times* said that Moi either would have to give in to some of his critics or face a revolution like those that had been taking place in Eastern Europe.

Koiga wa Wamwere, a Kenyan dissident who had once been a member of Parliament, took refuge in Norway; when he returned to Kenya in October, 1990, he was arrested. Norway's government sent an official protest, and Kenya then proceeded to cut off diplomatic relations with Norway.

In spite of Moi's stubbornness, the wave of protest in Kenya did not seem likely to die. The Kenyan people seemed in no mood to accept a repressive system that allowed them little freedom to speak their minds. Nevertheless, Moi remained firmly in power into the twenty-first century.

Richard A. Fredland

Green and Burnley Solve Mystery of Deep Earthquakes

> *Scientists discovered the process of fault slip, which causes earthquakes deep within the earth.*

What: Earth science
When: 1989-1992
Where: Riverside, California
Who:
HARRY W. GREEN II (1940-   ) and
PAMELA C. BURNLEY, American earth
    scientists

The Problem of Deep Earthquakes

The location at which an earthquake occurs beneath the earth's surface is known as its "focus." Estimates of depth of focus had been made as early as the 1755 Lisbon earthquake, but early estimates were crude and inaccurate. It was generally accepted by the scientific community, for lack of better evidence, that all earthquakes occurred at a shallow depth. This concept was not challenged until sometime after the advent of the earthquake instruments called seismographs in the 1880's. Such instruments could provide permanent records of ground movement at a seismograph station as earthquake waves passed through. H. H. Turner, who compiled travel-time data for earthquake waves, found that errors in arrival times of waves could be accounted for by assuming a deep focus for earthquakes; this suggestion was confirmed by analysis of arrival times of earthquakes in Japan. Subsequent analyses would show that a small number of earthquakes occurred as deep as 680 kilometers beneath the earth's surface.

Deep earthquakes pose a physical problem. Shallow earthquakes occur as forces build up to the point where rock breaks in a brittle fashion and a slip occurs on a fault surface within the rock. This process releases earthquake waves. Hundreds of kilometers deep within the earth, however, pressures are too high for brittle failure to occur. At such pressures and depths, rocks flow rather than break. This problem was recognized soon after the existence of deep-focus earthquakes was confirmed.

A clue to the origin of deep-focus earthquakes came from the suggestion that, at great depths in the earth under high pressure, minerals collapse to form denser mineral compounds. This collapse, called a phase change, could, scientists realized, somehow be related to earthquakes. The supporting link was the fact that deep earthquakes began and ended at the same depths at which phase changes were postulated; the exact process, however, was unclear.

Phase Changes

Deep earthquakes occur in thin, slab-shaped regions that extend into the earth beneath the oceans and continents. These slabs are relatively cold and brittle rocks consisting of the outer fifty to one hundred kilometers of the earth; they sink slowly into the earth at no more than a few inches per year as they eventually heat up and disappear. The slabs consist mostly of the iron- and magnesium-rich mineral compound olivine. At the pressures that exist at depths of 300 to 350 kilometers, olivine collapses and changes into the denser mineral compound spinel.

Harry W. Green and Pamela C. Burnley were the first to demonstrate in the laboratory that faulting can be triggered by the olivine-to-spinel phase change. Using experimental laboratory equipment, the two scientists were able to subject olivine samples to the same pressures and temperatures existing at depths of 300 kilometers and more. The changes that led to faulting in this environment were much different than those in the brittle faulting process occurring at shallow depths.

The shallow faulting process begins during compression by the formation of large numbers of microscopic cracks. The number of microcracks increases gradually as force increases, until the material begins to lose strength. The microcracks then extend and join, and a fault begins to form.

The process leading to faulting under high pressure was shown to be quite different. During high-pressure experiments on olivine, microcracks did not form. Instead, microscopic lens-shaped pockets of spinel began to appear. The spinel pockets were filled with tiny microcrystals. The material in the lens-shaped packets was much weaker than the surrounding olivine and had the ability to flow by sliding along grain boundaries. It was suggested that these lens-shaped packets served the same function as cracks in the process of brittle failure: That is, they are zones of weakness in the material. Since they are physically different, however, the term "anticrack" was applied to them.

As compressive stress increases, the anticracks grow in number and density. At some point, the density becomes critical, and the anticracks begin formation of a fault surface. At the same time, the anticracks dump their spinel microcrystals into the fault zone. The fault zone thus becomes lubricated by the sliding along the grain boundaries of the spinel crystals. Green and Burnley have also shown that faulting by this method releases earthquake waves, similar to those released during shallow brittle faulting.

The anticrack mechanism can therefore explain the process by which earthquakes and faulting can occur in the high-pressure environment of deep-focus earthquakes. The next step was to explain why earthquakes do not occur beyond about 690 kilometers beneath the earth's surface.

Phase changes are governed by both temperature and pressure. By the time depths of more than three hundred kilometers have been reached, however, both temperatures and pressures have increased to the point that the olivine-to-spinel conversion has occurred; however, the rock involved consists initially of colder rock composed largely of olivine. The rock heats up slowly from the outside; the interior thus remains colder than the phase-conversion temperature at depths greater than 350 kilometers. As the slab interior warms up, the phase transition occurs, releasing earthquake energy and producing fault slip. By the time the slab reaches about 690 kilometers, however, any remaining olivine, as well as any spinel, breaks down to two other phases in a process that occurs so slowly that earthquake energy cannot be released.

Consequences

The existence of deep earthquakes was not recognized until the 1920's. It was decades before the importance of phase changes of mineral compounds would be connected to deep earthquakes, providing a potential source of energy for deep tremors. Through the experimental work of Green and Burnley, the exact process of how the phase change of olivine to spinel provides this energy has been outlined in detail. This has resulted in a new concept, the anticrack process, to explain deep-seated, high-pressure fault slip and earthquakes. The process further explains the depth range observed for deep earthquakes.

There is another important implication of the anticrack hypothesis. If the hypothesis is correct, then the cessation of earthquakes at approximately 690 kilometers is not caused by the inability of down-going slabs of rocks to penetrate into the mantle below, but rather by the absorption of such slabs by phase change. There is thus no reason to suppose that the material of the upper and lower mantle cannot mix, as heat is transferred outward by solid convective flow of material. This has been an important constraint on ideas proposed to explain how heat is transferred outward in the earth to supply energy to drive rock plates at the earth's surface.

David S. Brumbaugh

U.S. Fighter Planes Down Two Libyan Jets

> *Tensions between the United States and Libya led to a clash between two U.S. fighter planes and two Libyan MiG-23's over international waters off the coast of Libya.*

What: International relations
When: January 4, 1989
Where: Over the Mediterranean Sea
Who:
RONALD REAGAN (1911-   ), president of the United States from 1981 to 1989
MUAMMAR AL-QADDAFI (1942-   ), president of Libya from 1977

Tensions Rising

Following World War II, Libya, a land of North Africa, became the first country to gain independence through the United Nations. The Kingdom of Libya was created in December, 1951. In September, 1969, a group of young army officers overthrew the Libyan government; the leader of the group was twenty-seven-year-old Colonel Muammar al-Qaddafi.

In 1977, Qaddafi became president of Libya and commander in chief of the country's armed forces. Qaddafi had particular ideas about how Libya should be governed; he called his ideas *jamahiriya*, or Islamic socialism. The president explained his *jamahiriya* policies in a collection of essays published as *The Green Book*.

In the following years, Qaddafi used huge profits from Libya's sales of oil to buy weapons, to build housing, roads, and communications networks, and to improve education. Soon Libya had become one of the major military powers of North Africa.

In 1983, Libya began sending aid to "revolutionary" groups in Arab countries. For some time Qaddafi had criticized the United States and Egypt for their policies toward Israel; the Libyan congress even called for armed attacks against Israel and its supporters.

In January, 1986, the United States decided to punish Libya economically for several Libyan-sponsored terrorist attacks in Western Europe. A few months later, however, terrorists bombed a West German discotheque that was often visited by American servicemen. The United States first gave Libya a warning, then staged a nighttime air attack on military targets in Tripoli, the Libyan capital. Qaddafi's home was heavily damaged in the raid.

In late 1988, the United States accused Libya of building a factory for chemical weapons at Rabta, in the desert south of Tripoli. Libya insisted that the accusation was untrue, that the factory actually produced medical drugs. American officials, however, said that they had evidence that the factory was preparing to produce mustard gas and other poisonous gases that could be used in chemical warfare.

Confrontation in the Air

On January 4, 1989, two Navy combat planes, F-14 Tomcats, took off from the aircraft carrier USS *John F. Kennedy*. According to Navy reports, the *John F. Kennedy* was going through normal training exercises off the southwest coast of Crete, in the Mediterranean Sea. As the F-14's flew on patrol about seventy miles (110 kilometers) north of the coast of Libya, American radar picked up two MiG-23 Floggers—Libyan fighter planes—taking off from a Libyan air base at al-Bumbah. The radar sightings were passed on to the crews of the F-14's.

Once the MiG's were airborne, they turned and headed straight for the U.S. Navy jets. Trying to avoid problems, the F-14 pilots descended to four thousand feet (one thousand meters) from about fifteen thousand feet (five thousand meters). When the MiG's kept coming toward them, the F-14's tried several other maneuvers in order

2296

to evade them. Each time the American jets tried to move away, however, the Libyan fighters would turn and continue to chase the F-14's.

Once the MiG's had come within fourteen miles of the U.S. planes, the commander of the lead F-14 decided that his aircraft was in danger. He then armed his weapons and maneuvered his plane so that it would be in position to fire on one of the MiG's. The F-14's chased the Libyan jets at near supersonic speeds over the Mediterranean.

Finally, both F-14 pilots fired their weapons at the MiG's. One MiG was hit and destroyed by a Sparrow missile, and the second was downed by a Sidewinder. The American fighter crews sighted two parachutes, which showed that the Libyan pilots had been able to jump before their planes went down.

Consequences

There had been several other military clashes between Libya and the United States during the administration of Ronald Reagan. In 1981, two U.S. F-14's shot down two Libyan SU-22's, and in 1986, several Libyan naval ships were sunk by U.S. naval forces after American aircraft were fired upon by the ships.

The United States insisted that there was no connection between the controversy over the supposed chemical-weapons factory and the January, 1989, conflict between American and Libyan fighter planes. Immediately after the incident, however, Colonel Qaddafi accused the United States of practicing terrorism, and he asked the United Nations Security Council to come together in an emergency meeting. There were reports that defenses around Qaddafi's home and headquarters in Tripoli were being strengthened.

At the United Nations, the Libyan delegate accused the United States of a premeditated attack against unarmed Libyan aircraft that had simply been gathering information. On January 6, however, the chief U.S. delegate to the United Nations, Vernon Walters, displayed photographs which, he said, proved that the Libyan aircraft had been armed.

Francis Poole

Japan's Emperor Hirohito Dies and Is Succeeded by Akihito

> *The death of Emperor Hirohito of Japan in 1989 ended the Showa era, and the ascension of the Crown Prince Akihito to the throne ushered in the Heisei era.*

What: Government
When: January 7, 1989
Where: Imperial Palace, Tokyo, Japan
Who:
HIROHITO (1901-1989), 124th emperor of Japan
TSUGU AKIHITO (1933-   ), 125th emperor of Japan

Death of the Emperor of Japan

On Friday, January 7, Hirohito, the 124th emperor of Japan and the longest-ruling head of the oldest royal dynasty in the world, died at the age of eighty-seven. He was the last of the World War II leaders, having presided over the most tumultuous era of Japan's modern history. Arguably more an observer than an actor, Hirohito's reign saw the nation of Japan embrace nationalistic militarism, conquer much of East and Southeast Asia, wage war on the Allied Powers, suffer the world's first atomic bombing, and rise in just four decades to become a vibrant economic power in the world.

When Shoichi Fujimora, the grand steward of the Imperial Household Agency, announced the emperor's death at 7:55 A.M., he revealed for the first time that Hirohito had been suffering from cancer of the duodenum since September, 1987. Adhering to the common practice in Japan, the emperor was never told that he had cancer.

The official time of mourning was set at six days for governmental agencies and two days for businesses. During this time, radio and television stations canceled advertising and replaced popular music with quiet, dignified classical music. Store clerks wore solemn black clothing, and black cloth draped over window displays to hide them from view. Black armbands were issued to employees, and business activity was kept to a minimum. Japanese flags were flown from offices and shops, most draped with black bunting at the end of the poles. The national airline carrier, Japan Airlines, canceled in-flight films and provided somber outfits for stewardesses.

As was customary, the funeral was not held until forty-eight days after the emperor's death, on February 24. Unlike the funeral of his father, the Taisho Emperor, sixty-two years earlier, the

Emperor Akihito formally ascends to the chrysanthemum throne in November, 1990.

2298

Showa Emperor's funeral did not contain a ceremonious parade of officials in military uniforms. It also contained fewer Shinto rituals because after World War II, the emperor had renounced his status as a Shinto deity.

Emperor Hirohito was an ardent marine biologist and wrote several authoritative books in the field. He was also credited with discovering previously unknown species of tiny sea creatures in the waters of Japan's Sagami Bay.

Akihito Becomes Emperor

On the day of Emperor Hirohito's death, a state ceremony was held to usher in the Heisei era and bestow the imperial regalia on the new emperor Akihito. As Akihito stood on a white rug in front of a white and gold throne, Imperial Household Agency chamberlains bowed twice, then held up a sword and a jewel wrapped in embroidered cloth and tied with a purple cord. These items are symbolic of Amaterasu, the sun goddess and mythical founder of Japan. Then another chamberlain took the Imperial seal and the state seal, each wrapped in purple cloth stamped in gold with the chrysanthemum crest, and placed them before Akihito. The new emperor bowed, then stood as Prime Minister Noboru Takeshita and other government officials bowed deeply before him. Emperor Akihito and the chamberlains then exited the hall. Akihito became the first emperor to be installed since Japan was transformed into a constitutional democracy at the end of World War II, although the multipart installation was not viewed as complete until November 12, 1990.

Born in Tokyo on December 23, 1933, Akihito, the elder son of Emperor Hirohito and Empress Nagako was educated among commoners at the elite Gakushuin school, unlike previous princes. During his years there, he also had an American tutor, Elizabeth Gray Vining, who helped prepare the crown prince for his future duties in a world that would have to include a broad, international worldview. Nevertheless, like his predecessors, Akihito was raised from the age of three by court-appointed tutors rather than by his parents, and until after World War II, he had very limited contact with those outside the palace. In 1959, Akihito married Michiko Shoda, the daughter of the president of a flour company and the first commoner to enter the Japanese imperial family.

Consequences

The fact that the Crown Prince Akihito was educated among commoners—albeit elite commoners—and that he was allowed to marry a commoner provided signs of other ways in which the Emperor Akihito and the Heisei era would differ greatly from the Showa era of his father. For one thing, Akihito was not considered to be a god when he ascended the Chrysanthemum Throne. He spoke English and was well-traveled. He and the Empress Michiko broke with tradition and raised their three children themselves at the palace.

These departures from tradition suggested that Akihito's reign as emperor would be very different from those of his predecessors. Royal family watchers predicted that his reign would be more democratic and less formal. When asked what he would have done with his life if he had been born an average Japanese person, Akihito replied that not ever having had that experience, he could not imagine being able to freely choose another way of life.

Akihito's belief that the Japanese imperial family should be more open to and in tune with the citizens of Japan has manifested itself in many ways, even in details of his everyday life. On one occasion, he became uncomfortable about speeding through the traffic with sirens blaring and lights flashing and ordered that his motorcade stop at traffic lights so as not to inconvenience the "ordinary people."

Victoria Price

Exxon Valdez Spills Crude Oil Off Alaska

> *The environmental damage to the Gulf of Alaska caused by the spillage of more than eleven million gallons of crude oil from the Exxon Valdez focused attention on the fragile Alaskan environment and raised questions about corporate responsibility, oil transport technology, and legal liability.*

What: Environment; Disasters
When: March 24, 1989
Where: Valdez, Alaska
Who:
JOSEPH HAZELWOOD (1946-   ), captain
  of the *Exxon Valdez*

Alaskan Oil

The breaching of the supertanker *Exxon Valdez* on the rocks of Bligh Reef outside the port of Valdez, Alaska, and the loss of its entire cargo of eleven million gallons of crude oil caused the world to focus its attention on the value of preserving the environment versus humanity's need for petroleum products. The players in the events that followed—Alaskan tribal communities, large corporations, the federal government, attorneys and courts, and environmental organizations—made the accident a model for the handling of environmental disasters.

In 1968, Atlantic Richfield and Humble Oil discovered the last great oil field in the United States, in the forbidding landscape of Alaska's North Slope, at a place called Prudhoe Bay. The ever-growing demand of American car owners for fuel for their personal transportation, combined with the use of oil for transport, heating, electricity generation, and manufacturing power, made the question of future supply of this essential ingredient a focus of American business. The land, which belonged to the state of Alaska, was offered for leasing in September of 1969, and a variety of U.S. oil companies won the right to extract oil at various points in the Prudhoe Bay area.

Prudhoe Bay posed one problem that had not existed in earlier oil development: Access to it by water was possible for only half of the year, as the

ocean around the bay froze during the other six months. Unwilling to see their ability to exploit the new mineral wealth thus constrained, the oil companies proposed the construction of a pipeline from Prudhoe Bay south across the middle of Alaska to the open-water port of Valdez, on the Gulf of Alaska. However, because Alaska is mostly pristine wild land, occupied by a few scattered tribes of Native Americans and many forms of wildlife, the scheme was bitterly opposed by environmentalists. An act of Congress, in 1973, brought the legal wrangling to an end and permitted the Alyeska Pipeline Service Company, formed by a group of oil companies, to begin construction.

The Alaskan pipeline, the largest such construction project in history up to the mid-1970's, required major engineering innovations. The 800-mile (1,287-kilometer) pipeline crossed two mountain ranges, the Brooks and the Alaska, and had to contend with temperatures falling as low as 70 degrees below zero Fahrenheit (57 degrees below zero Celsius) in the wintertime. In addition, it was part of the natural range of many varieties of wildlife, including the caribou, a species of deer. Elaborate provisions were made to minimize the impact on wildlife, including special corridors for wildlife to cross the pipeline. Transmission of crude oil began in 1977.

The Oil Spill

After all the concerns expressed by environmentalists, the environmental accident that happened occurred not on the pipeline, but in the ocean after the oil had been transferred from the pipeline to a tanker. The pipeline brings 2.2 million barrels a day of North Slope oil to Valdez, where fleets of tankers, mostly owned by the large oil companies, load the oil into their cargo holds. The risk is well known, as there have been other tanker disasters, notably the *Amoco Cadiz*, which grounded on the coast of France in 1978 and spilled some 1.5 million gallons of oil into the Atlantic Ocean, and the *Torrey Canyon*, which grounded on the south coast of England in 1967 and lost a little more than 800,000 barrels of oil. The *Exxon Valdez* disaster was on a vastly larger scale, although the amount spilled has since been exceeded by the *Braer*, which went aground off the Shetland Islands in 1993 and dumped 26 million gallons into the North Atlantic.

On the night of March 24, 1989, Joseph Hazelwood, captain of the *Exxon Valdez*, returned to his vessel, completed the loading process, and set the vessel in motion. He then, according to subsequent investigation, turned the navigation of the vessel over to a junior, inexperienced officer. Navigation in Prince William Sound is tricky, and the inexperienced officer was unable to avoid striking Bligh Reef, which penetrated the hull of the vessel. The cargo began to leak into the water of the sound, and subsequently into the Gulf of Alaska. By the time all the oil had run out, it had soiled 1,300 miles (2,091 kilometers) of beaches. Some 5,000 sea otters died, as did 300 harbor seals, 22 killer whales, 150 bald eagles, and an estimated 250,000 birds, chiefly waterfowl. The local population, partly Native American, makes its living largely from the fishery, which had to shut down for several years. Both environmentalists and the general public were outraged.

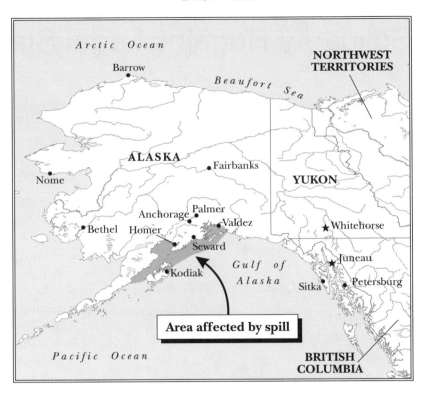

Area affected by spill

Consequences

Exxon immediately began a cleanup program, involving both attempts to capture the oil floating on top of the water and to remove it from beaches with high-pressure hoses. Some otters and birds were rescued and cleaned up, but subsequently, most of these also died. Between March of 1989 and 1991, Exxon spent $2.5 billion on cleanup and scientific evaluation, and in a trial in 1991, it asserted that according to its scientists, Prince William Sound had been environmentally restored. A settlement was negotiated, in which Exxon paid $900 million to finance further environmental monitoring and additional restoration efforts to a trustee council, which also handled the payment of compensation to affected communities. A 1994 legal judgment assessed Exxon $5 billion in punitive damages; for the rest of the twentieth century, the company continued to contest this judgment in court. However, in 1999, Hazelwood, who had been judged negligent in his handling of the vessel, began to perform the community services imposed on him as part of his sentence.

The *Exxon Valdez* spill led to legislation requiring all tankers to have double hulls from the year 2015 onward. It focused attention on the environmental risks of oil exploitation, especially in Alaska. It also made all companies much more sensitive to the costs that could result from environmental accidents.

Nancy M. Gordon

Solidarity Regains Legal Status in Poland

Seven years after being suppressed by the communist government, the independent trade union Solidarity became legal in Poland.

What: Labor; Political reform
When: April, 1989
Where: Warsaw, Poland
Who:
LECH WAŁĘSA (1943-   ), leader of Solidarity
TADEUSZ MAZOWIECKI (1927-   ), a Catholic journalist and Solidarity adviser
BRONISLAW GEREMEK (1932-   ), a historian who became Wałęsa's chief political adviser
WOJCIECH JARUZELSKI (1923-   ), first secretary of the Polish Communist Party from 1981 to 1989
MIECZYSLAW RAKOWSKI (1926-   ), prime minister of Poland from 1988 to 1989
CZESLAW KISZCZAK (1925-   ), Polish minister of internal affairs
MIKHAIL GORBACHEV (1931-   ), first secretary of the Soviet Communist Party from 1985 to 1991

Solidarity Struggles

In the summer of 1980, there was a new force for reform at work in Poland: an independent, self-governing trade union called Solidarity. Led by Lech Wałęsa, Solidarity tried to promote social justice throughout the country—but its efforts came up against the power of the communist government, which had been installed by the Soviet Union after World War II.

In December, 1981, General Wojciech Jaruzelski, first secretary of the Polish Communist Party, imposed martial law with the help of General Czeslaw Kiszczak, minister of the interior and head of the police. Solidarity's leaders were imprisoned, and the union was outlawed.

Wałęsa and the others were eventually released; martial law was ended in July, 1983, but Poland had not really recovered from its troubles. Wałęsa won the Nobel Peace Prize in 1983, and many Polish people continued to support Solidarity. During the years that followed, Wałęsa continued to speak up for workers' rights.

Poland's new prime minister, Mieczyslaw Rakowski, proved to be adamantly opposed to Solidarity. He declared to Jaruzelski that he could keep order and stability in Poland without taking Solidarity into account.

Yet Poland's economic problems were very serious. Workers were discouraged, the factories were not producing well, prices were rising quickly, and there were more and more strikes. In November, 1987, the government asked the people to vote in a referendum: Should there be a new "government program for radical economic recovery," and should Poland begin a program for "democratizing political life"? Most of those who voted marked "yes," but too many citizens did not bother to vote at all, so that the vote was not considered valid. It was another embarrassment for the government.

Rising from the Ashes

During 1988, workers went on strike often, chanting, "There is no freedom without Solidarity." Finally, Interior Minister Kiszczak decided that the government must try to come to an agreement with Solidarity. This decision had the unspoken support of the Soviet Union's president, Mikhail Gorbachev. On August 30, 1988, Solidarity spokesman Tadeusz Mazowiecki announced that Wałęsa and Kiszczak would meet the next day to discuss the possibility of official negotiations.

These talks led to another announcement: Wałęsa would appear on national television to debate economic and labor issues with the spokes-man of the government union. This confronta-tion took place in November, and Wałęsa quickly gained the upper hand in the debate. He empha-sized that Solidarity needed to be legalized so that it could work to restore human rights in Po-land and help rebuild the economy.

Some of the government leaders responded by criticizing Solidarity sharply, but Kiszczak per-suaded Jaruzelski to accept Solidarity's return. After arguments in a Central Committee meet-ing in January, 1989, it was agreed that there would be more talks with Solidarity representa-tives.

These roundtable talks began on February 6. Wałęsa had the help of advisers such as Mazo-wiecki and Bronislaw Geremek, a historian. The Catholic Church did not have formal representa-tives at the talks, but leading Catholic laypeople were present to speak for the Church. The gov-ernment's delegation was headed by Kiszczak.

The former prisoners met with their former jailers in Warsaw and the nearby village of Magdalenka. Most of the issues were hard to re-solve, but everyone was willing to compromise. Three months of discussions ended on April 5 with two historic agreements.

The first agreement contained three primary points: Solidarity would be legalized, Rural Soli-darity would be established (extending trade-union independence into the countryside), and those who had been fired because of their union work would be given back their jobs.

The second agreement had to do with politi-cal reform. The government agreed to set up a parliamentary system, with representation by op-position parties. New elections would be held within two months. In the lower house of the par-liament, called the Sejm, 65 percent of the seats would be reserved for the Communist Party and its coalition partners. An upper house, the Sen-ate, would be established. In this body, all one hundred seats would be open to members of any party. Finally, a presidency was established.

An opposition newspaper was given permis-sion to compete with the government media, and the opposition was also given radio and tele-vision time. The government also promised to le-galize the Catholic Church, which had been ille-gal in Poland since the end of World War II.

Consequences

Elections were scheduled for June 4 and 18. The communist regime was sure that it would win, for Solidarity had very little time in which to put together a list of candidates. But to every-one's surprise, the union did manage to make a national campaign.

When the returns were in, it was clear that Solidarity had won a remarkable triumph: all 35 percent of the open seats in the Sejm, and ninety-nine of the hundred seats in the Senate. Geremek became leader of the Solidarity forces in parliament. When the presidential election was held, Jaruzelski won, but only by one vote. To form a government, he eventually had to ask for Solidarity's help. Mazowiecki became the new prime minister, heading the first noncommunist government in Eastern Europe since the late 1940's. The people of Poland began learning again what it was like to live in freedom.

China Crushes Prodemocracy Demonstrators in Tiananmen Square

> *When hundreds of thousands of intellectuals gathered in Tiananmen Square to protest official corruption and suppression of freedoms, the People's Liberation Army massacred the demonstrators.*

What: Civil rights and liberties

When: April-June, 1989

Where: Beijing, People's Republic of China

Who:

FANG LIZHI (1936-   ), an astrophysicist who lost his Party membership after encouraging reform

HU YAOBANG (1915-1989), general secretary of the Chinese Communist Party from 1980 to 1987

DENG XIAOPING (1904-1997), the dominant figure in Chinese politics after Mao Zedong's death in 1976

LI PENG (1928-   ), premier of the People's Republic of China from 1988

An Unenforced Constitution

In 1954 Mao Zedong, the leader of the Chinese Communist Party (CCP), implemented his nation's constitution. *The Constitution of the People's Republic of China* contains sections which ensure open dialogue between citizens and state officials. Article 35, for example, states that Chinese citizens enjoy freedom of speech, of the press, of assembly, of association, of procession, and of demonstration. Articles 37 and 38 promise that the freedom and dignity of all Chinese citizens are inviolable (protected from attack). Article 41 provides the right to criticize and make suggestions to the government with the assurance that no official may suppress such complaints, charges, and exposures, or retaliate against the citizens making them.

Even with these laws in place, however, the Chinese were not protected by their constitu-tion. Repeatedly, when citizens spoke out for reform, they were silenced—sometimes imprisoned or killed—by their government. In May, 1956, in his "speech of one hundred flowers," Mao invited Chinese intellectuals to express their complaints against the CCP openly. In response, hundreds of thousands of Chinese openly criticized the CCP and demanded freedom of speech, better working conditions, the right to form unions, and more respect for human rights from the government. By June, 1957, Mao had had enough of the criticism; he placed Deng Xiaoping in charge of handling the intellectuals who had attacked the CCP. Nearly two million were questioned, 100,000 were given serious sentences, and several million were assigned lesser punishments, such as working in the countryside for reeducation.

In 1986, the CCP showed a renewed level of tolerance to open expression. Deng officially initiated debates on political reform. CCP general secretary Hu Yaobang spoke out for debate as an avenue toward reform, and Fang Lizhi, an astrophysicist and vice president of Hefei University in Anhui, blamed the CCP leaders for China's social ills and criticized the CCP's use of propaganda. Both Hu and Fang became inspirational leaders of the prodemocracy movement.

Deng, angered by the outburst of dissent, once again halted the free expression he had earlier encouraged. He replaced Hu Yaobang with premier Zhao Ziyang, and stripped Fang Lizhi of his Party membership. Hu's death on April 15, 1989, gave the students a cause for which to demonstrate. Though demonstrations were banned in Tiananmen Square, more than 100,000 students gathered there to rally for democracy in honor of Hu. These demonstrations set off the climactic events of the "Beijing Spring."

Growing Dissent

Though Zhao was appointed to stand against the reform movement, he became a supporter of openness, and Party conservatives quickly linked him with the student unrest. The students, who remained active after the April 15 demonstrations, hoped to gain the attention of reporters for the international media, who were covering Soviet leader Mikhail Gorbachev's May 16 visit to China. Three days prior to Gorbachev's arrival, three thousand students in Tiananmen Square had begun a week-long hunger strike, successfully drawing attention to reform demands. Hundreds of bureaucrats, intellectuals, and workers rallied behind the hunger strikers, and by May 17, more than one million demonstrators in Beijing called for resignations from Deng and Premier Li Peng. Zhao made a personal visit to negotiate with the strikers, but they would not compromise their demands. On May 20, Li implemented martial law, which remained in effect for two weeks.

Some of the leaders of the student movement became concerned about the unruly nature of the events and began dissolving the hunger strike and refocusing the protest on local campuses. As a result, new, radical leaders took over. Within days, Beijing art students had molded white plaster and Styrofoam into a statue symbolic of Asian freedom and reform. They erected their thirty-foot-high "liberty statue" in Tiananmen Square, and it won the attention of international media.

On June 4, 1989, the People's Liberation Army (PLA) marched on Tiananmen Square, crushing to pieces the "Goddess of Freedom" and killing an undetermined number of prodemocracy demonstrators and bystanders—some said that the death toll reached the thousands. The following day, another two hundred civilians were massacred. On June 6, PLA troops turned on one another, reflecting

differences within the army leadership. At least seven officers publicly sided with the demonstrators.

On June 11, the Chinese press claimed that no students had been killed, but the Western media had much evidence to disprove that claim.

Consequences

In the days following the crackdown, at least twenty-seven demonstrators were publicly executed. Hundreds of students and intellectuals remained in jail until 1990. In 1991, when media attention turned away from China during the Persian Gulf War, the CCP quickly tried and convicted some of the most famous dissident leaders of the 1989 demonstrations.

On November 13, 1989, eighty-five-year-old Deng stepped down from his chairmanship of

In Tiananmen Square, Beijing, student leader Wang Dan calls for a citywide march.

the Central Military Commission after planting Jiang Zemin as his successor. Jiang openly embraced Deng's policies: His primary concern was to uphold the Communist Party's leadership role.

In the aftermath of the Tiananmen Square demonstrations, the CCP enforced numerous regulations to help stifle prodemocracy uprisings, among them: The size of the freshman university class of 1989 was cut in half; very few, if any, government-supported students could get permission to study social sciences abroad; a thorough reindoctrination program was instituted at every educational level; the universities were forced to add required classes on CCP history and political ideology, to detract students from other academic pursuits; and demonstrations were restricted. The right of freedom of expression took a serious step backward in China after the Tiananmen Square demonstrations.

Fraud Taints Panama's Presidential Election

General Manuel Antonio Noriega, Panama's brutal leader, declared a presidential election void after his favored candidates suffered a crushing defeat.

What: National politics
When: May 7, 1989
Where: Panama
Who:

MANUEL ANTONIO NORIEGA (1934-   ),
   military ruler of Panama from 1983 to
   1989
GUILLERMO ENDARA (1936-   ),
   president of Panama from 1990
GEORGE HERBERT WALKER BUSH
   (1924-   ), president of the United
   States from 1989 to 1993

From Friend to Foe

The Central American country of Panama has had a rocky political history. It has had its share of coups and rigged elections. In 1968, for example, General Omar Torrijos took power in a coup against Arnulfo Arias, who had been democratically elected. Torrijos ruled the country until he was killed in a plane crash in 1981.

In 1983, General Manuel Antonio Noriega became leader of the armed forces and the controlling force in Panama. In the election of 1984, many Panamanians accused the government of fraud when the military-backed candidate, Nicolás Ardito Barletta, became president. In 1985, when Noriega became dissatisfied with Barletta, the general fired him and put Vice President Eric Arturo Delvalle in his place.

Panama has been considered important to American interests since the beginning of the twentieth century. The Panama Canal, which was built under United States supervision and opened in 1914, provided the only quick route for ships passing between the Atlantic and Pacific oceans. The United States has a political and military interest in keeping the canal open, and the U.S. government has attempted to encourage stable, friendly governments in Panama.

In the first five years of Noriega's rule, he was considered a friend of the United States. He passed information to the Central Intelligence Agency (CIA) about the activities of Marxists and leftist guerrillas in Central and South America, and he was also considered useful in fighting trade in illegal drugs. Noriega was even praised by U.S. officials for his help in slowing down the flow of illegal narcotics.

But all of this changed in February, 1988, when federal grand juries in Miami indicted Noriega for drug trafficking and racketeering. From that point onward, Noriega was an embarrassment to the United States, and the administration of President Ronald Reagan began working hard to remove him from power.

Under pressure from the United States, President Delvalle tried to fire Noriega. Within hours, however, Delvalle was forced out of office by the National Assembly (which was controlled by Noriega). The United States responded by freezing Panama's holdings in the United States and imposing tough economic penalties.

In March, 1988, Noriega defeated a coup attempt against him, and there were large street protests against his rule in Panama City. Two months later, the United States almost succeeded in making a deal to get rid of Noriega: The United States agreed to drop drug charges against him if he would give up power and leave Panama. But the deal fell through.

A Stolen Election

On May 7, 1989, there was an election in Panama. General Noriega wanted the Panamanian people to support his favorite candidate, Carlos Duque. Noriega hoped that Duque's election would distract attention from the brutality and corruption that had made his government infamous.

2307

So Noriega carefully planned a vote fraud. Many of his supporters were given more than one ballot paper; in at least one precinct the number of votes cast was higher than the number of residents. Other voters were handed ballots that had already been marked; they were told to deposit these ballots with no questions asked. Members of Noriega's military, the Panamanian Defense Force, were allowed to vote in any precinct—which meant that they could vote several times. Noriega was so confident of winning, and of disguising the fraud, that he allowed an international team—including two former U.S. presidents, Gerald Ford and Jimmy Carter—to observe the election.

But Noriega had made some bad mistakes. The people of Panama had become so strongly opposed to his regime that even fraud could not gain victory for him. After the vote count began, it was soon clear that Noriega's candidate was losing, by a ratio of three to one, to an opposition candidate named Guillermo Endara.

Then Noriega's forces tried even more obvious vote fraud. Troops confiscated ballots and tally sheets and passed fake ones to the election tribunal in Panama City. Even Carter was turned away from the building where votes were being tallied. There was a worldwide outcry at the unfairness of this election.

For several days no one was sure what would happen. Then Noriega turned brutal, sending thugs into the streets to beat up the opposition candidates, including Guillermo Endara and vice-presidential candidates Guillermo Ford and Richard Arias Calderón. Journalists captured some of the beatings on film; when they were shown on television, millions of people around the world were horrified.

Finally, Noriega announced that the election was not valid, because it had been tainted by foreign interference. In response, the United States condemned Noriega, sent eighteen hundred more U.S. troops to join the ten thousand who were already stationed in Panama, and tightened the economic penalties against Panama. President George Bush asked the Panamanian military to overthrow Noriega.

Consequences

For a while, Noriega seemed so tough that no amount of pressure could bring him down. He survived another coup attempt by members of the Panamanian military in October, 1989, and the rebels were cruelly punished; in fact, it was reported that nearly one hundred of them were executed. After this, it seemed that Noriega's rule was safe.

Ultimately, however, Noriega's arrogance worked against him, and he underestimated the determination of his enemies in Washington. After Noriega said, in December, 1989, that Panama was in a state of war with the United States, President Bush sent a powerful military strike against Panama. Bush justified the invasion by saying that American lives and property were at risk. Within a few days Noriega surrendered to U.S. authorities, and he was taken to the United States to stand trial on drug charges. On April 9, 1992, a Florida jury found Noriega guilty of eight out of ten drug and racketeering charges. In July a federal judge sentenced Noriega to forty years in prison.

Bryan Aubrey

2308

Iran's Ayatollah Khomeini Dies

The death of Ayatollah Ruhollah Khomeini, Iran's supreme religious and political leader, was followed by a time of intense mourning and a change to more moderate leadership.

What: National politics; International relations
When: June 3, 1989
Where: Tehran, Iran
Who:
AYATOLLAH RUHOLLAH KHOMEINI (1902-1989), rahbar (religious leader) of Iran from 1979 to 1989
AYATOLLAH ALI KHAMENEI (1939-   ), rahbar of Iran from 1989

Leading an Islamic Revival

Ayatollah Ruhollah Khomeini was born in 1902, in the town of Khomein. Following the custom for religious leaders, he later adopted the name of his hometown.

Having studied Islamic theology and law at Arak, in the early 1920's he moved to the holy city of Qom. There he became a teacher, poet, and writer of religious books. In the 1930's Khomeini became known as an important scholar, and he earned the title "ayatollah" in the 1950's. It was not until the 1960's, however, that he became known as a Shiite leader opposed to the government of Shah Mohammad Reza Pahlavi. (The Shiites follow a strongly conservative version of Islam, while the shah disregarded many Islamic rules and was eager to make Iran a modern nation.)

Protesting the shah's modernization program, the ayatollah organized strikes, demonstrations, and boycotts. He believed that Western influences would weaken the fundamental teachings of Islam. After calling for the shah's overthrow in 1963, Khomeini spent months in jail and almost a year under house arrest. In late 1964, he was forced into exile.

After stopping briefly in Turkey, Khomeini settled in Najaf, a Shiite holy city in Iraq, and spent thirteen years there. During these years, he continued to work for the overthrow of the shah. By the early 1970's, he was calling for a new Islamic state to be ruled by religious leaders.

In 1978, the shah expelled Khomeini, who then moved to Paris and set up his headquarters there. Now he was able to take advantage of the international news media and other modern communications tools, such as cassette tapes and the telephone, to communicate with his followers in Iran.

A revolution was brewing back in Iran, and on January 16, 1979, the shah fled into exile. Khomeini received a noisy, joyful welcome when he returned to his country on February 1. After a national referendum, Khomeini was made supreme religious and political leader for life. Then he moved quickly to reverse the shah's reforms. Women's rights were restricted, music from Western countries was banned, and members of religious minorities were persecuted, along with political opponents.

The most important event in Iran during Khomeini's leadership was the Iran-Iraq War, which lasted from 1980 until 1988. While Khomeini was able to draw his people together to fight for their country, the war was dreadfully costly for Iran. Hundreds of thousands of Iranian soldiers were killed, and the economy was left in a shambles. In July, 1988, Khomeini agreed to a cease-fire sponsored by the United Nations.

Khomeini Is Mourned

Less than a year later, Khomeini suffered a heart attack twelve days after he had had surgery for internal bleeding. He died on June 3, 1989, in a hospital in Tehran, Iran's capital.

On June 5, Khomeini's body lay in state in a refrigerated glass coffin as hundreds of thousands of Iranians gathered to pay their respects. As the huge crowd pushed forward to get a glimpse of

2309

Iranians reach out to touch the body of Ayatollah Khomeini during funeral ceremonies in Tehran.

AP/Wide World Photos

the body, eight people were crushed to death and hundreds of others were injured.

Khomeini was buried on June 6 in the midst of a great outpouring of grief. Crowds of frantic mourners jostled the coffin, finally knocking the corpse to the ground. Foreign journalists were there to record the event, along with Iranian television.

The crowd in and around Tehran was estimated at between three and ten million. When it was obvious that Tehran streets were so clogged that the hearse could not pass through, Khomeini's body was moved to a helicopter and flown to the burial site on the southwest outskirts of the city. Then grief-stricken mourners overwhelmed the guards who stood around the grave site. People tore at the wooden coffin, trying to grab pieces of the burial cloth or any other memento they could get. Many people were injured;

some frenzied mourners actually dove into the empty grave, while others fell or jumped on top of them.

Eventually, the guards were able to remove Khomeini's corpse from its burial site. It was once again placed aboard a helicopter and flown away, while the mourners were told that the funeral would be delayed until the next day. Later, after the crowd had thinned out, the helicopter returned with Khomeini's body; this time the corpse was sealed in a metal box.

As Ahmed Khomeini (the ayatollah's son) watched, along with parliament leader Hashemi Rafsanjani and other Iranian leaders, another attempt was made to bury the ayatollah. Even then helicopters had to kick up dust and spray the crowds with water cannons before Khomeini could be laid to rest at last. He was survived by his second wife and four children.

Consequences

After Khomeini's death, the Assembly of Experts, made up of eighty-three Islamic religious leaders, voted to elect President Ali Khamenei as Iran's supreme religious leader. Khamenei was not considered to be as radical as Khomeini had been.

Around the world, there were mixed reactions to the news of Khomeini's death. Shiite Muslims in Lebanon and Syria mourned his death, but there were expressions of relief and optimism from other Arab countries. In the United States, President George Bush called on Iran to accept a more responsible role within the world community.

Francis Poole

Panamanian President Noriega Survives Bungled Coup

The October, 1989, coup against General Manuel Noriega failed, but ten weeks later the United States invaded Panama and captured the dictator, who was soon in a Florida jail, facing charges of drug smuggling.

What: Civil strife; National politics; Coups

When: October, 1989-January, 1990

Where: Panama

Who:

OMAR TORRIJOS (1929-1981), leader of Panama from 1972 to 1981

MANUEL NORIEGA (1934-   ), military dictator of Panama from 1983 to 1990

GUILLERMO ENDARA (1936-   ), a political leader

GEORGE HERBERT WALKER BUSH (1924-   ), president of the United States from 1989 to 1993

MAXWELL THURMAN (1931-   ), U.S. Army general in charge of Operation Just Cause

An Unscrupulous Dictator

Manuel Noriega, a Panamanian official, specialized in government intelligence (the gathering of important secret information). In the 1960's, Panama's military chief, Omar Torrijos, selected Noriega to set up the nation's intelligence organization. The United States military provided intelligence training for Noriega.

As Noriega developed Panama's intelligence organization, he used his new skills and international contacts for his personal advantage. He was not committed to any particular set of ideas or system of government. Rather, he was a broker or salesman of information. He worked with Cuba's communist leader, Fidel Castro, but he also worked with Castro's enemies in the Central Intelligence Agency (CIA).

Sometimes Noriega became involved in selling weapons or drugs, and he did not stop these activities when he became dictator of Panama. Noriega had a special relationship with the Medellín cartel, a group of men in Medellín, Colombia, who were responsible for manufacturing and exporting billions of dollars worth of cocaine. He allowed cartel pilots who carried cocaine from South America to refuel in Panama on their way to the United States.

Apparently the Medellín cartel paid Noriega very well for these landing privileges. At the same time, Noriega supplied information on rival drug traffickers to the U.S. government. This contradictory role of drug-traffic controller and intelligence source put him in a bind.

On May 21, 1984, under pressure from the United States, Noriega ordered his troops to raid the cartel's new drug-processing laboratory in the Panamanian jungle. Because of this raid he lost much of his influence with the drug cartel. At the same time, he was losing credibility in the United States; the Drug Enforcement Agency, congressional investigators, law-enforcement officials, and journalists stacked up considerable information about his involvement in the international drug trade.

Coup Attempt and Invasion

Noriega's international reputation came from his drug dealing, but the base of his power was the Panamanian military. Torrijos had tried to use Noriega's tough aggressiveness by appointing him to top positions in the military. As always, however, Noriega was more concerned to gain power for himself than he was for the well-being of his country.

A series of violent acts were committed against Noriega's political opponents, but no one could say for sure that Noriega was responsible. For example, Hugo Spadafora, an outspoken critic of Noriega, was tortured and murdered in 1985. Four years later, Guillermo Endara defeated Noriega's favorite candidate in the presidential election (Noriega preferred to rule through "puppet presidents"). The outraged dictator announced that the election did not count. Noriega supporters confronted Endara in the street and beat him in the head with a steel pipe. Endara survived, but Noriega was not willing for him to take his rightful place as president.

On October 3, 1989, some officers in the Panamanian Defense Forces tried to overthrow Noriega because of his corruption. Although the officers captured Noriega and held him for about five hours, they could not decide what to do next. Meanwhile, a special operations company loyal to Noriega was moving to the scene of the coup; the rebels were overpowered, and the dictator was released.

Noriega's return to power meant brutal deaths for the leaders of the coup. Two days later, the dictator boasted in a nationally televised speech that he had defeated the United States and its collaborators, just as Castro had turned back the CIA-sponsored invasion of Cuba in 1961.

American president George Bush was troubled by events in Panama. His war against the drug trade was a major priority for him in 1989, and Noriega had been indicted for smuggling drugs into the United States. On December 20, 1989, the United States launched Operation Just Cause, an invasion of Panama with the primary goal of capturing Noriega.

General Maxwell Thurman commanded forces that included paratroopers, amphibious assault units, attack helicopters, and jet fighters. Thurman's invasion overwhelmed the Panamanian Defense Forces within a few days. About twenty-five thousand U.S. military personnel were involved, and their weapons included two sophisticated Stealth fighter-bombers.

In spite of their superior technology, Thurman's forces did come upon some difficulties. Their attack on Noriega's military headquarters was effective, but this multistory building was located in a densely populated part of Panama City. After American attack helicopters and artillery set the building ablaze, the fire spread until almost two thousand surrounding homes were destroyed. Fifteen thousand people were left homeless, and a number of civilians were killed or injured.

For several days Noriega succeeded in hiding from the U.S. forces. Finally, he took refuge in the Papal Nunciature (the office of the pope's representative), where he stayed until January 3, 1990. After officials of the Catholic Church informed the Bush administration where Noriega was hiding, General Thurman decided to use

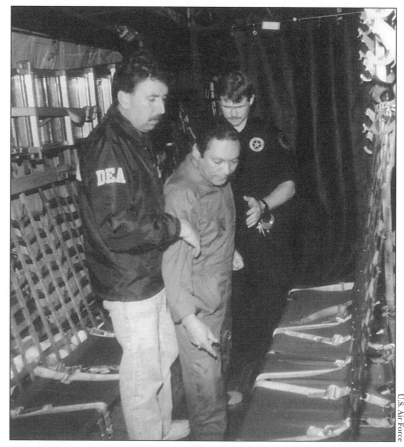

U.S. Drug Enforcement Agency members escort Manuel Noriega onto an Air Force plane, for a flight from Panama to the United States.

psychological weapons: He surrounded the Nunciature with large numbers of troops and played rock and roll music over powerful loudspeakers.

Noriega finally surrendered to the U.S. military. After an overnight airplane flight to Miami, he found himself in a Florida jail awaiting trial. On April 9, 1992, he was convicted of eight out of ten drug and racketeering charges, and in July he was sentenced to forty years in prison.

Consequences

Noriega's capture was a victory over brutal dictatorship and international lawlessness, but it also raised some controversial issues. Using military force to capture the citizen of one country for criminal charges in another country was a dangerous step to take, for in other circumstances it might bring greater bloodshed, even war. A gutted, burned-down area of Panama City remained as a sad illustration of the risks President Bush had taken in ordering the invasion. Also, Latin American countries remembered how the United States had used its military forces to get its way in their countries earlier in the twentieth century. They were unhappy to see the pattern repeated.

Kathleen S. Britton
John A. Britton

Berlin Wall Falls

The Berlin Wall stood as a barrier between East and West for twenty-eight years, until the thaw in the Cold War allowed the people to take down the wall and reunify Germany.

What: Political reform
When: November 9, 1989
Where: Berlin, East Germany
Who:

Mikhail Gorbachev (1931-   ), first secretary of the Soviet Communist Party from 1985 to 1991

Erich Honecker (1912-1994), first secretary of the East German Communist Party from 1971 to 1989

Egon Krenz (1937-   ), leader of East Germany in 1989

Hans Modrow (1928-   ), prime minister of East Germany from 1989 to 1990

Helmut Kohl (1930-   ), chancellor of West Germany from 1982 to 1990

East of the Wall

After World War II, relations between the Soviet Union and the Western powers broke down. In the Soviet Union and the countries in its sphere of influence, called "the Eastern Bloc," business and industry were taken over by the government, the Communist Party ruled, and anyone who criticized the government could be thrown into prison.

On August 13, 1961, Soviet leader Nikita Khrushchev ordered that a thirteen-foot-high wall of steel and concrete be built to divide the eastern half of Berlin, East Germany, from the western half, which had remained under the control of West Germany. Khrushchev wanted to stop unhappy East Germans from fleeing to the West.

The Berlin Wall created a country of political prisoners. East Germans who wanted to leave their country had to fill out many forms to try to get a visa; usually the visa was denied. Many of those who applied for visas came under the surveillance of the East German state police.

The standard of living in East Germany was better than that of most Eastern European countries, but poor compared to West Germany. There were often shortages of food and other necessities. Attempting to prove that East Germany was a strong, healthy nation, the government began an athletic training program. Many children were taken from their parents at an early age to be trained as future Olympic stars.

As more East Germans obtained television sets, they began to receive broadcasts from the West. These programs gave them a look at the kinds of products people in the West could buy. The East Germans became more and more unhappy with their own way of life.

Some people tried to find ways to go over or under the Berlin Wall so that they could make a new life in the West. The government ordered that anyone trying to escape be shot on sight, and a line of machine guns was placed along the top of the wall. Between 1961 and 1989, seventy-seven people were killed while trying to cross the wall, but about forty thousand escaped successfully.

The New Openness

In 1983, Erich Honecker, leader of East Germany's Communist Party, ordered the guards at the wall not to shoot at people who tried to escape. More changes were on the way after Mikhail Gorbachev came to power in the Soviet Union and began reforming the economy and political system of his country.

In September, 1989, Hungary opened its border with Austria, letting East German refugees leave the East without exit visas. About thirty thousand East Germans jumped at the chance. Within a few days, East Germans visiting Czechoslovakia and Poland began seeking refuge at the West German embassies in Prague and Warsaw.

2314

By the beginning of October, seven thousand East Germans had boarded trains from these cities headed for West Germany.

On October 9, nearly seventy thousand people marched in Leipzig, East Germany, demanding government reforms. Egon Krenz, chief of internal security, ordered the police not to interfere with the protesters.

Gorbachev had become impatient with Honecker, who was unwilling to move quickly on reforms. On October 18, Honecker was forced to resign as head of the country, and Krenz was chosen to replace him. East Germans were upset to hear that there had been corruption in Honecker's administration, and demonstrations continued in Dresden and other cities.

On October 31, travel restrictions to Czechoslovakia were lifted, and the next day thousands of East Germans rushed to the West German embassy in Prague. Between twenty and fifty thousand people fled the country that week. On November 7 and 8, the entire East German cabinet (Politburo) resigned; reform-minded politicians took their place.

On the evening of November 9, the East German government unexpectedly opened the borders to West Berlin and West Germany. Although those crossing the border were supposed to have a police permit, the government did not order the soldiers to stop people who had no permit.

Throughout the night, thousands streamed into West Berlin. The border guards, faced with lines of cars three miles long, quickly stamped the East Germans' papers and waved them through. Some people who wanted to avoid the long lines simply climbed over the wall—with the guards' help.

Once inside West Berlin, most East Germans headed for the banks, where they were given a money gift (about fifty-four U.S. dollars) as a welcome. Relatives and friends who had not seen one another in years embraced joyfully.

On November 12, a new crossing point was set up at Potsdamerplatz, as Walter Momper, Mayor of East Berlin, shook hands with Eberhardt Krack, Mayor of West Berlin. The official end to the Berlin Wall came on December 22, when West German chancellor Helmut Kohl and the

Workers tear down the Berlin Wall.

2315

new East German prime minister, Hans Modrow, together opened the Brandenburg Gate. Kohl became the first West German chancellor to set foot in East Berlin. The wall was torn down within a few days, and Germany was reunified as one country less than a year later, on October 3, 1990.

Consequences

The fall of the Berlin Wall was an important symbol of huge political changes. Germany became one country after more than forty years of division. The Cold War between the United States and the Soviet Union changed to friendship.

For individual Germans, life changed overnight. Many Easterners fled to West Germany only to find that everything was not rosy there. Under West German law, all East Germans who came to the West automatically became citizens and were entitled to medical care and other social services. With so many refugees flooding in at once, the West German government had a hard time meeting all the needs.

Early in 1990, the East German government made important reforms in the economy. Foreign investors were allowed to own shares in East German industries, which had formerly been owned and controlled completely by the state. The East German people were promised free elections and improved housing. Many of those who had left decided to return.

Jo-Ellen Lipman Boon

Chileans Vote Pinochet Out of Power

> *After seventeen years of military rule under General Augusto Pinochet, Chileans returned to democracy and elected Patricio Aylwin to the presidency.*

What: Political reform
When: December 14, 1989
Where: Chile
Who:
SALVADOR ALLENDE (1908-1973), president of Chile from 1970 to 1973
AUGUSTO PINOCHET (1915-   ), military ruler of Chile from 1973 to 1989
PATRICIO AYLWIN (1918-   ), president of Chile from 1989

The Pinochet Years

Chileans in the 1960's were proud of their many years of democratic rule. In 1970, a socialist named Salvador Allende was elected president in a race among three candidates that gave him just over one-third of the vote. Allende's administration began trying to redistribute wealth in Chile, but this quickly led to inflation and other economic problems.

Allende's opponents hoped to impeach him after the congressional elections of 1973, but they did not win enough seats. So the military plotted to overthrow him. On September 11, 1973, they staged a coup; Allende died in the national palace, probably by suicide.

A junta (group of military officers) took control of the country. The police rounded up Allende supporters and executed many of them—between five thousand and fifteen thousand. Many others escaped death by fleeing the country.

At first the officers said that they would hold power only long enough to restore order and get the country ready for new elections. But then General Augusto Pinochet, commander in chief of the Chilean army, began to seek more power for himself. Pinochet was named president in 1974, and in the next two years his National Intelligence Directorate (Dirección Nacional de Inteligencia, or DINA) tracked down persons who had spoken out against his rule. Several hundred of these people disappeared, never to be heard from again.

Pinochet was most concerned to get Chile's economy moving. Foreign investors had left Chile because of high inflation and nationalization of businesses during Allende's administration. To keep prices down, Pinochet kept workers' wages low, and he encouraged foreign investors to return. The economy did grow as a result—but only a few Chileans really benefited, because most found that they could barely make ends meet on their low wages.

Meanwhile, DINA continued its repression, even outside Chile. On September 21, 1976, with the help of Cuban exiles, DINA planted a car bomb in Washington, D.C., and killed Orlando Letelier, who had been foreign minister of Chile during Allende's presidency. Letelier's American aide, Ronni Moffit, was killed as well. Many Americans protested, and DINA was abolished in 1977.

Chileans Vote

In 1980 Pinochet produced a new constitution which promised that in 1988 Chilean citizens would be given the chance to approve or end Pinochet's government: They could vote "yes" or "no" on another term in office for the general.

In 1981-1982 Chile began suffering economic problems, and by 1983 the number of unemployed had grown from a 1973 total of 145,000 to more than 1 million. More and more people were unhappy with Pinochet's rule. In 1983, women in well-to-do neighborhoods in the capital, Santiago, took to the streets, banging on empty cooking pots to show that it had become much harder to feed their families. Leaders of the political parties, which had been banned, gathered quietly to discuss how to get rid of Pinochet.

The military responded with more repression, but the strikes and protests continued. The commanders in chief of the navy and air force met with the political-party leaders to talk about ousting Pinochet. In September, 1986, a leftist group tried to assassinate the general, and he imposed a new "state of siege."

Yet the plebiscite (popular vote) promised by the constitution was approaching. Leaders of the sixteen political parties banded together to urge Chileans to vote "no." The coalition of parties was led by the president of the Christian Democratic Party, Patricio Aylwin.

Preparing for the plebiscite, Chileans feared the worst—that Pinochet would use military forces to punish the people for voting against him. Meanwhile, Pinochet seemed confident that he would win.

In the October 5, 1988, plebiscite, 54.7 percent voted "no." Pinochet responded by promising to prepare for presidential elections, in which he would not be a candidate. Yet the constitution allowed him to stay in command of the armed forces for eight more years and to appoint people to Congress before he stepped down.

Now the Chilean people flung themselves into politics. Realizing that many still supported Pinochet's economic policies, the opposition leaders promised to keep his policies even after he was gone. The coalition that had called for "no" votes in the plebiscite now united behind one candidate—Aylwin.

Some socialist leaders who had supported Allende had spent the Pinochet years in Eastern European countries, where they had seen the weaknesses of communism. Returning to Chile, they announced that now they supported liberal democracy and capitalism.

Pinochet's supporters chose Hernan Buchi, who had been minister of finance from 1985 to 1988, as their candidate. They claimed that he would help Chile continue economic growth.

On December 14, 1989, a majority of Chilean voters cast their ballots for Patricio Aylwin; they also elected a Congress for the first time since the 1973 coup. The elections were peaceful, and the people of Chile were delighted that democracy had been restored to their land.

Consequences

New challenges lay ahead for the new president and Congress. First, Pinochet remained in charge of the army, and he and the other military leaders were not willing to admit that they had violated the human rights of Chileans.

Second, the coalition that had elected Aylwin was made up of many groups with very different ideas. The leaders were determined to keep working together, but agreeing on decisions would sometimes be quite difficult.

Finally, there was a serious economic challenge. Chile's gross national product (GNP) had been growing in the late 1980's, yet most workers earning the minimum wage could not support their families. The new government wanted to find a way to help the workers without alarming foreign investors with the threat of socialism. Chile's problems could no longer be blamed on Pinochet; they had become the responsibility of all the Chilean people.

Joan E. Meznar

Romanian President Ceausescu Is Overthrown

> *The democracy movement that swept through Eastern Europe in 1989 toppled Nicolae Ceausescu, the tyrannical dictator of Romania.*

What: Political reform; Coups
When: December 23, 1989
Where: Bucharest, Romania
Who:
NICOLAE CEAUSESCU (1918-1989), general secretary of the Romanian Communist Party from 1974 to 1989
ION ILIESCU (1930-   ), a Communist reformer who became prime minister of Romania in 1989
GHEORGHE GHEORGHIU-DEJ (1901-1965), a leader in the Romanian Communist Party

The Coming of Communism

As a result of the Balkan Wars and World War I, Romania gained Dobruja from Bulgaria, Transylvania from Austria, and Bessarabia from Russia. This brought a mixture of ethnic groups under Romanian rule—Hungarians, Russians, Serbs, Bulgarians, Gypsies, Ukrainians, and Turks. The new lands included many religious groups as well: Muslims, Catholics, Protestants, Orthodox Christians, and Jews.

Riots between ethnic groups became common. Hungarians in Transylvania and Jews in Bessarabia were especially unhappy with the way Romania treated them. Matters became worse in the 1930's, when Romania adopted fascism and became an ally of Nazi Germany.

The Romanian workers were quite poor and were not allowed to join free labor unions; they suffered greatly during the Great Depression. In response, many of them began turning to communism. Members of oppressed minorities, especially Hungarians and Jews, were also attracted to the Communist Party. Some of them became leaders in the Party and made their way to Moscow.

But it was another group of communist leaders, those who stayed in Romania to fight during World War II, that took over the government in 1944. Chief among these leaders was Gheorghe Gheorghiu-Dej, who headed the Romanian government until his death in 1965.

Ceausescu Rises and Falls

Nicolae Ceausescu, who had worked closely with Gheorghiu-Dej, managed to push aside his major rivals and take control of the government after Gheorghiu-Dej's death. Ceausescu was a nationalist who quickly became quite popular.

In the 1960's and 1970's, the economy grew quickly, and the Romanian people found that their standard of living was rising. Ceausescu was admired by other world leaders for his skill in governing. Gradually, however, his reputation changed.

Entering the 1980's, Romania's economy went into a depression, and its foreign debt grew. Poverty increased; food and energy were rationed by the government. But the people began noticing that Ceausescu did not seem to suffer as they did. In fact, he was enjoying great wealth and did not seem too concerned about hiding his extravagance. He had become like one of the old Romanian kings.

Ceausescu built up a large secret-police force, along with informers who would report dissent to the government. The liberal reforms vanished. Romanians found that they could not travel freely; those who criticized the government were often fired from their jobs or thrown in jail.

2319

Ceausescu's socialist reforms also brought unhappiness. He wanted to build huge complexes of factories and farms, and to do so he forced many peasants to move out of their cottages and into large dormitory-type buildings. Then the peasants' villages were bulldozed to the ground. Many historical monuments were also destroyed.

Ceausescu singled out Hungarians for mistreatment. Many of the people who had lived in the destroyed villages were Hungarian. Some of these people fled across the border into Hungary, where liberal reforms were already taking place.

Even in the late 1980's, when democratic reform was spreading across other parts of Eastern Europe, Ceausescu refused to change. He said the Hungarians who had fled were traitors to Romania and to socialism. He blamed Romania's problems on the Communist Party, but he did not offer to change the political system.

Early in December, 1989, newspapers and other media around the world carried stories of the mistreatment of the Reverend Laszlo Tokes, a Protestant Hungarian pastor who lived in Timisoara, Transylvania. Tokes had protested the government's discrimination against Hungarians, and in response the Romanian authorities had been making life hard for him. Three Hungarian journalists who came to visit him were expelled from the country.

On December 17, the Romanian police attempted to deport Tokes himself, but the townspeople of Timisoara, both Hungarians and Romanians, rose up in a protest that turned into a revolt. Thousands died in battles on December 19 and 20. Governments around the world condemned Ceausescu for his abuses, but he insisted angrily that the demonstrators were "fascist agitators" who were being egged on by Hungary.

On December 21, the demonstrations spread to the capital, Bucharest. Crowds jeered Ceausescu and shouted him down when he tried to make a speech at the university. The next day, he tried to flee the country. A group of dissident Communists, led by Ion Iliescu, formed a Com-

President Nicolae Ceausescu, with his wife, in about 1980.

mittee of National Salvation as a temporary government.

Ceausescu did not succeed in leaving Romania; he and his wife were captured on December 23. Two days later, a hasty court martial tried the Ceausescus on charges of genocide, destroying Romania's economy and spiritual values, and stealing more than one billion dollars, which they had deposited in foreign banks. The court found them guilty and ordered them shot the same day. The execution was filmed and then shown on national television as proof of the dictator's death.

Consequences

The fall of Ceausescu did not bring an end to the Communist Party's rule in Romania. Some reforms were introduced, including civil liberties such as freedom of speech, assembly, and religion. Yet students and intellectuals in Bucharest were unhappy, for they wanted all Communists out of power. The new government made plans to solve Romania's economic problems, but little came of these schemes.

When the government gave Hungarians the right to establish schools using the Hungarian language, Romanian nationalists became quite upset. A peaceful Hungarian march at Tîrgu-Mures, Transylvania, in the spring of 1990 was broken up by club-wielding Romanians, who killed eight people and injured hundreds more. Thousands of Hungarians again fled the country.

In May, 1990, Iliescu and his Socialist Party (the former Communists) won a landslide victory in an election. Observers said that the election was honest, but dissidents in Bucharest claimed there had been fraud. There was a violent demonstration in June, and Iliescu brought in miners to attack the protesters because he could not rely on the police. Clearly, Romania faced a struggle to regain stability and establish democracy.

Frederick B. Chary

Havel Is Elected President of Czechoslovakia

> *With the election of Václav Havel as president, Czechoslovakia changed from a communist state to a Western-style democracy.*

What: Political reform
When: December 29, 1989
Where: Czechoslovakia
Who:

GUSTÁV HUSÁK (1913-1991), president of Czechoslovakia from 1975 to 1989

LADISLAV ADAMEC (1926-), prime minister of Czechoslovakia from 1988 to 1989

VÁCLAV HAVEL (1936-), a playwright who became president of Czechoslovakia in 1989

ALEXANDER DUBČEK (1921-1992), a political reformer

MIKHAIL GORBACHEV (1931-), first secretary of the Soviet Communist Party from 1985 to 1991

A Time of Reform

When Mikhail Gorbachev became leader of the Soviet Union in 1985, he quickly began reforming the political system and economy through what he called *glasnost* (openness) and *perestroika* (restructuring). The changes he brought to the Soviet Union eventually spilled over into Eastern Europe, which had been dominated by the Soviets since the end of World War II.

The year 1989 was a time of many changes in the communist countries of Eastern Europe. Dissidents gained the courage to speak openly of the need for political reform. Hungary, Poland, and East Germany all left communism behind and embraced democracy. In the last weeks of 1989, Czechoslovakia followed their example.

On November 17, students led a demonstration in Prague, the capital of Czechoslovakia; this protest was brutally ended by the police. Thousands of Czechs became so inspired by the students' action, and so angry at the government, that they poured out into the streets in the days that followed.

Václav Havel, a playwright who had often spoken up against the government, joined other dissident leaders to form Civic Forum. This coalition organized protests and helped guide the prodemocracy movement.

On November 27, a two-hour strike was honored by millions of Czech workers; the government got the message that students and intellectuals were not the only people who supported Havel and Civic Forum. The next day, Prime Minister Ladislav Adamec announced that the Communist Party would give up its dominance of Czechoslovakia, even though the country's constitution guaranteed the Communists that right.

By December 10, President Gustáv Husák, a strict Communist, had resigned. A new cabinet was formed, with Communists in the minority for the first time since 1948.

Havel Is Elected

Now Czechs wondered whether Civic Forum would be allowed to take over the government peacefully and quickly, and who should be named the new leaders of the country. There were rumors that the military might be plotting a coup on December 18. The rumors turned out to be false, but they added to the tension people were feeling in this time of political change.

There were negotiations and compromises among the political leaders. Adamec negotiated with Civic Forum and hoped to keep a place in the government for a reformed Communist Party; he resigned as prime minister on December 7. By the middle of December, Alexander Dubček, who had led the reform movement of 1968 known as the Prague Spring, was supporting Havel for president; in exchange, Civic Forum backed Dubček as the new head of the Federal Assembly (parliament).

Dubček was named to that post on December 28. Though the Communist Party still had a majority in the Federal Assembly, the assembly elected Havel president of Czechoslovakia on December 29. So both Havel and Dubček were leaders of the new Czechoslovakia.

The transfer of power had been peaceful; some called it "the velvet revolution." People around the world admired the Czechs for their success in organizing huge but orderly demonstrations to bring political reform. Civic Forum leaders had spoken often of discipline and responsibility, and this approach had paid off.

Consequences

As president, Havel did not lose any of his courage or honesty. He traveled to both East and West Germany in early January, 1990, and then found himself caught up in the first controversy of his presidency. Adamec, still a leader in the Communist Party, quoted Havel as saying that Czechs owed an apology to the Germans who had been expelled from the Czech Sudetenland at the end of World War II. Czech nationalists were displeased at this statement.

Actually, Havel had made this remark a month before his election as president. The comment showed that he was a humanitarian who understood his country's history and wanted to make amends for past wrongs. In the end, though, Havel did apologize for the comment—although it was determined that no restitution was necessary, and the controversy died down.

On January 15, 1990, Havel's government began negotiating to remove the nearly seventy-five thousand Soviet troops that had occupied Czechoslovakia since 1968. Five days later, Pope John Paul II announced that he would visit Czechoslovakia in late April. This would be the pope's first visit to Eastern Europe outside Poland, his homeland. It was clear that the Roman Catholic Church would enjoy a good relationship with Czechoslovakia's new government.

Havel even tried to persuade leaders of the United States and the Soviet Union to hold their next summit meeting—scheduled for early summer, 1990—in Prague. He did not succeed, but he had demonstrated his willingness to take his place as a world leader.

At first Havel announced that he intended to stay in office only until new elections could be held for the Federal Assembly. Eventually, he was reelected to a two-year term in Czechoslovakia's first free elections since the 1940's. He remained president until the summer of 1992, when he resigned on July 18, following a Slovak vote for sovereignty. The idealistic first president of a postcommunist Czechoslovakia was philosophical about the potential split up of his nation. *Time* magazine (August 3, 1992) quoted him as saying: "I do not place the highest value on the state, but rather on man and humanity. . . . We cannot be surprised that now, when the straitjacket of Communism has been torn off, all countries [including discrete ethnic groups such as the Slovaks and the Czechs] wish to establish their independence." Havel also reaffirmed his intention to remain active in public life.

William D. Bowman

Miloš Fikejz

Václav Havel.

Political Reforms Come to Albania

After decades of isolation under a communist government, Albania made it easier for foreigners to visit the country; this was the first step in a process of liberalization.

What: Political reform
When: 1990
Where: Tirana, Albania
Who:
ENVER HOXHA (1908-1985), first secretary of the Albanian Communist Party from 1941 to 1985
RAMIZ ALIA (1925-   ), Hoxha's successor as first secretary of the Albanian Communist Party
ISMAIL KADARE (1936-   ), a nationalist writer

The Hoxha Years

Modern Albania was formed just before World War I as a result of conflict among the Balkan states and the Great Powers. The people of this new country all spoke the same language, but they were divided by religion; there were two different Muslim groups, Orthodox Christians, and Roman Catholics. Between the world wars, Albania changed from a democratic republic to a dictatorship, whose leader proclaimed himself Zog I, king of Albania.

In 1939, Italian dictator Benito Mussolini took over Albania and made it part of Italy. Two Albanian groups began resisting the fascists: a democratic organization called Balli Kombetar and a communist group called the Albanian Party of Labor. Enver Hoxha organized the communists into a powerful force.

After World War II, Hoxha became ruler of the People's Socialist Republic of Albania. Hoxha admired Soviet dictator Joseph Stalin; when a more liberal leader, Nikita Khrushchev, came to power in the Soviet Union, Hoxha was displeased, and in 1962 he broke off all ties with the Soviets.

Hoxha was determined to keep Albania a pure Stalinist state—which meant totalitarianism. The dictator was glorified, and strict censorship ensured that all criticism was silenced. People were executed for breaking even the smallest laws. Private ownership was severely limited; for example, Albanians were not allowed to own cars. And Hoxha tried to prevent foreign ideas from entering the country. All public worship was banned in Albania.

Though Hoxha avoided relations with the two superpowers, the United States and the Soviet Union, he did not completely isolate his country. Albania did have diplomatic relations with some other communist nations, a few Western countries, and many nations in the Third World (Africa, Asia, and Latin America).

Hoxha wanted Albania to have a self-sufficient economy. The country was rich in resources, but since these riches had never been used to benefit the people, Albania's population was one of the poorest in Europe. Hoxha modernized the country gradually, raising living standards slowly without piling up a foreign debt.

A Time of Change

In the 1970's, Hoxha gradually pulled back from the limelight and turned most of his power over to Ramiz Alia. Alia made new diplomatic and business contacts, though still not with the superpowers. Some foreigners—even Albanian Americans—were allowed into the country on tours. After Hoxha died in 1985 and Alia took full control, there was even more liberalization.

The changes that swept Eastern Europe in 1989 affected Albania as well. Students began demonstrating for liberalization. At first Alia tried to stop them; then he tried to keep the communist system going while making compromises with the growing reform movement.

In March, 1990, the Albanian government introduced its own reform program, including economic changes, free elections in factories and on collective farms, and open debate on some government policies, such as education. The government also reached out abroad, trying to end its isolation. For the first time, ordinary Albanian citizens were able to place direct telephone calls to the West. In May, Alia restored the people's right to practice their religion openly and to travel abroad. In the months that followed, Alia opened up other political and civil rights and made it easier for foreigners, including Americans and Soviets, to visit the country.

Yet protests continued, especially among educated Albanians, who were impatient to see a complete end to communism. In October, 1990, while on a visit to France, the well-known Albanian writer Ismail Kadare announced that he would stay in the West in protest against his nation's government. His action inspired even larger demonstrations in Tirana, the Albanian capital, and other cities.

In the meantime, the Albanian government kept trying to establish ties with foreign governments, especially those of the United States and the Soviet Union. In August, 1990, Albania restored diplomatic relations with the Soviet Union, and in March, 1991, with the United States. The next June, Secretary of State James Baker became the highest-ranking American official ever to visit Albania.

Consequences

Albania was divided into two camps: those who supported Alia's slow reforms and those who wanted to move away from communism as rapidly as possible. The debate turned violent; though the government gave in to some of the reformers' demands, demonstrations and riots continued. Statues of Stalin and Hoxha were knocked down, and demonstrators frequently came into conflict with the police.

In November, 1990, Alia called for the constitution to be revised; the Communist Party's powers would be cut back, and the official policy of atheism would be ended. He promised free elections and steps toward a free-market economy. The parliament passed laws to allow foreign investors into Albania.

The government freed political prisoners, and a new law in February, 1991, made the courts independent. Opposition political parties and newspapers were allowed. Yet the communists could still rely on the support of people in the countryside, who did not trust the reform-minded intellectuals and city dwellers.

Albania's economic struggles did not end with liberalization. Many Albanians were eager to leave the country to find a better life. In 1990 and 1991, many Albanians of Greek background and some other Albanians fled south into Greece, until Greece was forced to close its border. Other Albanians crammed themselves onto overloaded ships headed for southern Italy,

2325

where an Albanian community had lived for centuries. The Italian government sent some of these refugees back and refused to feed the others in the hope that they would return of their own accord. But many Italians protested, so that the authorities gave in and allowed some of the refugees to stay.

Albanian elections in the spring of 1991 kept Alia and the communists in power. International observers believed that the elections were mostly fair, but some Albanians accused the communists of fraud. There were more violent demonstrations. In June, Alia agreed to bring the opposition into a coalition government to work out a solution to the country's conflicts. Alia was the only communist leader in Eastern Europe who was able to bring democracy to his country and still remain in power a year after the changes.

Ralph L. Langenheim, Jr.

South African Government Frees Mandela

> *African nationalist leader Nelson Mandela, who had been sentenced to life imprisonment in 1964, was released from prison and began to negotiate with the ruling white-dominated Nationalist Party to end apartheid in South Africa.*

What: Civil rights and liberties; Political reform
When: February 11, 1990
Where: Cape Town, South Africa
Who:
NELSON MANDELA (1918-   ), deputy president of the African National Congress (ANC)
WALTER SISULU (1912-   ), a leader of the ANC
FREDERIK WILLEM DE KLERK (1936-   ), president of South Africa from 1989 to 1994
PIETER WILLEM BOTHA (1916-   ), prime minister of South Africa from 1978 to 1989

Struggle for Justice

When the Nationalist Party came to power in South Africa in 1948, the years of apartheid began. Europeans in southern Africa had always treated the native Africans badly, but apartheid (an Afrikaans word meaning "apartness") was a new, strict system of racial segregation. Under the Population Registration Act of 1950, South Africans were divided among four racial groups: white, colored, Indian, and African. The pass laws required blacks to carry special passports at all times. Marriage between people of different races was forbidden, and the Bantu Education Act of 1953 established segregated, inferior schools for black children.

Yet various groups continued to work for democracy and civil rights in South Africa. One of the oldest was the African National Congress (ANC), which was founded in 1912. In 1944, Nelson Mandela, Oliver Tambo, and Walter Sisulu formed the ANC Youth League to bring new energy into the struggle against discrimination. Cooperating with the South African Communist Party, the South African Indian Congress, and the South African Coloured People's Organization, the ANC organized special days of protest and "stay-at-home" demonstrations and strikes. The ANC claimed that South Africa belongs to all who live in it, black and white.

In 1955, Mandela, along with 155 other activists, was arrested and charged with treason. The treason trial lasted five years, during which Mandela managed to stay active in the ANC.

Opponents to apartheid began a campaign against the pass laws, and at Sharpeville (a black area near Johannesburg), the government made its response: Dozens of protesters were killed, and 186 were wounded. A state of emergency was declared, and the ANC was banned. Mandela went into hiding.

On December 16, 1961, the ANC founded a subgroup called Umkhonto we Sizwe, or Spear of the Nation. As commander in chief of this military organization, Mandela helped plan its strategy: sabotage and destruction of property rather than violence against people.

In August, 1962, Mandela was arrested, tried, and sentenced to five years in prison for launching an illegal strike and leaving the country without a valid passport. In July of the following year, he was tried for sabotage along with Sisulu and eight other activists, and in 1964 he was sentenced to life in prison. Over the years, some Nationalist Party leaders offered him freedom in exchange for renouncing violence and promising to stay in his "homeland." He always refused.

Walking into Freedom

By 1980, apartheid was getting wobbly. Nations all around South Africa had achieved independence from European domination, and most governments around the world criticized South Africa for its mistreatment of black citizens. Whites in South Africa, who made up only a small minority of the population, were divided: Some wanted reform, while others demanded even stricter controls on blacks.

Pieter Willem Botha, who became prime minister in 1978, decided on gradual reforms. A three-chamber parliament giving representation to whites, coloreds, and Indians was established in 1983; the pass laws were repealed in 1986; and Africans were allowed to vote in elections to local councils in 1988. Most anti-apartheid organizations, including the underground ANC, refused to support the reforms; they wanted to see apartheid completely done away with.

After Botha suffered a stroke in February, 1989, Frederik Willem de Klerk became the leader of the Nationalist Party and the country. He released Sisulu and several other antiapartheid activists in October, 1989, and on February 2, 1990, de Klerk announced that the ANC and thirty other antiapartheid organizations were no longer banned.

On February 11, 1990, Mandela was released from prison; the event was broadcast around the world. The next day he gave a rousing speech at Cape Town City Hall, asking fellow South Africans to continue the armed struggle. Mandela also asked foreign nations to keep economic pressure on the South African government until apartheid was abolished.

Consequences

Although Mandela's release did not immediately end apartheid, it was an important step toward reform. After all the years of working underground, the ANC was still highly respected among the black people of South Africa; Mandela was welcomed as a hero and a major spokesman for blacks.

Talks between the ANC and the Nationalist government began in May, 1990. Soon afterward, several important apartheid laws were repealed, including the Group Areas Act (1950) and the 1913 and 1936 Native Land Acts, which had reserved 87 percent of the land for whites. Although extremist whites protested, the government seemed committed to forming an assembly to write a new constitution and holding free elections in 1994.

There were two major obstacles to an agreement. First, the ANC stood firmly for one person, one vote in South Africa—which would mean rule by the black majority—while the government wanted to protect minority rights. Sec-

AP/Wide World Photos

Nelson Mandela (left) and his wife, Winnie, the day after his release.

ond, the government wanted to repeal the Native Land Acts without compensating the millions of Africans whose lands were taken, while the ANC insisted that those who were evicted should be paid.

Meanwhile, violence in the black areas continued; between 1987 and 1991, about five thousand Africans died in township violence. ANC supporters were often attacked by members of the Inkatha Freedom Party, a more conservative black group headed by the Zulu chief Gatsha Buthelezi. Leaders of the ANC accused the government of supporting Inkatha and encouraging the violence. March 17, 1992, became a turning point in South Africa.

In a referendum, the white minority overwhelmingly approved de Klerk's efforts to establish majority rule—South Africans had finally "closed the book" on apartheid.

Catherine Scott

Sandinistas Lose Nicaraguan Elections

The Sandinistas (Frente Sandinista de Liberación Nacional), a revolutionary party that had overthrown the Somoza family dictatorship in 1979, held open elections in February, 1990, and lost to a coalition of more conservative parties.

What: National politics
When: February 25, 1990
Where: Nicaragua
Who:
DANIEL ORTEGA (1945-   ), president of Nicaragua from 1984 to 1990
VIOLETA BARRIOS DE CHAMORRO (1929-   ), president of Nicaragua from 1990
GEORGE HERBERT WALKER BUSH (1924-   ), president of the United States from 1989 to 1993

The Sandinista Years

In July, 1979, the Sandinista Front for National Liberation (FSLN, or Sandinistas) rescued Nicaragua from its long years of domination by the Somoza dictators. Over the next few years, Daniel Ortega became the main leader of the Sandinistas.

Ortega's rise to power did not come overnight. He had spent his whole life fighting for Nicaragua's freedom, and had been jailed by the Somozas for seven years. Though he was a hardworking soldier and leader, he was awkward in public; his speeches were long, rambling, and uninspiring.

In the first years after the revolution, Ortega was just one among equals. In 1979 the FSLN passed a bylaw that said no attention would be called to any particular member of the nine-person National Directorate that governed the party and the nation. The Sandinistas wanted group leadership, rather than domination by one person.

Yet gradually Ortega became more comfortable in public, and he was respected for helping fellow members of the National Directorate to compromise on important decisions. He led Nicaragua into cooperation with the Central American Peace Plan designed by Oscar Arias, president of Costa Rica.

Keeping its promise, the FSLN held open elections in November, 1984. American president Ronald Reagan considered the Sandinistas communists and did not trust them; his administration convinced some Nicaraguan political parties to stay out of the election campaign. Yet international observers traveled to Nicaragua to watch the elections, and they agreed that the FSLN did not use fraud to gain its 67 percent victory.

In 1987, the Nicaraguan government proclaimed a new constitution that called for regular elections. A new election was called for 1990.

The 1990 Campaign

During the 1980's, a long struggle against the Contras, a conservative rebel group, caused much suffering among the people of Nicaragua. The U.S. Central Intelligence Agency (CIA) provided money, weapons, and training for the Contras, for Reagan was determined to get the Sandinistas out. The FSLN's health and education programs did not work, because most of the government's money went into the war. The fighting left thirty thousand dead, and hundreds of thousands of Nicaraguans fled the country. Wages fell more than 70 percent between 1978 and 1990. As if all this were not enough, Hurricane Joan swept the coast and left many people homeless.

By the time of the 1990 elections, the Nicaraguan people were tired and beaten down. Many of them believed that a new government might make things better; it could hardly make them worse. The people were especially upset that government leaders were rewarded with houses, cars, private clubs, and tax privileges. Clearly the Sandinistas were not as corrupt as the Somozas had been, but people suspected that some of the country's economic problems had been caused by inefficiency and mismanagement.

During the election campaign, Ortega and other FSLN candidates emphasized that the Sandinistas had been courageous in their battle against the Americans' efforts to control Nicaragua. Yet realizing that the people were tired of war, Ortega tried to show that his government could bring peace. In the week before the election, the Sandinista newspapers were filled with stories saying that the new American president, George Bush, would recognize a Sandinista victory, and that the Contras were about to stop the fighting. The FSLN's slogan was "*Ganamos, todo será mejor*" (if we win, everything will be better).

It has been estimated that the FSLN spent more than twenty million dollars on the campaign. Trucks delivered tens of thousands of T-shirts, hats, red and black scarves, and posters to campaign headquarters across Nicaragua. In villages and in working-class neighborhoods, Sandinista slogans were painted on countless houses and buildings.

Yet this hard campaigning worked against the FSLN in the end. The government had been asking the people to sacrifice for the good of the country; people were so poor that few of them could afford meat or medicine. Now hats and shirts were being given away by the truckload. To the people this seemed hypocritical.

Still, the FSLN's main opponent faced problems as well. The Unión Nicaragüense Opositora (UNO), a coalition led by Violeta Barrios de Chamorro, included some Contra leaders. During the campaign, Contra attacks were blamed for a number of deaths in various parts of the country. Also, Chamorro edited a newspaper, *La Prensa*, which was known to have received money from the CIA during the war years.

Chamorro did not have much political experience, but her message was simple and direct: She could heal Nicaragua's divisions. She also promised that upon her election the Contras would be disbanded, Nicaragua's relationship with the United States would be restored, the draft would be abolished, the government bureaucracy would be cut back, and foreign investment would help rebuild the economy.

On February 25, 1990, voting was mostly quiet and orderly. There were more than three thousand observers throughout the country, including teams sent from the United Nations and the Organization of American States. Also, poll watchers from all political parties checked computerized registration lists to make sure there was no fraud.

The Sandinistas lost the election; the Nicaraguan people had sent the message that they could not keep sacrificing, even for patriotic causes. Chamorro received more than 53 percent of the vote and was soon inaugurated as Nicaragua's new president.

Consequences

In spite of its failures, the FSLN had succeeded in establishing real democracy in Nicaragua. Even the hated draft had been an equalizer: The idea that everyone in Nicaragua had to serve, not only the poor, was a new idea.

As the new president, Chamorro faced an enormous challenge: preserving the democratic reforms the FSLN had brought, bringing peace and reconciliation to a land that had been at war with itself for almost twenty years, and finding ways to mend the economy. The hard-fought election campaign had been easy by comparison.

Allen Wells

Soviet Troops Withdraw from Czechoslovakia

When Soviet troops left Czechoslovakia, that nation was finally free from outside domination.

What: International relations
When: February 26, 1990
Where: Frenstat, Czechoslovakia
Who:

GUSTÁV HUSÁK (1913-1991), president of Czechoslovakia from 1975 to 1989

MILOS JAKES (1922-   ), first secretary of the Czechoslovak Communist Party from 1987 to 1989

LADISLAV ADAMEC (1926-   ), prime minister of Czechoslovakia from 1988 to 1989

VÁCLAV HAVEL (1936-   ), president of Czechoslovakia from 1989

ALEXANDER DUBČEK (1921-1992), a reform-minded politician

A Movement Emerges

In 1955, Czechoslovakia joined the Warsaw Pact Treaty Organization, a mutual-defense organization dominated by the Soviet Union. The pact allowed member nations to hold joint military maneuvers on one another's territory.

In 1968, reformers within the Czechoslovak Communist Party managed to take control of the government. Led by Party Secretary Alexander Dubček and Prime Minister Oldrich Cernik, the new government ended strict censorship and allowed citizens to travel more freely. The Party's Central Committee discussed allowing freedom of the press and of assembly. This time of reform was called the Prague Spring.

To stop the reforms, the Soviet Union, along with four other Warsaw Pact countries, invaded Czechoslovakia and seized control. Soviet tanks rolled into Prague, the capital, on August 21, 1968, and thousands of young Czechs took to the streets in protest. They took down street signs to confuse the soldiers; in some parts of the city, electricity, gas, and water were shut down.

Nevertheless, on October 16 the Soviets forced Prime Minister Cernik to sign a treaty allowing Soviet troops to be temporarily stationed within Czechoslovakia. The treaty could only be changed by agreement on both sides. For the next twenty years, the Soviets kept approximately 73,500 troops in Czechoslovakia.

Gustáv Husák was chosen to replace Dubček as head of the Party, and he was named president as well. The government now made it illegal to form noncommunist political groups, and censorship was restored. Husák tried to satisfy the people by providing more consumer goods for them.

Yet the resistance continued. In early 1969, students named Jan Palach and Jan Zajic became national heroes when they publicly set themselves on fire to protest the Soviet occupation. To mark the first anniversary of the Soviet invasion, residents of Prague boycotted public transportation for the day. A group of intellectuals, led by playwright Václav Havel, also sent the federal parliament a declaration in which they protested the Soviet invasion, censorship, and the Communist Party's cruel repression.

The Soviets Depart

Havel had grown up in a wealthy family; after the communists seized power in 1948, his family's property was confiscated, and he was not allowed to finish his professional education. Yet he worked hard to educate himself, and by 1968 he had become well known as a poet and playwright, both in his country and abroad. Some of his works were banned in Czechoslovakia, but he continued to send his writings to publishing companies in Western Europe.

Meanwhile, Husák's government kept up its policies of repression—even of popular music. Because rock music was seen as a challenge to the government's power, many recordings were banned. By 1983, thirty-five "new wave" bands were not allowed to perform in Czechoslovakia. Yet young Czechs and Slovaks illegally passed around homemade copies of the banned tapes.

When Mikhail Gorbachev began reforming the Soviet economy and political system in the mid-1980's, many in Czechoslovakia hoped that their government would follow suit. In 1987, Milos Jakes replaced Husák as head of the Communist Party, but the new first secretary was unwilling to share power with noncommunists.

Dissidents in Czechoslovakia became more and more active in 1989. Student activists found that many writers, artists, and actors in Prague were their allies in the antigovernment movement. On November 19, 1989, a number of dissidents gathered at the Magic Lantern Theater and formed Civic Forum, a prodemocracy organization. The next day, 200,000 people crowded into the main street of the capital in an antigovernment rally. Jakes threatened to take action against them, but Prime Minister Ladislav Adamec promised the Civic Forum leaders that the government would not impose martial law.

On November 24, Alexander Dubček traveled from Bratislava to Prague to join in the demonstrations. Dubček also met with leaders of the Communist Party. Later that same evening, Jakes and all the other main leaders of the Party resigned; Jakes was replaced by Karol Urbanek.

Three days later, Civic Forum called a day-long nationwide strike. Everyone was surprised to see that workers across the country honored the strike; this proved that Civic Forum's cause was extremely popular.

In early December, Adamec announced that a new government was being formed, and that five cabinet positions would be given to noncommunists. But Civic Forum did not accept this compromise, and the next day Adamec was forced to resign. The next few weeks saw the creation of a new, noncommunist government in Czechoslovakia. Havel became president, and Marian Calfa was named the new prime minister. Calfa promised to hold free elections in June, 1990.

Workers remove a statue of Vladimir Lenin from the central square in Zilina, Czechoslovakia.

2333

Within days, the Havel government began asking the Soviet Union to take its troops out of Czechoslovakia—in fact, Deputy Foreign Minister Evzen Vacek insisted that the soldiers must be gone before the midyear elections. On February 26, 1990, Soviet leader Gorbachev and President Havel signed an agreement that the Soviet troops would be withdrawn gradually over sixteen months. By May 31, 1990, more than half of the 73,500 Soviet troops were out of the country; the last of them left on June 30, 1991.

Consequences

Havel's success in persuading the Soviet Union to withdraw its troops from Czechoslovakia gave the Czechoslovak people new faith in the power of democratic action. This new confidence allowed Czechoslovakia to move quickly into democracy.

On June 8-9, 1990, Czechoslovakia held its first free election in forty-one years. Twenty-three political parties participated, but most Parliament seats were won by five major parties. Civic Forum won 48 percent of the vote and became the largest party in Parliament.

The Soviets left some problems behind. Czechoslovakia's ministry of the environment estimated that the Soviet forces had polluted five thousand to eight thousand square miles of land with diesel oil, toxic chemicals, live ammunition, land mines, and chemical weapons. The Czechoslovaks said that it would take $125 million to clean up the pollution.

Supreme Court Overturns Law Banning Flag Burning

The U.S. Supreme Court held that the Flag Protection Act of 1989 was unconstitutional because it violated flag burners' freedom of speech as guaranteed in the First Amendment.

What: Civil rights and liberties; Law
When: June 11, 1990
Where: Washington, D.C.
Who:
WILLIAM J. BRENNAN (1906-1997), associate justice of the U.S. Supreme Court
JOHN PAUL STEVENS (1920-), associate justice of the U.S. Supreme Court

Desecration or Protest?

In 1989, the U.S. Supreme Court held unconstitutional a Texas law that prohibited desecration of the flag in a way that the person committing the act knew would be offensive to others. In *Texas v. Johnson*, the decision was close: The Court split 5-4. One of the issues in the case was whether a state government has sufficient protective interest in the flag of the United States to override the free speech rights of demonstrators. The majority held that state governments do not. The decision in *Texas v. Johnson* engendered considerable public criticism. Resolutions of protest were quickly adopted in both the House of Representatives and the Senate. The issue quickly became whether Congress should propose a constitutional amendment or attempt to deal with flag desecration by statute. In the hope that a new federal law might be viewed more favorably by the Supreme Court, Congress passed the Flag Protection Act of 1989 by overwhelming majorities.

The assumption of some of the more reluctant backers of the act was that the constitutional amendment would be by far the greater evil. Such an amendment would, for the first time, have altered an original provision of the Bill of Rights.

The Flag Protection Act criminalized the conduct of anyone who "knowingly mutilates, defaces, physically defiles, burns, maintains on the floor or ground, or tramples upon" a U.S. flag, except conduct related to the disposal of a "worn or soiled" flag. The new law was challenged almost immediately in two cases. The challenge took the form of public flag burning—in one of the two cases, on the steps of the U.S. Capitol Building, directly in front of several Federal Bureau of Investigation agents. The second case involved a group in Seattle, Washington, who were protesting the act's passage. Both district courts before which the demonstrators were tried held the new law unconstitutional under the First Amendment on the strength of the precedent set by *Texas v. Johnson*. The government then asked the U.S. Supreme Court to bypass the Courts of Appeals and grant direct review of the lower court decisions. The Supreme Court granted the motion and heard the cases, combined as *United States v. Eichman*, during its 1989 term.

The Decision

The Court ruled in *United States v. Eichman* on June 11, 1990. Associate Justice William J. Brennan wrote the Court's opinion for the majority of five. His opinion pointed out that the government's interest in protecting the physical integrity of the flag relates to its status as a symbol of the nation. If a person destroys a privately owned flag, the act does not affect the symbol itself, but it does have communicative impact. Under the freedom of speech portions of the First Amendment, the government is not free to per-

2335

mit the expression of ideas about the flag and nation of which it approves and criminalize the expression of other ideas. Of course, the government may create a national symbol, promote it, and encourage its respectful treatment. However, even though desecration of such a symbol is deeply offensive to many people, the same may be said of the expression of other opinions that the court has protected. Brennan cited such examples as virulent ethnic and religious epithets, vulgar repudiation of the draft, and scurrilous caricatures. In closing, he wrote, "If there is a bedrock principle underlying the First Amendment, it is that the government may not prohibit the expression of ideas simply because society finds the idea itself offensive or disagreeable. Punishing desecration of the flag dilutes the very freedom that makes this emblem so revered, and worth revering." Justice Brennan's opinion was joined by Justices Thurgood Marshall, Harry A. Blackmun, Antonin E. Scalia, and Anthony M. Kennedy.

The dissenting opinion was written by Justice John Paul Stevens, who was joined by Chief Justice William H. Rehnquist and Justices Byron White and Sandra Day O'Connor. Stevens argued that some methods of expression may be prohibited under certain circumstances. If the prohibition is supported by a legitimate societal interest that is not connected to the suppression of a speaker's ideas, if there is no interference with a speaker's ability to promote his ideas in other ways, if allowing a speaker to choose his method of expression conflicts with a legitimate social interest such as the preservation of the flag, then the government may constitutionally restrict the method of expression. He cited the example of the statute prohibiting draft-card burning that the Court had upheld in *United States v. O'Brien* in 1968.

Consequences

The decision in *United States v. Eichman* brought about a renewed attempt to pass a constitutional

During this August, 2000, protest, a demonstrator throws an American flag into the burning pile outside the Democratic National Convention at the Staples Center in Los Angeles.

AP/Wide World Photos

amendment giving the federal government the power to criminalize flag desecration. A proposed amendment reached the floor of both houses in 1990 but fell thirty-four votes short of passage in the House and nine votes short in the Senate. A two-thirds majority is necessary to pass an amendment. In 1995, a new version of the proposed amendment did better. It actually passed in the House but fell three votes short in the Senate. Renewed proposals for a flag desecration amendment circulated in later sessions of Congress, but by 2000, interest in an amendment had waned, probably because few instances of flag burning in political protests had taken place and even fewer had come to the attention of the public.

The legal result of *Texas v. Johnson* and *United States v. Eichman* is that neither the states nor the federal government may constitutionally punish people who desecrate the flag.

Robert Jacobs

Algeria Holds First Free Multiparty Elections

Local elections in June, 1990, were the first allowed by the ruling Front for National Liberation since Algeria had gained independence from France in 1962.

What: Political reform
When: June 12, 1990
Where: Algeria
Who:
ABASSI MADANI (1931-   ), leader of the Islamic Salvation Front, a Muslim fundamentalist party
ALI BELHAJ (1954-   ), another Islamic Salvation Front leader
CHADLI BENJEDID (1927-   ), president of Algeria from 1979 to 1992

The One-Party Years

France occupied Algeria in 1830 and declared it to be part of France in 1848. The French colonists lived mostly along the Mediterranean coast, while the native Muslims (Arabs and Berbers) were concentrated in the interior of Algeria.

In November, 1954, a Muslim nationalist group known as the Front for National Liberation (FLN) began an armed struggle for independence. After eight years of fighting, which took the lives of more than one million Algerians, the FLN negotiated an agreement with French president Charles de Gaulle. Algeria became independent under the FLN's domination.

Ahmad Ben Bella, one of the leaders of the revolution, was chosen president of Algeria. Under his administration, foreign investments in Algeria were taken over by the government (nationalized). In 1965, Ben Bella was overthrown by his defense minister, Colonel Houari Boumedienne; Boumedienne continued one-party socialist rule by the FLN until his death in December, 1978. He was succeeded by Colonel Chadli Benjedid, who was known for his skill at compromising.

Though most Algerians were Muslim, Algerian government and society had become more and more separate from religious influence. This secularization had been begun by the French. Algeria's industry grew quickly in the 1960's and 1970's, but by the 1980's there were food shortages and high unemployment, and many people were unhappy with Benjedid and the FLN.

In October, 1988, Algerians rioted to protest the scarcity of food; thousands of young protesters were wounded, and at least one hundred were killed. Many people called on the FLN to reform the political system and allow other political parties to compete. Benjedid's government responded with a new constitution on February 23, 1989, opening the way to a multiparty system.

The Vote and Its Meaning

On June 12, 1990, Algeria had its first free municipal (city) elections. Eleven political parties participated, including the FLN; the Islamic Salvation Front (FIS), a Muslim fundamentalist group led by Abassi Madani, a professor of philosophy; and a secular party, Rally for Culture and Democracy (RCD).

The election returns were remarkable: In Algeria's four largest cities—Algiers, Oran, Constantine, and Annaba, the FIS won most of the municipal seats. Overall, the FIS received 65 percent of the vote and won 55 percent of the fifteen thousand municipal posts. It won representation in thirty-two of the forty-eight provinces.

Probably this support for the FIS expressed the growing admiration many Arab Muslims had for Islamic fundamentalist leaders in countries such as Iran, Jordan, Turkey, Tunisia, and Morocco. Yet Algeria's economy was at least as important for those who voted. When asked why they had supported the FIS, many voters answered that they had done so as a protest against the FLN.

2338

On June 25, 1990, *The New York Times* reported that many Algerians had used their vote to protest low salaries, high inflation, and the fact that job promotions usually went to those who were connected to the FLN. Jobs and apartments were scarce in Algeria; young Algerians sometimes found that their only work option was moving to France to sweep streets in Paris or Marseilles. Out of frustration, many of these young people had joined the FIS and voted for Islam.

Regardless of how they had voted, everyone in Algeria seemed to be excited about being given the chance to vote at all. To them it seemed as if a revolution had come.

From the beginning, the FIS was rather vague about its goals. In an agreement with shopkeepers and other business leaders, Madani supported a free market economy rather than socialism. Both Madani and his second-in-command, Ali Belhaj, stated that women should mostly be devoted to caring for their families, and that those who needed jobs should choose traditional careers such as nursing and teaching. Madani and Belhaj were strongly opposed to public drinking, all kinds of dancing, and radio and television programs that were not religious.

In the summer of 1990, when Madani urged the government to schedule national and parliamentary elections, the FLN regime did not make a promise either way. Yet the FLN leaders—who still controlled the parliament, the cabinet, the army, and the media—expressed some concern about the rise of Islamic fundamentalism; they said they would make sure that Algeria was not taken over by a religious government.

Consequences

Though the elections had been an important step toward democracy in Algeria, many governments around the world were unhappy with the FIS victory. About four million North Africans had immigrated to France and other nations of Western Europe in the late 1980's, and Europeans feared that now they would be flooded with Algerians who opposed the FIS. It seemed that democracy had brought something very different in Algeria from what it had produced in Eastern Europe and Latin America: Though the FIS had benefited from democratic elections, it was not committed to democracy and freedom.

Parliamentary elections were scheduled in Algeria for June 27, 1991—one year after the historic municipal elections. More political organizations arose to challenge both the FLN and the FIS. Demonstrations broke out; in two months of rioting, forty people were killed and more than three hundred were wounded. The army rounded up at least seven hundred political opponents and arrested them. Martial law was then imposed on June 5, and the scheduled elections did not take place.

Madani and Belhaj openly questioned the need for martial law and called for the elections to be rescheduled immediately. On Sunday, June 30, the FIS headquarters in Algiers was surrounded by the National Guard, and both FIS leaders were arrested. The military charged Madani and Belhaj with "armed conspiracy against the security of the state" and said that they would face trial. No one was sure what this would mean for the future of democracy in Algeria.

Michael M. Laskier

Iraq Occupies Kuwait

On August 2, 1990, Iraq invaded its southern neighbor, Kuwait, and set the stage for the Gulf War, in which the United States led a coalition of nations against Iraq.

What: War; Political aggression
When: August 2, 1990
Where: Iraq and Kuwait
Who:
SADDAM HUSSEIN (1937-   ), president of Iraq from 1979
GEORGE HERBERT WALKER BUSH (1924-   ), president of the United States from 1989 to 1993

Hussein Seeks Power

During the 1980's, Iraqi president Saddam Hussein worked systematically to get rid of all of his political rivals. For example, he crushed the Kurdish minority, using chemical weapons; more than sixty thousand Kurds fled as refugees into Iran and Turkey. More than half a million others were forced to move away from the border so that Iraq could take tighter control of the frontier areas. Hussein also kept independent-minded people from taking important political posts; he surrounded himself with advisers who told him what he wanted to hear.

After Iraq's victory in the long Iran-Iraq War, Iraq was recognized as the most important military power in the Persian Gulf region. At the end of the conflict, Iraq's military included more than one million men. Defense industries were growing, especially in production of chemical weapons and development of missiles.

Hussein dreamed of leading the entire Arab world. With the collapse of communism in Eastern Europe and growing chaos in the Soviet Union, he thought, the time was ripe for Iraq to draw the Arab states into a position of world leadership. Hussein began to make critical statements about Israel, the Arabs' traditional enemy.

On April 2, 1990, Hussein announced that if Israel should attack Iraq (as it had in 1981, when the Iraqi Osirak nuclear reactor was destroyed), Iraq would respond with a chemical attack on Israel. Later, Hussein said that Iraq would share its chemical weapons with any other Arab state that feared an attack from Israel. He also stated that Iraq supported the Palestinian cause. These bold promises won him friends throughout the Arab countries.

Though Hussein seemed successful in his quest for power, his nation was facing some severe problems. The war against Iran left Iraq with a debt of approximately $80 billion; $30 billion to $35 billion of this amount was short-term loans owed to the Western powers, and the rest was owed to the oil-producing Arab states of the Persian Gulf. Rebuilding areas damaged by war, continuing to produce weapons, and importing food and other goods from abroad all required money.

The Dispute with Kuwait

Iraq hoped to make more money from oil to take care of these problems, but between January and June, 1990, the price of oil fell from twenty dollars to fourteen dollars a barrel. Many Iraqis blamed the West; the fall in oil prices, they said, was part of a conspiracy against Iraq.

Hussein also accused Kuwait and the United Arab Emirates of cheating on the production quotas set by OPEC (the Organization of Petroleum Exporting Countries). Also, Kuwait refused to forgive the debt Iraq owed it.

Iraq also had some territorial disputes with Kuwait. Before giving up its claim to all Kuwaiti territory in 1963, Iraq had tried to control the entire region. After 1963, some conflicts continued. Iraq claimed the entire Khor Abd Allah estuary along the border, while Kuwait insisted that each country owned half of the waterway. Iraq also wanted to take control of Bubiyan and Warbah islands at the mouth of the estuary.

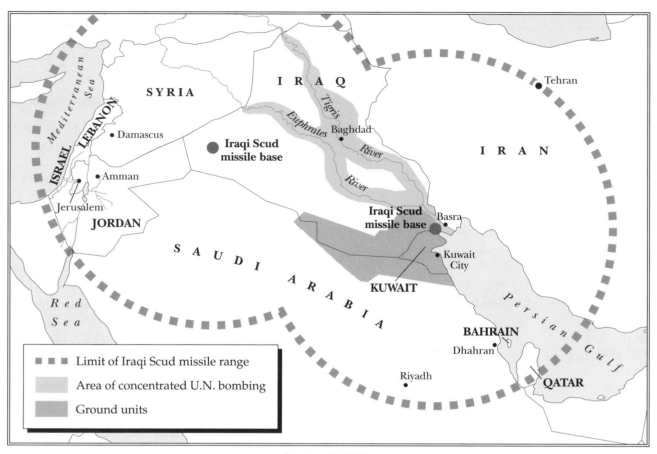

Persian Gulf War.

Another disputed territory was the Rumailah oil field, which straddled the border between Iraq and Kuwait. Iraq wanted control of the entire field; Hussein accused Kuwait of pumping from this field and selling the oil at high prices during the Iran-Iraq War. Iraq demanded that Kuwait give up the field and pay $2 billion for the oil, but Kuwait rejected these claims.

Across the Arab world, many people resented Kuwait and the other wealthy, conservative monarchies of the Persian Gulf. So Hussein hoped to win support in his struggle against Kuwait.

Tensions between Iraq and Kuwait rose quickly in late May, 1990. In mid-July, Hussein threatened action, and he sent Iraqi forces into position along the Kuwaiti border.

Suddenly, on August 2, 1990, Iraq invaded Kuwait. Within hours the entire country was overrun; most of the ruling family escaped to Saudi Arabia. At first the Iraqis claimed that an opposition group within Kuwait had invited them into the country, but the Iraqis could not persuade any Kuwaitis to form a new pro-Iraq government. Soon Iraq was simply claiming Kuwait as an annexed territory. The northern portion of Kuwait was made part of Basra Province, while the remainder of Kuwait became the nineteenth Iraqi province.

Consequences

After the invasion, the Iraqis looted Kuwait of any movable item: furniture, cars, gold, factory equipment, and even treasures from the Kuwaiti Museum. About half of the citizens of Kuwait, along with many Asian and Palestinian workers, fled the country. The Iraqis ruthlessly crushed any dissenters among those who remained. There were many arrests, tortures, and executions; also, many Kuwaiti males between eighteen and forty-five were taken back to Iraq to be held as hostages.

Immediately after the invasion, United States president George Bush ordered an economic embargo of Iraq. The Western European states

2341

and Japan also imposed an embargo, and on August 6, 1990, the United Nations Security Council ordered a world embargo against Hussein's government.

The next day, President Bush ordered U.S. military forces to Saudi Arabia, beginning what was called Operation Desert Shield. A number of Arab nations also sent forces. On August 25, the Security Council decided that force could be used to enforce the embargo against Iraq.

Meanwhile, Iraq closed its borders and those of Kuwait, so that foreign citizens could not leave. On August 28, foreign women and children were allowed to leave, but the foreign men were still held as hostages.

On November 29, the Security Council set January 15, 1991, as the deadline for Iraq to withdraw from Kuwait. On December 6, Iraq decided to release the foreign hostages, but the Iraqis stayed in Kuwait. On January 12, 1991, the U.S.

Congress authorized President Bush to use military force to drive the Iraqis from Kuwait.

Operation Desert Storm began on January 17 at 12:50 A.M. Over the next five and one-half weeks, bombers from the United States and other nations pounded military targets in Iraq. Ground forces were sent into the area on February 23. Saddam Hussein ordered Iraqi forces to withdraw from Kuwait on February 25.

The war left Kuwait, Iraq, and the whole region in chaos. The Iraqis had lit fires in Kuwait's oil fields and dumped a large quantity of oil into the Persian Gulf; they also left mines and other explosives behind them in Kuwait. Large portions of Iraq were left without clean water or medical supplies. Iraqi missiles fired into Saudi Arabia and Israel caused considerable damage as well. The Kurds in Iraq tried again to rise up and assert their independence after the war, but Hussein was successful in smashing them.

Cambodians Accept U.N. Peace Plan

> *After years of conflict, opposing parties in the Cambodian civil war agreed to guidelines for peace suggested by the permanent members of the United Nations Security Council.*

What: Civil war; International relations
When: September 10, 1990
Where: Jakarta, Indonesia, and Cambodia
Who:
NORODOM SIHANOUK (1922-   ), the former Cambodian monarch, head of the National United Front
SON SANN (1911-   ), leader of the Vietnamese-backed People's National Liberation Front government
HENG SAMRIN (1934-   ), leader of the Kampuchean People's Revolutionary Party
KHIEU SAMPHAN (1930-   ), leader of the Party of Democratic Kampuchea (Khmer Rouge)

A Bloody History

Twentieth century Cambodia has been a bloody land. Cambodia gained full independence from France in 1953. Norodom Sihanouk, a leader in the independence movement, became the country's king. He later took the title "Prince" and started his own political party.

Sihanouk was overthrown in 1970 by General Lon Nol, and he fled into exile. Almost immediately a communist group, known as the Khmer Rouge, began to fight against Lon Nol, and four years later it defeated him. When the Khmer Rouge took over, they forced people to leave the capital city, Phnom Penh, and work the land in the countryside. Hundreds of thousands of Cambodians died under the harsh rule of the Khmer Rouge, led by Pol Pot.

In 1979, Vietnam invaded Cambodia after a dispute over territory, and the Khmer Rouge leaders were forced to flee into Thailand. Thousands of Cambodian refugees fled into Thailand too. Finally, the world began to hear about the torture and brutality of the Khmer Rouge government. Most countries, though horrified at the stories, refused to recognize Vietnam's invasion as legal; they also refused to recognize the Cambodian government of Heng Samrin, who was placed in power by the Vietnamese.

Instead, most governments, including China, the United States, France, and Great Britain, supported a coalition of Cambodian factions that continued to oppose Vietnam's occupation. Sihanouk joined forces with Son Sann, of the non-communist People's National Liberation Front, and with the Khmer Rouge; these groups stood united to oppose the Heng Samrin government. Together these opposition groups were recognized by the United Nations (U.N.) as the legitimate government of Cambodia, even though they operated outside the country.

For ten years, occasional fighting took place near the border of Cambodia and Thailand. Thousands of Cambodian refugees were eventually settled in Europe and the United States, but many more continued to live along the unsafe border in camps controlled by Sihanouk, Son Sann, and the Khmer Rouge; they hoped to return home someday.

Hope for Peace

In 1989, Vietnam finally pulled all of its troops out of Cambodia. With the Vietnamese out, the most powerful members of the United Nations Security Council, called the "Big Five"—the United States, the Soviet Union, Great Britain, France, and China—began to explore ways to make peace among the Cambodian factions. The big questions were how to form a government between groups that had fought bitterly against one another, and how to prevent the Khmer Rouge, which had the largest opposition army, from grabbing power and starting the bloodshed all over again.

2343

In 1990, the Big Five met six times to talk over these problems and design a peace plan. A new spirit of cooperation between the United States and the Soviet Union made negotiations for peace in Cambodia easier. This cooperation included China, which had been a longtime supporter of both the Khmer Rouge and Prince Sihanouk. In August, representatives of the Big Five finished their peace plan in New York.

France and Indonesia then called a meeting in Jakarta, Indonesia, to present the plan to the Cambodians. After two days of discussions, representatives from all the major factions in Cambodia agreed to the plan. Their decision was announced on September 10, 1990.

The plan had five major principles. One principle called upon all Cambodians to agree to elections under the supervision of the United Nations. But these elections could not take place until the Cambodian rebel groups left their bases across the border in Thailand and returned home—and they could return home only if everyone agreed to stop fighting and lay down their arms.

To encourage this, two other principles called for a transition government to be created in Cambodia, and for military arrangements to ensure safety for all Cambodians. The plan's fourth and fifth principles called for protection of human rights and for international guarantees to make sure that the bloodshed that had taken place in the 1970's would not be repeated.

The Big Five called upon the Cambodian parties to create a national council, composed of representatives of the four major factions. The Cambodians quickly responded by forming the Supreme National Council (SNC). Included as representatives on the SNC were Prince Sihanouk, head of the National United Front; Son Sann, leader of the People's National Liberation Front; Khieu Samphan, leader of the Party of Democratic Kampuchea (formerly known as the Khmer Rouge); and Heng Samrin, leader of the Kampuchean People's Revolutionary Party. By creating the SNC, the various factions had provided a transitional government to pave the way for achieving the other goals in the peace plan.

Consequences

The peace plan finally bore fruit in October, 1991, when all the Cambodian factions signed a peace treaty. The United Nations was given major responsibility for preparing Cambodia for peaceful elections, and it began bringing a peacekeeping force into Cambodia from Australia in November, 1991.

Countries in the Southeast Asia region, including Vietnam, Thailand, Malaysia, and Indonesia, strongly supported the peace treaty. The treaty's success would depend on the continued cooperation of the Cambodian factions in the new transition government, as well as strong backing from the Big Five. Around the world, people were hopeful that the years of civil war in Cambodia had finally come to an end.

Incidents in Cambodia, however, tempered this hope. In May, 1990, when Minister of Communications, Transport, and Posts Ung Phan, in the spirit of the pending peace plan, suggested the creation of a new political party, he was accused of subverting the Cambodian government and jailed. Ung was released from prison after seventeen months, and in mid-January, 1992, he successfully organized Cambodia's first new political party, the Free Democratic Society Party. In late January, Ung survived a shooting attempt on his life.

The circumstances surrounding the shooting were similar to those surrounding the murder that had occurred the previous week of Tea Bun Long, a government official who had spoken out against government corruption. Some reports blamed the government for the shootings and convinced many Cambodians that the leaders of the transitional government were unwilling to give up their power or create a stable government. They and their families and friends were taking advantage of the instability and becoming rich while many people suffered. The Cambodian people mistrusted these leaders' guidance: They had been directly responsible for Cambodia's widespread poverty and for the decade of genocide that took the lives of more than a million Cambodians.

Bryan Aubrey

First Patient Undergoes Gene Therapy

A team of scientists infused genetically altered cells into a girl who had been classified as incurable, helping her to recover from a debilitating illness.

What: Medicine; Health; Genetics
When: September 14, 1990
Where: Bethesda, Maryland
Who:
W. FRENCH ANDERSON (1936-), a geneticist who helped to perform the first gene-therapy trial
R. MICHAEL BLAESE (1939-), a geneticist who helped to perform the first gene-therapy trial
RANDY A. HOCK (1952-) and DUSTY MILLER, researchers who helped to produce the first successful viral vectors
STEVEN A. ROSENBERG (1940-), the chief of surgery at the National Institutes of Health
DINKO VALERIO, a Dutch researcher who completed the first successful bone-marrow transfer experiments

Background

Until the late twentieth century, genetic disorders could only be approached from the dead-end angle of treating the symptoms. The mutant genes that caused the defects were unreachable, often leaving afflicted patients helpless. In the early 1970's, however, new findings sparked interest in gene therapy, a process by which defective genes could be replaced. This process involved several steps, each posing a challenge to contemporary technology.

Before a particular genetic disorder could be treated, the gene containing it first had to be isolated. This was often difficult, however, because few of the tens of thousands of human genes had been identified. Once a mutated gene had been isolated, moreover, healthy copies of the same gene had to be made to insert into cells within the patient. The choice of recipient cells was also important, as it determined the effectiveness and longevity of the cure. Finally, the healthy new genes had to be delivered into the target cells.

Because viruses reproduce by inserting genetic material into host cells, scientists focused their research on means by which viruses deliver such material via agents called "vectors." They theorized that a disabled virus could deliver healthy genes to replace a mutation. The first successful viral vector was produced in 1984 by Randy Hock and Dusty Miller at the Fred Hutchinson Cancer Research Center.

Adenosine deaminase (ADA) deficiency, a rare fatal disease that disables the immune system, was the target of the first genetic experiments in the mid-1980's. Research on other disorders, such as cystic fibrosis and tumors, began shortly after. These tests mainly involved inserting concentrated cultures of genetically altered cloned cells into a patient to temporarily replace cells the body lacked. This gene transfer method, however, lacked potency and longevity. Bone marrow cells were next targeted because they actually produce the body's defective blood cells, but they proved difficult to distinguish from other types of cells.

The Achievement

Before any experiment could be attempted on humans in the United States, it had to gain approval from many health organizations such as the Food and Drug Administration (FDA) and the National Institutes of Health (NIH). The benefits of the therapy and the knowledge gained from it had to be weighed carefully against potential risks. Scientists were still learning the viral genetic insertion process, and these agencies were concerned that vectors might alter healthy genes along with defective ones. After many rejections, however, the first gene therapy trial was finally approved on July 30, 1990.

The theory of gene therapy was first practically applied on September 14, 1990, at the National Institutes of Health in Bethesda, Maryland. A four-year-old girl suffering from ADA deficiency was infused with cells altered to produce the ADA enzyme that was needed to destroy harmful chemicals disabling her immune system. Six years earlier, a twelve-year-old boy afflicted with ADA deficiency had died after being exposed to ordinary germs. Over the next year and a half, the girl received additional injections. Tests showed healthy production of ADA from the newly cloned cells. Nearly two years after her first infusion, the patient returned to normal life.

Still, further infusions of corrected cells every several months were necessary to maintain her health. Experiments by Dinko Valerio in 1996 gave new hope for permanent gene replacement. He successfully identified the ADA gene in blood cells produced by transplanted bone marrow cells in rats and rhesus monkeys. Experiments on humans were delayed, however, as an ethics committee at the Dutch National Health Council reviewed the proposal.

Consequences

Approval of the first gene therapy experiment opened an international gateway to researchers seeking genetic cures for formerly incurable diseases. France, Germany, Italy, and the Netherlands soon began to perform gene therapy with the help of NIH scientists. In January, 1996, the British Department of Health declared that there were no major ethical objections to gene therapy, leading British researchers seeking a cure for cystic fibrosis to introduce a new line of experiments involving synthetic chromosomes as vectors.

In the United States, dozens of patients have been treated for ADA deficiency, and genetic-therapy clinics opened throughout the country. Gene therapy theoretically offered a cure to every genetic disorder, from cancer to AIDs to asthma. Future technology may also allow babies to be diagnosed with birth defects and cured even before they are born. These possibilities, however, raise more ethical questions about the extent to which such technology should be employed: For example, it may one day be possible to engineer humans with specific features, including eye color, brain type, and athletic ability.

John Powell

Biologists Initiate Human Genome Project

The Human Genome Project is intended to extract all the instructions hidden within the DNA molecules that make and maintain the human body.

What: Biology; Genetics
When: October 1, 1990
Where: United States
Who:

JAMES DEWEY WATSON (1928-   ), an American molecular biologist and 1962 recipient of the Nobel Prize in Physiology or Medicine

ROBERT LOUIS SINSHEIMER (1920-   ), an American biochemist

WALTER GILBERT (1932-   ), an American molecular biologist and 1980 recipient of the Nobel Prize in Chemistry

The Chemistry of Heredity Revealed

Since color blindness was linked to the male X chromosome in 1911, biologists have attempted to unravel the chemical secrets of heredity. It was suggested in 1944 that a complex molecule called deoxyribonucleic acid (DNA) played an important role in heredity. Normal human cells contain twenty-three pairs of chromosomes, each of which is composed of a very long DNA molecule that is elaborately wound into a bundle. DNA is formed by joining together molecular units called "nucleotides," which come in four different varieties that are identified as A, T, C, and G.

In 1953, the field of molecular biology was born when James Dewey Watson and Francis Crick worked out the three-dimensional structure, or "double helix," of DNA, which stores hereditary information written using a four-letter alphabet. In the 1960's, the use of this alphabet in heredity, the genetic code, was unraveled. Genes are sections of a chemical message inside the DNA molecule that control inherited traits ranging from hair and eye color to diseases such as cystic fibrosis or Down syndrome. In a gene, three consecutive nucleotides are grouped together, spelling a "codon," which represents a particular amino acid unit of a protein. The order of these words forms the "sentences" of a recipe telling cells how to make particular proteins such as those that color the hair and eyes.

Biology's Holy Grail Sought

In the 1970's, techniques allowing hundreds of genes to be assigned to specific chromosomes were developed, although their exact locations could not be determined. In 1977, researchers discovered a way to orient themselves in the DNA molecule using spelling differences in the DNA's text as markers for genetic diseases. Since members of chromosome pairs (except the sex-linked ones) are twinlike, with one chromosome being inherited from the mother and the other from the father, they can be compared for spelling differences. If a marker is regularly inherited along with a disease, the gene for the disease must be close to the marker on the DNA strand, and careful study of the code near it allows the gene to be located.

Throughout the 1980's, scientists sought to construct regional maps of chromosomes while searching for disease genes by breaking chromosomal DNA into smaller fragments called RFLPs (pronounced "rif lips") with restriction enzymes and studying the fragments' codes. For example, five years of intensive searching at a cost of $120 million using this technique located the gene for cystic fibrosis, a disease afflicting one in 2,000 newborn children.

In 1985, Robert Louis Sinsheimer proposed a project, causing much controversy within the biological community, to map the entire human genome (the complete human genetic structure) and determine the exact order of all the letters in its chemical code. Walter Gilbert, a Nobel laureate for his research in DNA sequencing and a staunch project supporter, proclaimed the project "the Holy Grail of Biology" because it would make it possible to understand everything human. Such an undertaking would be enormous, since nucleic acid molecules are made up of about three billion words, enough to fill 360 books each the size of an encyclopedia volume.

Some biologists opposed the project on the grounds that the technology to complete the job in a reasonable amount of time was not available. After all, in the three decades since the determination of the structure of DNA, only a small fraction of the human genome had been deciphered. Of an estimated 50,000 to 100,000 genes, about 4,450 had been identified, with only about 1,500 located on the chromosomes. Moreover, tracking down the location of a specific gene is easy when compared to determining the sequence of code letters within the DNA chain. Deciphering the entire genome would take centuries, because the technology of the early 1980's allowed only about 10,000 letters per year to be sequenced. The development in the mid-1980's by Leroy Hood of a device capable of sequencing 16,000 letters per day, however, did much to silence this opposition.

By 1988, support for the project was broad, and James Watson was chosen to oversee its implementation by the National Institutes of Health. The Human Genome Project officially began on October 1, 1990, with a strategy to map and interpret important regions of the genome first, and then, as more sophisticated technology developed, read the entire code. It was expected that within the first five years, 50 percent of the genome would be mapped; within 10 years the mapping would be completed, with about 10 percent of the genome being sequenced; and within 15 to 20 years all sequencing would be completed and all genes found. In actuality, the project moved much more rapidly. In mid-2000, completion of the first draft of the human genome was announced, to much fanfare.

Consequences

Total sequencing of the DNA code would yield the standard human genome—that is, the genome found in a normal healthy human being. Benefits would be gained even in the initial mapping phase, as the defective genes responsible for the thousands of inherited physical and behavioral disorders are identified and diagnostic tests are developed to detect them. At 50 percent completion, most of the major disease genes and those for other characteristics should have been located. The understanding that is gained concerning growth, development, and health would result in new medical therapies. The ability to predict, based on a genetic profile, an individual's vulnerability to genetic disorders would allow preventive lifestyle measures to be taken, thus minimizing risk factors. The development of new drug therapies to treat diseases or even prevent their onset would be possible.

Complete sequencing would allow scientists to understand genetically complicated traits and disorders resulting from multiple gene action. The development of gene therapy, a process by which actual instructions in the human genome are altered to eliminate genetic defects, could be accelerated. Although it is now anticipated that translation of the "encyclopedia of life" will not be completed until 2003, it will provide material for biological and medical research for centuries.

National Institutes of Health

A depiction of a cell and a chain of deoxyribonucleic acid, or DNA.

Arlene R. Courtney

East and West Germany Reunite

The East German populace revolted against their oppressive government, and the two Germanies, which had been separated since 1949, reunited socially, politically, and economically.

What: Political reform
When: October 3, 1990
Where: East Germany and West Germany
Who:
KONRAD ADENAUER (1876-1967), chancellor of West Germany from 1949 to 1963
ERICH HONECKER (1912-1994), communist leader of East Germany from 1971 to 1989
HELMUT KOHL (1930-   ), chancellor of West Germany from 1982 to 1990, and chancellor of reunited Germany from 1990 to 1998

A Nation Divided

The end of World War II saw deep division drawn among the nations of Europe. With Nazism and fascism overrun and with a need for massive rebuilding, the United States, the Soviet Union, England, and France moved in to assert their influence over the damaged countries and ensure that tensions would not again escalate into a world war. The four nations, however, could not agree on what to do with Germany.

In 1949, two German states emerged: the Federal Republic of Germany, or West Germany, governed by a democratic and pro-Western administration, and the German Democratic Republic, or East Germany, under the stranglehold of the Soviet Union's communist regime. After 1955, both Germanies joined rival military pacts. West Germany became a member of the North Atlantic Treaty Organization (NATO), and East Germany became a member of the Warsaw Pact. They were caught on opposite sides of the Cold War.

When Germany split in two, relatives and friends who had previously been citizens of the same country now found themselves barred from each other. Many who were stuck in East Germany crossed to the West in search of freedom and better living conditions. These refugees were an embarrassment to the communist East German government, which reacted by closing nearly seventy border crossings. In 1961, the infamous Berlin Wall was built as a barrier between East and West. It was guarded by armed soldiers, and many who tried to cross the Wall were killed or captured and held as political prisoners.

The first chancellor of West Germany, Konrad Adenauer, refused to recognize the legitimacy of East Germany as an independent state. His country's constitution accepted only one German citizenship for all Germans. For many years there were no official diplomatic relations between the two governments. Only after socialist Willy Brandt became chancellor of West Germany in 1969 did the two nations establish closer contact. In 1972, the two Germanies signed a Basic Pact that regulated official contacts between them. The following year they were both admitted into the United Nations.

After 1972, the two states increased economic and political contacts. Even during tense periods of increased East-West conflict, they attempted to continue a policy of détente. Despite official West German commitment to German unification, as late as October, 1989, two-thirds of West Germans did not expect German reunification to occur in the twentieth century. Furthermore, the East German leader Erich Honecker pursued a policy aimed at creating an independent East German identity. He ruled according to hard-line communist policies and was disliked by many East Germans. The East German regime depended on the support of Moscow for its survival. Large Soviet forces stationed in East Germany provided protection for the communist leadership in the event of social disturbances.

2349

Toward Unification

The emergence of Mikhail Gorbachev as a new, liberal Soviet leader in 1985 drastically altered conditions. The Brezhnev Doctrine, which had promised Soviet military assistance to endangered communist regimes, was abandoned. In May, 1989, Hungary began dismantling its border fortifications with Austria, allowing thousands of East Germans to flee to the West. By the summer of 1989, East Germans were flooding into West German embassies in Budapest, Prague, and Warsaw. Finally, in early October, the East German government allowed thousands of its citizens to emigrate. These concessions by the East German leaders, however, were not enough to prevent more mass demonstrations and calls for reform in East Germany.

On October 7, 1989, Honecker officially celebrated East Germany's fortieth anniversary. A few days later, he was ousted from office and replaced by his former protégé, Egon Krenz. Krenz promised to bring reforms, but the East German people were not satisfied with promises. On October 30, more than 200,000 people demonstrating in Leipzig demanded the legalization of New Forum, an opposition coalition group.

In a desperate attempt to stop the flow of refugees to West Germany, the East Germans officially opened the Berlin Wall in November, 1989. Communist leaders expected that the East Germans would choose not to flee if they were given access to the West.

After only a few months as leader, Krenz resigned in response to continuing demonstrations against communist rule. He was replaced by the reformist communist boss of Dresden, Hans Modrow. Modrow wanted to reform the political system of East Germany, but he did not favor unification with West Germany. While he resisted, the East German masses increasingly called for union and continued to pour into West Germany.

In late November, West German chancellor Helmut Kohl announced his ten-point plan for gradual German unification. The issue was not

Schoolchildren from East and West Germany hoist the German flag in front of the Reichstag building in Berlin to celebrate the reunification of Germany.

simply an internal German question. The powers that had defeated Germany in World War II were concerned that a reunited Germany might eventually cause trouble again. By December, 1989, U.S. president George Bush and U.S. secretary of state James Baker had begun to support Kohl's efforts with the stipulation that Germany had to remain a member of NATO. On February 10, 1990, Kohl visited Moscow and received Gorbachev's approval that the two German states could define their internal issues of unification. At Ottawa, Canada, a "Two-plus-Four" system was established that allowed the two German states to work out the details of German unification while they negotiated the external aspects of German unification with France, England, the United States, and the Soviet Union.

On March 18, 1990, the East Germans, in their first democratic election, voted in favor of the pro-Western Christian Democratic coalition. The new East German prime minister, Lothar de Maizière, supported Kohl's efforts to reunite the two German states without delay. By July 1, the Germans had established a monetary union. In a meeting with Gorbachev on July 15-16, at Stavropol in the Soviet Union, Kohl was able to remove the last obstacle to German unification by obtaining Gorbachev's approval that a united Germany could remain a member of NATO. On August 31, the two Germanies signed a treaty that called for unification by October 3, 1990.

Consequences

The external Two-plus-Four negotiations were completed on September 12, 1990, in Moscow. The united Germany promised to reduce its military and to ban nuclear, chemical, and biological weapons from its arsenal. The new Germany would also officially recognize the Oder-Neisse border with Poland and provide massive financial assistance to the Soviets.

On October 3, 1990, the two German states were officially united, and on December 2, the first general election was held in the new nation. Chancellor Kohl's coalition government emerged victorious. Although numerous political, economic, and social problems remained to be resolved, the new Germany had become firmly anchored in a democratic framework and closely tied to the European community.

Johnpeter Horst Grill

Thatcher Resigns as British Prime Minister

Margaret Thatcher, the leader of Great Britain's government and head of the Conservative Party, resigned after more than eleven years in office, the longest tenure of a British prime minister in the twentieth century.

What: Government
When: November 22, 1990
Where: London, England
Who:
MARGARET THATCHER (1925-   ), prime minister from 1979 to 1990
JOHN MAJOR (1943-   ), Thatcher's successor as party leader and prime minister from 1990 to 1997
MICHAEL HESELTINE (1933-   ), cabinet member and party leader candidate
DOUGLAS RICHARD HURD (1930-   ), cabinet member

The Thatcher Leadership

On November 22, 1990, Margaret Thatcher announced her intention to resign as prime minister. It was the end of an era in British history. Elected as leader of the Conservative Party in 1975, she became Britain's prime minister following the 1979 national elections. Her party won national elections in 1983 and 1987. Her leadership and policies in the period after 1987 became more controversial, however, not only within the nation itself but also among her senior government staff and the Conservative Party membership in the British House of Commons.

Thatcher led the nation during a period of economic and social transition. Articulate and self-confident, she urged a rejection of the socialist policies and attitudes of the pre-1979 period. This included an effort to encourage more private ownership. The transfer of many government programs and enterprises to the private sector promoted a capitalistic, free-enterprise environment. However, it also alienated many who criticized her for reducing government services and employment opportunities.

A second issue involved Britain's membership in the European Community (EC), an association of twelve European states. One goal created greater trade between member states by the reduction and elimination of tariffs. Decisions were to be made by EC committees representing all members. By the 1980's, the EC moved toward common fiscal and monetary policies that affected each member. Thatcher favored cooperation within the EC, but she criticized policies that she argued either complicated or reduced British conditions and decision-making authority. The situation became serious by 1987-1988, as several senior cabinet officials believed her critical statements had negative effects on Britain's future role in European economic and political affairs. Changes in Thatcher's government occurred in 1988-1989 as several key officials resigned, were transferred to other positions, or fired.

A third factor was her personality. Intelligent, articulate, and strong-willed, Thatcher was supremely confident in her ability to lead the nation. To many, however, her self-assurance was seen as arrogance, especially in light of problems that continued to affect the British economy and society: unemployment, inflation, and social discontent. The fact she had served as prime minister for more than a decade, longer than any other British leader in the twentieth century, also gave critics an additional reason to call for a change of national leadership.

The 1990 Controversy

What brought the issue to a head was the community charge. This plan, sometimes referred to as a poll tax, sought to spread local taxation to all inhabitants rather than only to businesses and property owners. The plan, passed in 1988, was implemented in the spring of 1990. It was a com-

2352

plex scheme, and many criticized its uneven effects. Conservative Party politicians worried about its impact on the next parliamentary election. Thatcher's strong commitment to the policy increased concerns about her effectiveness. Economic news also became worrisome in 1990, as unemployment increased for the first time in nearly four years.

By early fall, several longtime Thatcher associates were gone from the cabinet, either through dismissal or by resignation. On November 1, Geoffrey Howe, the deputy prime minister and a longtime Thatcher supporter, abruptly resigned and publicly criticized her leadership. This led to further division within the Conservative Party. On November 14, Michael Heseltine announced his intention to seek election as party leader. He had served in earlier Thatcher cabinets as secretary of the environment (1979-1983) and secretary of defense (1983-1986). He used the community charge issue as a central focus of his opposition of Thatcher.

Under the British parliamentary system, the person selected to lead the largest party in the House of Commons becomes the nation's prime minster. The Conservative Party's leader was elected by its senior party leadership and its party members in the House of Commons. If Heseltine succeeded in his bid to become party leader, he would become the next head of the government. Intensive discussions took place before the crucial vote on November 20. Thatcher's supporters included Foreign Secretary Douglas Richard

Hurd and Chancellor of the Exchequer John Major. However, others seemed ready for a change. The November 20 vote gave Thatcher a majority of the votes (204-152), but she fell four votes short of the required margin of victory on the first ballot: a 15 percent majority over the next candidate.

Thatcher announced her determination to go into the second round, predicting victory. However, erosion of party support quickly became evident. Reluctantly, on November 22, she announced her resignation as party leader and prime minister. In the second vote for party leader, Major emerged the winner. He replaced Thatcher as prime minister on November 28.

Consequences

The Thatcher years were a dynamic and active period of British politics in the late twentieth century. In both domestic and foreign policy issues, Thatcher made her mark on her nation and Europe. Her leadership sometimes was controversial because she sought to guide Britain in directions that encouraged change, resilience, talent, and potential success. Thatcher's record reflects a mixture of achievements and shortcomings in reaching the desired goals. The effects of her leadership and policies continued after she left office in 1990. Thatcher remains one of the most interesting and important political leaders of the twentieth century.

Taylor Stults

Wałęsa Is Elected President of Poland

Election of Lech Wałęsa, a long-time opponent of communist rule, symbolized the end of domination of Poland by the Soviet Union.

What: National politics; Political reform
When: December 9, 1990
Where: Poland
Who:
LECH WAŁĘSA (1943-), leader of Solidarity labor union and president of Poland from 1990 to 1995
WOJCIECH WITOLD JARUZELSKI (1923-), military officer and president of Poland from 1989 to 1990
TADEUSZ MAZOWIECKI (1927-), first noncommunist prime minister of Poland and candidate for president

The Rise of Solidarity

During World War II, the military forces of the Soviet Union swept across Poland, driving back the forces of Nazi Germany. After the war, the Soviet Union remained in control of Poland, imposing communist rule under a series of puppet government leaders. A one-party dictatorship was maintained, and individual freedoms were sharply limited. Poland's economy was forced into a program of centrally controlled decisions. No independent labor unions were permitted.

Soviet domination was weakened by a series of upheavals, chiefly provoked by economic crises. In 1970, government proposals to raise food prices and to close the Gdańsk shipyards led to extensive protest demonstrations. One of the leaders was Lech Wałęsa, who worked at the Gdańsk shipyard as an electrician. He lost his job there for criticizing the government in 1976. In 1980, worsening economic conditions were again punctuated by government efforts to raise food prices. A series of protest strikes occurred across the country. In August, sit-down strikes took place in the shipyards at Gdańsk and elsewhere. Although Wałęsa no longer worked at the Gdańsk shipyard, he was able to make his way into its facilities and became one of the leaders of the strike.

From this experience emerged the organization called Solidarity, drawing not only on economic grievances but also on opposition to foreign domination and the desire to restore individual freedom. In September, 1980, Wałęsa was elected chair of Solidarity's National Coordinating Committee. By the end of 1980, Solidarity had enlisted nearly ten million members, in a country in which the total working population numbered around seventeen million.

In February, 1981, former defense minister Wojciech Witold Jaruzelski became head of the communist government. In December, 1981, he acted to suppress the labor unrest by declaring martial law and arresting thousands of Solidarity activists, including Wałęsa, who was imprisoned for eleven months. Wałęsa had become a symbol of the Polish people's struggle for freedom and was awarded the Nobel Prize for Peace in October, 1983.

Martial law was lifted in July, 1983, but Solidarity was not recognized as a legal entity. Its leaders were harassed, but it continued to press for free-market economic reforms and for genuine democracy. Wałęsa insisted that Solidarity maintain a nonviolent approach. The communist government also generally refrained from using violence against the dissidents. The government was trying to win approval and economic assistance from the West. However, the Polish economy was not doing well: Real national income in 1989 was still below that of 1979, and prices were rising rapidly.

The End of Communist Rule

In January, 1989, the government reluctantly granted legal recognition to Solidarity as a labor union. It also acknowledged Solidarity's political strength by agreeing to participate in a series of round table political meetings from February through April. Independent newspapers began to appear. In June, Solidarity candidates swept the free seats in a parliamentary election; the communists retained a majority of guaranteed seats.

In July, 1989, it became clear that the Soviet Union would not use military force to put down the Polish moves toward democracy and independence. In November, Hungary declared itself an independent republic. In November, the Berlin Wall fell, and Germany moved rapidly toward reunification.

With Solidarity cooperation, General Jaruzelski was chosen president (a relatively powerless figurehead position at that time) by parliamentary vote in July, 1989. In August, he appointed the first noncommunist prime minister, Tadeusz Mazowiecki. However, Wałęsa pressed for further reforms.

In January, 1990, the Polish Communist Party voted to disband, and its members formed two new parties. The parliament voted to reshape the office of the president to give it more power and make it elective by the public rather than by the parliament. A new election was scheduled. In September, 1990, Wałęsa announced his candidacy. In October, Jaruzelski agreed to relinquish the presidency when the results of the election were determined.

Besides Wałęsa, there were five other candidates, including Prime Minister Mazowiecki. In the first round of voting on November 25, 1990, Wałęsa received the largest part of the vote, about 40 percent, while Mazowiecki received only 18 percent. However, a relative unknown, Stanislaw Tyminski, received 23 percent, enough to force a runoff. Tyminski had lived in Canada for twenty-one years and was supported by many of the old-line communists. In the runoff December 9, Wałęsa received 73 percent of the vote. He was sworn in as president on December 22, 1990.

Consequences

The Polish elections occurred in the midst of the turbulent period in which communist regimes were overthrown throughout Eastern Europe and the Soviet Union itself ceased to exist. In rapid succession, the Baltic States, Czechoslovakia, Hungary, Bulgaria, Romania, and East Germany freed themselves from Soviet domination and communist control. The activities of Wałęsa and Solidarity are generally regarded as having started the collapse of Soviet communism.

Although Wałęsa's economic ideas were not very clear, he retained Leszek Balcerowicz as deputy prime minister with primary responsibility for economic policy. Balcerowicz had begun a courageous but controversial plan to liberalize the economy, removing many government controls and privatizing some state enterprises. In the short run, these measures led to both inflation and unemployment, but they put Poland on the road to long-term growth and stability. However, Wałęsa's own personality shortcomings and the difficulties attending the economic transition led to his not being reelected when his term expired. He lost a close runoff to Aleksander Kwasniewski, a former communist, in November, 1995. Solidarity faded into relative insignificance.

Lech Wałęsa in 1983.

The Nobel Foundation

Paul B. Trescott

Ethnic Strife Plagues Soviet Union

As the Soviet people gained new freedom to speak up for themselves, open conflict broke out between competing ethnic groups.

What: Civil strife; Ethnic conflict
When: 1991
Where: Soviet Union
Who:

MIKHAIL GORBACHEV (1931-   ), first secretary of the Soviet Communist Party from 1985 to 1991

BORIS YELTSIN (1931-   ), president of the Russian Republic from 1991 to 1999

Longing for Independence

For many decades, the ethnic groups living around the edges of Russia struggled for self-determination. These groups represented a wide variety of religions, languages, and traditions, and they wanted to be free to govern themselves and express their culture. During the reigns of Czar Alexander III (1881-1894) and Czar Nicholas II (1894-1917), the Russian government tried to "Russify" those who were different. When the communists came to power in 1918-1919, they promised to give more freedom to ethnic groups, but these promises were soon broken as dictator Joseph Stalin began his program of "Sovietization."

Those who followed Stalin as leaders of the Soviet Union continued his policy of dominating the ethnic populations. Sometimes there were emotional, even violent, confrontations between local Communist Party leaders and minority groups, but the central government kept working to keep control.

Things began changing, however, when Mikhail Gorbachev became first secretary of the Soviet Communist Party in 1985. Gorbachev believed that the Soviet Union needed to be reformed; he introduced new policies that he called *glasnost* (openness) and *perestroika* (restructuring). He was ready to change the Soviet economy to make it more like Western systems, and he realized that the Soviet people needed the opportunity to express their political ideas and to vote freely.

Gorbachev probably did not realize how much turmoil he was setting loose. As he allowed the "satellite" nations of Eastern Europe to begin choosing their own destinies, ethnic groups in republics within the Soviet Union began hoping for the same privileges. To these groups, the republics were their national homelands.

Meanwhile, conservative leaders feared that the Communist Party would be destroyed and the central government would fall apart. It was primarily because of these fears that some of them tried in August, 1991, to remove Gorbachev from the presidency. The coup failed, and people across the Soviet Union responded with a great tide of emotion in support of greater freedom and democracy. Boris Yeltsin, president of the Russian Republic, became a hero to many Soviets for his strong stand against the plotters.

Demands for Freedom

When Gorbachev was restored to the presidency after three days of uncertainty, he had little choice. People were demanding an end to Communist Party rule; he had hoped to move the Party out of the government gradually, but now it had to happen quickly.

Soon after he returned to the Kremlin, Gorbachev agreed to recognize the independence of the Baltic States (Lithuania, Latvia, and Estonia). These three small nations had become part of the Soviet Union as a result of the Nazi-Soviet Pact of 1939, and Baltic dissidents had always insisted that the annexation was illegal. In September, 1991, representatives of these states were officially welcomed into the General Assembly of the United Nations.

2356

Nationalism also rose up within the more traditional Soviet republics. The states of the Caucasus Mountains (Armenia, Azerbaijan, and Georgia), Ukraine, Kazakhstan, Uzbekistan, and Moldova all expressed strong desires to be free of Moscow's control.

Gorbachev gave the republics more control over their internal affairs, yet he tried hard to find a way to keep them somehow within the Soviet Union. His task appeared impossible. The Communist Party, which had been the glue that held the Soviet Union together, was in disarray. If the central government lost control, the huge Russian Republic would become dominant, and its leader, Yeltsin, would have more power than anyone in the Kremlin.

In Armenia and Azerbaijan, greater political freedom was likely to lead to greater violence. The problem there was not only hostility toward the central government but also hatred between the Azerbaijanis, who follow Islam, and the Armenians, who are Orthodox Christians. Armenia claimed rights over Nagorno-Karabakh, an area within Azerbaijan where Armenians made up the majority. Conflicts over this region brought considerable bloodshed.

Fighting between Armenians and Azerbaijanis led Gorbachev to send troops to stop the violence in January, 1990. In September, 1991, Russian president Yeltsin tried to negotiate a settlement of the Nagorno-Karabakh problem. Leaders of Armenia and Azerbaijan realized then that the central government had become so weak that the Red Army was no longer available to control outbreaks of violence.

In April, 1990, Gorbachev warned that uncontrolled nationalism could lead the Soviet Union's ethnic minorities into civil war. The world watched with a mixture of pleasure and alarm as the people of the Soviet republics tried to find their way to freedom.

Consequences

Gorbachev's attempts to hold the Soviet Union together as one nation did not succeed. By the end of 1991, the union had collapsed. As many had predicted, Yeltsin became the most important political leader. Leaders of the republics met to discuss their future, and most of them agreed to join the new Commonwealth of Independent States, governing themselves but sharing responsibility for foreign policy and military forces. The Soviet superpower was no more.

The conflict between Armenia and Azerbaijan raged on. Economic problems were so severe that many people were going hungry in Soviet cities during the winter of 1991-1992. Clearly, the road to independence would be a rocky one.

Kendall W. Brown

Hostages Are Freed from Confinement in Lebanon

> *Various shadowy groups of kidnappers, who might have taken orders from Iran and Syria, released the last American hostages who had been held in Lebanon since the 1980's.*

What: International relations
When: 1991
Where: Lebanon, mostly Beirut
Who:
TERRY A. ANDERSON (1947-   ), chief Middle East correspondent of the Associated Press
WILLIAM F. BUCKLEY (1929-1985), head of the U.S. Central Intelligence Agency (CIA) station in Beirut
WILLIAM HIGGINS (1945-1989), a U.S. Marine Corps officer who had been serving in a United Nations force in Lebanon
AYATOLLAH RUHOLLAH KHOMEINI (1902-1989), rahbar (religious leader) of Iran from 1979 to 1989
JAVIER PÉREZ DE CUÉLLAR (1920-   ), secretary general of the United Nations from 1982 to 1991

War by Other Means

Hostage taking, like other kinds of terrorism, is a way for weak governments or private groups to make themselves heard in nations with military might. Terrorist groups take hostages as bargaining chips in order to demand money, political power, or other benefits. Usually the group believes that it has an important grievance or complaint against the country where the hostage is a citizen.

The nation of Iran started off the wave of hostage taking in 1979, at the time of the Islamic Revolution. A group of students and Shiite clergymen, encouraged by the government of Ayatollah Ruhollah Khomeini, held fifty-two people in the American embassy in Tehran for 444 days.

Khomeini was angry because the United States had supported Shah Mohammad Reza Pahlavi, Iran's former ruler.

Beginning in 1982, forty-eight Westerners were kidnapped by various groups in Lebanon. They included twenty Americans, twelve French, seven British, five Germans, two Swedes, one Irishman, and one Italian. A few non-Westerners were captured as well.

The kidnappers' groups had a variety of names: Islamic Jihad (Holy War), Islamic Jihad for the Liberation of Palestine, Revolutionary Justice Organization, Organization of the Oppressed on Earth, Amal, Organization of Islamic Dawn, Arab Commando Cells, Revolutionary Organization of Socialist Muslims, and others. Several of them were linked to the Hezbollah (Party of God), an organization backed by Iran.

There were economic or religious frustrations behind the actions of some of the hostage takers. Shiite Muslims (a fundamentalist branch of Islam) believed that they were being treated unfairly in some Arab nations, especially those that had a majority of Sunnis (Muslims who are more moderate in their views). Even where the Shiites were a majority among Muslims (for example, in Lebanon), they were usually poorer than Sunnis and members of other religious groups.

After bomb attacks in Kuwait in 1983, fifteen Shiites were taken prisoner. Shiite groups then took some Western hostages to trade for these prisoners. The fifteen Shiites were finally freed—perhaps by mistake—after Iraq invaded Kuwait in August, 1990.

Terry Anderson, a leading journalist with the Associated Press in Beirut, was among those kidnapped in the mid-1980's after four Iranians were kidnapped (and probably killed) by Christian militiamen from Lebanon.

The Captives Are Freed

Some of the hostages did not survive their ordeal. American hostages William F. Buckley, Peter Kilburn, and Lieutenant Colonel William Higgins were executed, as were two Britons. Various reasons were given for the killings: an Israeli air raid on the headquarters of the Palestine Liberation Organization in Tunis in 1985; American air raids on Libya in 1986; the abduction of a Lebanese Shiite leader, Sheikh Abdel Karim Obeid, by Israeli commandos in 1989; the shooting down of an Iranian airliner by the USS *Vincennes* on July 3, 1988; and Israeli air strikes against Palestinian guerrilla bases in Lebanon.

The hostages who were not killed were sometimes threatened with execution. Some of them were physically mistreated, and a few suffered permanent injuries.

The United States and other governments whose citizens were taken hostage vowed not to bargain with the terrorist kidnappers. In the Iran-Contra scandal of the mid-1980's, Ronald Reagan's administration secretly sold weapons to Iran, hoping that this would bring the release of hostages. Publicly, however, U.S. government officials were still promising never to bargain with terrorists.

Terry Waite, a clergyman from the Church of England, was sent by the British government to try to win release for the hostages. Waite himself was kidnapped, however, and was held for years.

In 1990 and 1991, many of the hostages were released—sometimes in an agreed-upon exchange of prisoners. For example, American hostage Jesse Turner, who had worked at the American University of Beirut, was released by the Islamic Jihad on October 21, 1991, after Israel freed one Arab prisoner and the Israeli-backed Southern Lebanese army freed fourteen others.

It seemed that after the Ayatollah Khomeini's death in 1989, leaders in Iran wanted to give themselves a new, more moderate image. They hoped to improve relations with nations of the West—and to begin receiving aid from the West. At the same time, the Soviet Union was begin-

Former hostage Terry Anderson arrives in Wiesbaden, Germany, where he is met by his sister, Peggy Say.

AP/Wide World Photos

2359

ning to dissolve. Syria had always received considerable military aid and political support from the Soviets, so now it began trying to befriend the United States instead.

There were other reasons for the release of hostages. The Gulf War of early 1991 put the United States, Syria, Iran, and Israel on the same side (though for different reasons) against Saddam Hussein of Iraq. Also in 1991, every major Arab country or group finally agreed to sit down and negotiate for peace with Israel.

In the United States and around the world, there was great celebration after the longest-held Western hostage, Terry Anderson, was released on December 4, 1991.

Consequences

The United Nations, its secretary general, Javier Pérez de Cuéllar, and a special U.N. representative named Giandomenico Picco had been involved in persuading the hostage takers to release their captives. Their success added to the growing prestige of the United Nations.

Another result of the release of hostages was that the United States and Iran began taking quiet steps toward a better relationship. The two countries' diplomatic relations had been broken after Americans were taken hostage at the Tehran embassy in 1979, but now officials began to hope that relations could be restored.

Peter B. Heller

Scientists Identify Crater Associated with Dinosaur Extinction

---

Geologists discovered a huge crater, probably formed by the collision of an asteroid or comet, that may have played a key part in the extinction of the dinosaurs.

What: Earth science
When: 1991
Where: The Yucatán Peninsula, Mexico
Who:

WALTER ALVAREZ (1940-), an American geologist who led the research team that suggested that a meteorite collision caused the dinosaurs' extinction

LUIS W. ALVAREZ (1911-1988), a Nobel Prize-winning American physicist and Walter's father, who helped to construct the collision-extinction hypothesis

GLEN T. PENFIELD, an American geophysicist whose team suggested that the Chicxulub crater was the result of a collision with a meteorite

ALAN R. HILDEBRAND, a Canadian planetary scientist who led the search for evidence that the Chicxulub crater was associated with the extinction of the dinosaurs

Death of the Dinosaurs

The extinction of dinosaurs about sixty-five million years ago has intrigued and challenged scientists ever since the first dinosaur fossils were discovered. Several hypotheses have been proposed to explain their disappearance. In 1980, Walter Alvarez and his coauthors suggested a new, interesting, and somewhat worrisome hypothesis—that the dinosaur extinction occurred as a result of a collision between the earth and an asteroid approximately six miles (ten kilometers) in diameter.

Models suggested that such an impact should have blasted out a crater at least 120 miles (200 kilometers) in diameter. According to the theory, the dust thrown into the atmosphere surrounded the earth for months, reflecting large amounts of sunlight. The earth cooled, and photosynthesis by plants was reduced. All animals depend on the food produced by photosynthesis, because they either eat plants or eat plant-eating animals. As a result of these and other disturbances caused by the impact, dinosaurs became extinct.

Evidence for the hypothesis centered on the Alvarez team's discovery of a sixty-five-million-year-old layer of clay that was rich in iridium. Iridium is rare in the earth's crust but is common in meteorites. First discovered in Italy, the clay layer was also found at a number of locations widely scattered around the earth. Alvarez explained its widespread occurrence as the result of the settling global dust cloud that had been formed by the impact of the iridium-rich meteorite.

Announcement of the hypothesis triggered a search for an impact crater of the appropriate size and age. Various craters were suggested as possibilities, but none fit expectations in all respects. For example, the Manson Crater in northern Iowa was a candidate for a time, although its small size was a problem; careful dating later showed it to be too old. A sixty-five-million-year-old crater might be deeply buried under more recent deposits; in addition, if the collision occurred in the ocean, the crater might be completely destroyed by plate tectonics. Despite the inherent difficulty of finding such a structure, skeptics embraced the crater's absence as evidence against an impact.

A Big Hole

In the 1950's, the Mexican oil company Pemex began drilling exploratory wells on the

Yucatán Peninsula near the Mayan village of Chicxulub. As information from the various wells accumulated, geologists realized that they were drilling into a geologically unique structure more than a half mile (about one kilometer) beneath the surface. Glen T. Penfield and others came to believe that the structure was an impact crater. The presence of shock metamorphism—rock particles fractured in a way that occurs only under extreme conditions, such as those that accompany a meteorite collision—was one form of evidence. Another was the presence of microtectites, tiny particles of melted rock solidified into a glassy state. Such particles are formed only under intense heat and pressure, such as occur during impacts.

Measurements and calculations indicated that the crater was approximately 110 miles (180 kilometers) in diameter, large enough to have been formed by the impact of a meteorite six miles in diameter. Aging studies indicated that the crater was formed approximately sixty-five million years ago, about the time that the dinosaurs became extinct. In 1991, Alan R. Hildebrand, Penfield, and their colleagues suggested that the Chicxulub structure was the crater formed by the meteorite that ended the reign of the dinosaurs. The meteorite was first assumed to be derived from an asteroid; later, a comet. Whatever its origin, here was the crater the Alvarez group had been looking for.

Consequences

The discovery of a crater of the correct size and date did not convince everyone that the impact caused the dinosaur extinction. Volcanic activity, climatic change caused by plate tectonics and continental drift, and combinations of these with the impact were alternative suggestions. Arguments centered on the pattern of the dinosaur and other extinctions shown by the fossil record. Two general viewpoints developed based on the argument that all the dinosaurs and other species should have become extinct at about the same time—at impact and shortly thereafter—if the impact caused the extinctions. The catastrophic argument claimed that this was exactly what happened: Most extinctions occurred at, and just after, the impact. In contrast, the gradualist view held that extinctions occurred over millions of years, beginning long before the impact, and that therefore the impact could not have been the single cause. Both sides argued from the fossil record, but the record is so incomplete that no definitive answer was available.

Many gradualists agreed that the Chicxulub structure suggests an impact at the time of the dinosaur extinctions and that the impact had some effect on those extinctions. One theory holds that the dinosaurs were declining before the collision, as a result of climatic and other changes, and the collision simply hastened the disappearance of the last members.

The general acceptance of a huge asteroid or comet's collision with the earth has also had an effect on scientists' view of the planet's future. What are the chances that another large meteorite will collide with the earth, and what would be the results? Some scientists believe that the chance of such a collision is exceptionally small at any point in time, but that one is inevitable some time in the future. The results would be catastrophic.

No comets or asteroids are known to be on a collision course with the earth, but there is no guarantee that scientists would know very far in advance if one was. In May of 1996, two small asteroids passed close to the earth without being discovered until it would have been too late had they been on a collision course. Scientists are working on techniques to search for such comets and asteroids and to nudge any that are discovered onto another course.

Carl W. Hoagstrom

Civil War Rages in Yugoslavia

> *In the midst of growing ethnic and political strife, Yugoslavia dissolved into warring ethnic factions.*

What: Civil war; Political independence
When: 1991-1992
Where: Yugoslavia
Who:

TITO (JOSIP BROZ, 1892-1980), marshal and prime minister of Yugoslavia from 1945 to 1953, and president of Yugoslavia from 1953 to 1980

SLOBODAN MILOŠEVIĆ (1941-   ), president of the Serbian Republic from 1990 to 1997

FRANJO TUDJMAN (1922-1999), president of the Republic of Croatia from 1990 to 1999

ANTE MARKOVIC (1924-   ), prime minister of the Yugoslavian Federation

MILAN KUCAN (1941-   ), president of the Republic of Slovenia from 1990

Yugoslavia Without Yugoslavs

Composed of six republics—Serbia (which also included the semiautonomous regions of Kosovo and Vojvodina), Croatia, Slovenia, Bosnia-Hercegovina, Montenegro, and Macedonia—Yugoslavia came into existence in 1931 and remained united after World War II because of the dictatorship of the Communist Party and the strong personality of Tito, who had led the resistance against the Nazis.

As ruler of Yugoslavia from 1945 to 1980, Tito created a federation in which each of the republics kept its own culture and identity. Instead of letting the Soviet Union control the country's policies, Tito insisted on Yugoslav independence. He allowed enough economic and political freedom to make Yugoslavia the most prosperous of the Eastern Bloc countries.

After Tito died in 1980, however, the ethnic, religious, and political divisions within Yugoslavia slowly began to tear the nation apart. Yugoslavia's ethnic problems were especially serious. According to the census of 1981, only 5 percent of the population considered themselves Yugoslavs. The rest called themselves Serbs, Slovenes, Albanians, Croats, (Bosnian) Muslims, and Macedonians.

Some of these groups barely tolerated one another. To the Serbs, for example, Croats were traitors and butchers who had sided with the Nazis during World War II. (The Croat fascist organization Ustashe had killed thousands of Yugoslavs; it had cooperated with the Germans in the hope of creating an independent Croatia.) Serbia also resented the growing number of Albanians in Kosovo, who outnumbered the Serbs there by nine to one.

Meanwhile, Croats, Albanians, and Slovenes resented Serbia. They believed that the Serbs and their leader, Slobodan Milošević, wanted to dominate all of Yugoslavia. In fact, Milošević talked of creating a Greater Serbia that would unite all Serbs in one territory.

Religious differences added to the hostility. Most Croats and Slovenes were Roman Catholic, whereas Serbians were predominantly Eastern Orthodox. Parts of Yugoslavia had once been part of the Ottoman Empire, and some inhabitants of those areas had converted to Islam. Having refused to convert, the Serbs despised the people who had become Muslims.

The Albanians were Muslims, and the Serbs were afraid that if they made Kosovo independent of Serbia, and perhaps joined it to Albania, the Serbs would lose a number of their most sacred Orthodox monasteries. Soon after Tito's death, Albanian nationalists began calling for independence from Serbia. Between 1987 and 1990, Milošević was ruthless in crushing this movement. This made Croats and Slovenes even more afraid that Serbia intended to dominate Yugoslavia.

Serb troops and civilians walk by a body in the fallen city of Vukovar, Croatia.

The Federation Splinters

Yugoslavia's constitution and political system contained no solution to the tensions. Each republic could veto the central parliament's decisions, so that any one republic could block change. This meant that Prime Minister Ante Markovic was almost powerless to stop the nation from falling apart.

As a wave of democracy swept over Eastern Europe in response to Mikhail Gorbachev's reforms in the Soviet Union, communists were no longer able to monopolize power in Yugoslavia. Instead, they too began to split along ethnic lines. Those favoring greater democracy and self-government mostly sided with Slovenia and Croatia. Serbian communists such as Milošević supported a strong central government and Yugoslav unity; they were willing to let Slovenia secede, but not Croatia, because it had a minority population of Serbs.

President Franjo Tudjman of Croatia and President Milan Kucan of Slovenia tried to take independence slowly; they hoped to come to an agreement with the central government. When they became convinced, however, that Serbia would not give Croatia and Slovenia enough freedom within the federation, they decided to move forward.

Elections in the two republics in the winter and spring of 1990-1991 showed that voters overwhelmingly favored independence. On June 25, 1991, Croatia and Slovenia declared their independence from Yugoslavia.

Still, Tudjman and Kucan tried to keep peace. Neither claimed to be seceding from Yugoslavia; instead, they used the word "dissociation." Realizing that Serbs living in Croatia were not in favor of independence, the Croat government issued a decree to protect the rights of its ethnic minorities. Both Slovenia and Croatia also announced that they were willing to be part of a confederation with the other Yugoslav republics if they were allowed to run their own internal affairs.

2364

AP/Wide World Photos

In spite of this caution, Yugoslavs and other Europeans recognized that the country was about to fall into civil war. The Slovene newspaper *Delo* described the situation as "a spontaneous and uncontrolled dying of the state," made worse by ethnic hatreds.

Encouraged by Milošević, some Serbs in Croatia armed themselves, formed into paramilitary groups, and began to make terrorist attacks. Others migrated to Serbia, claiming that they feared persecution by the Croatian majority. Meanwhile, Serbia insisted that if Yugoslavia fell apart, new boundaries would have to be drawn to permit all Serbs to live in the same territory.

Consequences

The Yugoslav People's Army and the Serbs quickly responded to Slovenia and Croatia's declaration of sovereignty. On June 27, 1991, the Serbian-dominated army moved against Slovenia. Yet Slovenia was well armed, and its resistance was fierce. Finally the Yugoslav officers decided to draw back, leaving behind some units that had been cut off and surrounded.

Unable to force Slovenia back into the federation, the federal army turned its attention to Croatia. Chief of Staff General Blagoje Adzic, a Serb whose family had been murdered by the Croat Ustashe during World War II, announced on July 2 that the army would crush the rebellious republics, even if it meant total war. The next day, armored forces left Belgrade, Serbia, for Croatia.

The army soon occupied Serbian regions of Croatia, with the help of Serb irregulars. Meanwhile, the Croats held back, believing that European nations or the United States would soon help them come to a settlement with the Yugoslav government. When this help did not appear, however, the Croats began to fight back in ear-

nest. The Serbs bombarded the ancient city of Dubrovnik. Numerous cease-fires negotiated by the European Community (EC) dissolved in the wake of ethnic hatred. Not even a United Nations peacekeeping force was able to keep thousands from being killed or wounded.

By early 1992, both sides showed signs of exhaustion, and EC nations helped bring a temporary halt to the fighting by taking applications for separate statehood from the warring republics. By May, 1992, all the Yugoslav republics except Serbia and Macedonia had seceded. Fighting broke out again, this time in Bosnia-Herzegovina. Serbian nationalists and those who remained in the federal army occupied large parts of Bosnia-Herzegovina, claiming to be protecting the Serbian minority that lived there. They besieged Sarajevo, which faced surrender or starvation.

The United Nations talked of sending a peacekeeping force to Yugoslavia, but negotiators failed in attempt after attempt to establish a truce that would enable U.N. troops to take up positions. The United Nations did impose economic sanctions against Serbia, but such tactics offered no prospect of a quick resolution to the crisis. Meanwhile, Serbian forces instituted programs of "ethnic cleansing" in both Bosnia-Herzegovina and the Albanian-populated region of Kosovo which resembled the Nazi concentration camps of World War II.

By the fall of 1992, tens of thousands of people had died in the continuing violence. Some analysts predicted that Yugoslavia would remain embroiled in decades of ethnic, religious, and ideological violence and destruction. In the meantime, U.N. officials continued to meet over the political and human fate of a destroyed nation.

Kendall W. Brown

Soviet Union Tries to Suppress Baltic Separatism

> *In January, 1991, the Soviet Union used armed force to seize key buildings in Lithuania and Latvia and prevent these republics from proclaiming their independence from the Soviet Union.*

What: Civil strife
When: January 11 and 20, 1991
Where: Vilnius, Lithuania, and Riga, Latvia
Who:
MIKHAIL GORBACHEV (1931-   ), first secretary of the Soviet Communist Party from 1985 to 1991
VYTAUTUS LANDSBERGIS (1932-   ), president of Lithuania from 1990 to 1992
ANATOLIJS GORBUNOVS (1942-   ), president of Latvia from 1990 to 1993

A Centuries-Old Battleground

The three Baltic nations, Lithuania, Latvia, and Estonia, cover an area that has been a battleground for centuries. In the Middle Ages, Lithuania ruled a mighty empire for one hundred years, but after that time it came under the control of other nations. Latvia and Estonia had never been independent before the twentieth century. The area was considered very valuable because it lay at the head of the Baltic Sea. For centuries the powers of northern Europe waged war to control these lands. Russia did not subdue the area until the eighteenth century.

After the chaos of World War I (1914-1918) and the Russian Revolution (1917-1918), the Allies and the new Soviet government recognized all three Baltic republics as independent. Still there were struggles. For example, Poland seized the city of Vilnius, Lithuania, and it remained in Polish territory until 1939.

In August, 1939, Nazi dictator Adolf Hitler and Soviet dictator Joseph Stalin secretly agreed that the Soviet Union would be allowed to take over the Baltic area. The next year, Soviet troops invaded the Baltic states and made them a part of the Soviet Union by force. The Soviet government expelled hundreds of anticommunist politicians and ordinary citizens from the region, and others were imprisoned or executed.

In 1941, Hitler broke his treaty with Stalin and invaded the Soviet Union; the Baltic states were made part of the German empire. Preferring fascism to communism, many Baltic people cooperated with the Nazis. When the war was over, however, the Soviet Union took back the Baltic states, though the United States and other countries protested.

Lithuania, Latvia, and Estonia remained under Moscow's control for the next forty-five years. Tens of thousands of citizens were tried and imprisoned for collaborating with the Germans or for opposing the communists; some were executed. Many managed to flee to Western countries, where they continued to dream of freedom for their homelands.

Assault on Independence

When Mikhail Gorbachev became leader of the Soviet Union in 1985, independence was finally a realistic goal for the Baltic republics. Even members of the Communist Party in these states sided openly with the independence movements.

Gorbachev allowed free, competitive elections to be held throughout the Soviet Union in 1989. Vytautus Landsbergis in Lithuania, Anatolijs Gorbunovs in Latvia, and Arnold Rüütel in Estonia campaigned on the independence platform, and all three of them won.

Once in power, these three presidents argued that their countries should never have been made part of the Soviet Union. In treaties signed in 1920 and 1921, the Soviet government had agreed to respect their independence, and the Stalin-Hitler Pact had been illegal and wrong.

In 1990, the presidents, cabinets, and parliaments of all three countries declared independence, but Moscow refused to recognize the proclamation. Most other nations preferred to wait; the United States, for example, wished to support Gorbachev, whose leadership was being challenged by conservative communists.

Still, the Baltic people were determined. Those who were called up for Soviet military service refused to report for duty. Lithuania and Latvia set up their own military and police. By January, 1991, Soviet police had begun to guard Communist Party buildings in Vilnius, Lithuania, and Riga, Latvia, so that the dissident governments could not take them over. (Estonia's government was seen as more willing to cooperate with Moscow.)

On January 10, 1991, Lithuanians favoring Soviet rule led strikes to protest the independence resolutions—even though a huge majority of Lithuanians wanted independence. Soviet paratroopers, backed by the army's tanks and machine guns, seized control of the Lithuanian army buildings and the press and television centers. Shouting insults at the invading soldiers, Lithuanian citizens gathered around their government's parliament building to prevent the troops from entering. The forces killed fifteen civilians, and more than one hundred were wounded.

Gorbachev called on the Lithuanian government to end its defiance of Moscow; he said that secession had to come through constitutional processes, and that the rights of non-Lithuanian minorities had to be protected. To replace the Lithuanian government, the Soviet authorities set up a National Salvation Committee made up of Communist Party members and members of the Polish and Russian minorities in Lithuania.

Yet many members of minorities in Lithuania sided with the independence movement; even in Moscow, Soviet citizens demonstrated in front of the Kremlin (the central building of the Soviet

Latvians guard barricaded government buildings in Riga against possible attack by Soviet troops, who were trying to end the republic's drive for independence.

2367

government) to show their support. The Russian people were experiencing their own wave of nationalistic feelings, as were other ethnic groups living in the Soviet capital. Some members of the parliament joined in the Moscow demonstrations.

On January 19, the Kremlin set up an all-Latvian Committee of National Salvation similar to the one that had been established in Lithuania. The next day, Soviet special forces attacked the Latvian Ministry of the Interior and raided the Latvian parliament, seizing the building in a ninety-minute assault. Five people died.

Consequences

For seven months, the confrontation between the Baltic states and the Kremlin remained at a standoff: The republics insisted on independence, while Moscow demanded that they use "constitutional procedures" to apply for withdrawal from the Soviet Union.

Then, in August, 1991, conservative Party members tried to overthrow Gorbachev. Although the coup failed, Gorbachev quickly lost prestige, while Russian leader Boris Yeltsin gained popularity. Yeltsin was much more sympathetic to the Baltic states' desire for independence, and on September 6, 1991, Moscow finally recognized all three republics as independent from the Soviet Union. By this time many other countries, including the United States, had already done so.

Frederick B. Chary

South Africa Begins Dismantling Apartheid

> *At the opening session of the South African parliament in 1991, President F. W. de Klerk announced plans to repeal the basic laws of racial separation and domination.*

What: Civil rights and liberties
When: February 1, 1991
Where: Cape Town, South Africa
Who:
FREDERIK WILLEM DE KLERK (1936-   ),
    president of South Africa from 1989 to
    1994
NELSON MANDELA (1918-   ), a leader of
    the African National Congress
MANGOSUTHU GATSHA BUTHELEZI
    (1928-   ), a Zulu chief and head of
    the Inkatha Party

A Historic Announcement

At the opening session of the racially segregated parliament of South Africa in 1991, President Frederik W. de Klerk delivered a dramatic speech. He promised to support legislation that would scrap the remaining laws of apartheid (South Africa's system of racial segregation, or "apartness"). The most important of these laws was the Population Registration Act of 1950, which divided South Africans into four racial categories: whites, blacks, Indians, and coloreds (people of mixed race). Other apartheid laws included the Native Lands Act (1913) and the Native Land and Trust Act (1936), which reserved 87 percent of South Africa's land for the white minority. The Group Areas Act (1950) segregated residential areas.

De Klerk also said that he would propose a new law to let communities work out integrated government for themselves. Finally, he said that a multiparty conference should be called to discuss writing a new constitution for the country. Because de Klerk was the leader of the National Party, which had a clear majority of votes in Parliament, it was evident that the proposed laws would be passed quite easily.

Before this time, the government had insisted that the Population Registration Act could not be repealed as long as South Africa was bound by the constitution of 1983, which set up parliamentary representation for whites, coloreds, and Indians. Now, however, de Klerk explained that the law could be repealed temporarily, until a new constitution was written. Babies born after the repeal would not be classified by race, but other South Africans would keep their classification until Parliament voted itself out of existence to make way for a new government.

In his address, de Klerk did not mention three homeland laws: the Bantu Self-Government Act (1959), the Bantu Homelands Citizenship Act (1970), and the Status Acts. The first two laws had given limited self-government to ten tribal homelands—desert regions of South Africa where all blacks were assigned as citizens. The Status Acts had made four of the homelands (Transkei, Ciskei, Bophuthatswana, and Venda) independent, so that their citizens were no longer citizens of South Africa. This issue could only be settled with a new constitution.

Reactions to the Speech

Right-wing Afrikaners (South Africans of Dutch background) were horrified; during de Klerk's speech they shouted, "Traitor to the nation!" and "Hangman of the Afrikaner!" Andries Truernicht and the forty other Conservative Party representatives angrily marched out of Parliament.

On the other hand, South Africa's Democratic Party, which opposed apartheid, was pleased with the proposals. In the United States, the administration of President George Bush declared that its policy toward South Africa was working; one State Department official said, "It's the equivalent of the fall of the Berlin Wall."

Yet de Klerk's proposals did not go far enough

to please black organizations in South Africa, especially the African National Congress (ANC). Nelson Mandela and other ANC leaders pointed out that de Klerk was unwilling to allow elections to choose those who would write the new constitution; this meant that whites would have greater representation in negotiations over the constitution. "We still do not have the vote," declared ANC leader Walter Sisulu, "and this is what our people demand today, to vote for a constituent assembly."

The ANC had other complaints. De Klerk had offered no specific plans to bring forty thousand political refugees back to South Africa, or to release political prisoners in the independent homelands. He also did not suggest repealing the security laws that had allowed the police to detain thousands of people for political reasons. Throughout South Africa, hundreds of thousands of blacks marched to demand greater reforms.

Three days before President de Klerk's speech, Mandela had met with his primary black rival, Zulu chief Mangosuthu Buthelezi, for the first time in twenty-eight years. After an eight-hour discussion, they had announced a peace pact for Natal Province, where violent fighting between ANC supporters and members of Buthelezi's Inkatha Party had killed four thousand people in five years. Unfortunately, the peace pact did not solve the problem; violence between the two groups soon sprang up again. Eventually, the ANC announced that it had evidence that the government had encouraged Inkatha to attack ANC members.

At the annual summit of the Organization of African Unity (OAU), June 3-5, 1991, there was a long debate about South Africa. Finally, the organization stated that its members would stop economic sanctions against South Africa if its new laws made "profound and irreversible changes toward the abolition of apartheid."

On June 5, the South African parliament voted to repeal the Land Acts of 1913 and 1936 and the Group Areas Act. Now all South Africans had the legal right to buy property and live where they pleased. On June 17, Parliament repealed the Population Registration Act; the white House of Assembly voted 129 to 38, with 11 abstaining. This meant that South Africans would no longer be assigned to racial categories at birth.

Consequences

Early in July, 1991, the U.S. State Department informed President Bush that South Africa was definitely moving to end apartheid, according to the requirements of the Comprehensive Anti-Apartheid Act of 1986. On July 10, the president officially ended trade and investment sanctions against South Africa. The ANC and many other antiapartheid groups were displeased, for they believed that the United States should wait until apartheid was completely done away with. But several countries, such as Finland and Japan, soon announced that they were following the American example. The International Olympic Committee also decided that South Africa would now be allowed to participate in the Olympics (it had been barred from the games for twenty-one years).

Many difficulties remained for the South Africans. The ANC was determined that the new constitution must provide "one person-one vote" democracy, while de Klerk and the Nationalists wanted to protect "minority rights" by allowing the whites to keep veto power.

Negotiations for writing the constitution began late in 1991. On March 17, 1992, the black Africans came one step closer to their demands for one person-one vote when the white majority, voting in a referendum, approved de Klerk's efforts to establish majority rule.

Thomas Tandy Lewis

East European Nations Dissolve Warsaw Pact

The Soviet-dominated military alliance that included most of Eastern Europe dissolved with the fall of communism and changes in Soviet president Mikhail Gorbachev's strategic priorities.

What: International relations; Military defense; Political reform

When: July 1, 1991

Where: Prague, Czechoslovakia

Who:

MIKHAIL GORBACHEV (1931-), president of the Soviet Union from 1988 to 1991

VLADIMIR ZHIRINOVSKY (1946-), a hard-core Soviet nationalist who opposed Gorbachev

End of an Era

On July 1, 1991, the Soviet Union and its allies dissolved the political structures of the Warsaw Treaty Organization at a meeting in Czechoslovakia. This organization, commonly known as the Warsaw Pact, had existed since 1955. The military elements of the alliance had already been shut down on April 1, 1991. The decision to end the alliance had been made on February 25, 1991, at a meeting in Budapest, Hungary, attended by the Soviet ministers of both defense and foreign affairs. The breakup came, ironically, at the climax of the Persian Gulf War, in which U.S. prestige and power were at a peak.

Throughout most of the Cold War (roughly 1946-1991), the Warsaw Pact had squared off against the U.S.-led North Atlantic Treaty Organization (NATO) in a divided Europe. However, the fall of the communist governments of Eastern Europe in 1989 and the reunification of Germany the next year combined with Mikhail Gorbachev's "new thinking" in Soviet foreign policy to make the Warsaw Pact untenable.

The Soviet Union had created the Warsaw Pact at the end of the first decade of the Cold War as a counterweight to NATO (formed in 1949), as an obstacle to the reunification of West and East Germany, and as a tool for legitimizing the presence of Soviet troops in the countries of Eastern Europe. The official purpose of the pact was to protect Eastern Europe from an attack by the United States or its allies. The pact's members, besides the Soviet Union, were Poland, Hungary, Romania, Czechoslovakia, Bulgaria, and East Germany. Socialist Yugoslavia was never a member. Like NATO, the members of the Warsaw Pact all had a common official ideological orientation in politics and economics. Unlike the United States in NATO, however, the Soviet Union had far greater degree of actual military, economic, and political control over the Warsaw Pact.

The Warsaw Pact's military leaders and political representatives met annually to coordinate policies. Obviously, the Warsaw Pact nations were never called upon to fight a major war in Europe, but the pact did militarily assault one of its own member states in 1968. In that year, reform Communists had begun relaxing strict Soviet-style controls in Czechoslovakia in a movement known as the Prague Spring. The Soviet Union, assisted by East German, Polish, Bulgarian, and Hungarian forces, crushed the movement in August. The official Soviet justification for the invasion, later known as the Brezhnev Doctrine, made no mention of the Warsaw Pact, because the pact guaranteed the independence and sovereignty of its members. Soviet claims that the West was using covert and cultural means to attack communism in Czechoslovakia rang hollow.

The Warsaw Pact had not been consulted in the 1956 Soviet invasion of Hungary. Later, the organization survived a partial defection of Romania; it successfully pressured East Germany at various times but had to expel Albania, which had drifted into China's camp in the 1960's. The Warsaw Pact's 1.5 million soldiers allowed the Soviet Union to concentrate its power elsewhere, a fact that was especially important by the late 1960's, when conflicts with China crescendoed in Siberia.

End of the Pact

After 1985, Soviet president Gorbachev made radical changes in Soviet military strategy and in the nature of the pact. He announced major unilateral cuts in Soviet troop levels and in defense spending, both because he personally believed that world wars and nuclear exchanges were unwinnable and because his country could not afford to continue the arms race with the United States. By 1987, Gorbachev had replaced most of the top Soviet military leadership. Moscow was no longer going to prevent reforms in Eastern Europe, and it intended to forge new security policies based on disarmament and pan-European cooperation. In 1989, new noncommunist governments came to power from Bulgaria to Poland.

By 1990, Western observers were predicting the imminent demise of the Warsaw Pact, using terms such as "failed" and "shattered bloc" to describe its state. After Gorbachev's brief crackdown on nationalist resistance in the Baltic states in January, 1991, the East European members of the Warsaw Pact began to bolt. In February, Poland unilaterally announced its exit from the Pact; Hungary followed suit and went even further, saying that it would immediately seek cooperation with NATO.

Consequences

Gorbachev hoped that NATO would dissolve at the same time as the Warsaw Pact, or at least that the Warsaw Pact could be preserved as a purely political alliance. When neither of these things occurred, the Soviet Union began concluding new bilateral treaties with each of the East European states. By mid-1991, Soviet troops had been pulled out of Czechoslovakia and Hungary; by 1994, they had left Poland and East Germany as well. Tricky issues of pollution cleanup,

compensation for military buildings, and closer ties between the Soviet Union's Baltic republics and the East European states troubled the new relationships. The Soviets' complicated task of shoring up defenses along their western border was made even more complicated when the Soviet Union broke up altogether in December, 1991; the reach of the government in Moscow by then extended only to the borders of Russia.

The fall of the communist governments in 1989 and then the dissolution of the Warsaw Pact gave much ammunition to Gorbachev's political opponents. They, especially the hard-core nationalist Vladimir Zhirinovsky, blamed him for squandering "Russia's" empire. All Russians experienced a sense of loss of fraternal countries with whom they felt in ideological solidarity, especially since the friendly communist governments in those states were officially portrayed as a lasting legacy of the great suffering and ultimate victory of the Soviet Union in World War II.

Eastern Europe shrank in importance in Russian thinking. The Russian government was worried about its international prestige in relation to that of the United States and about its minority populations in "the near abroad," as the successor states to the Soviet Union (such as Ukraine, Latvia, Moldova, and Kazakhstan) were called. Russia's main concern with its former allies was that they not move too fast to join the European Union or NATO. Eastern Europe embarked on its famous, but still unfinished, transition to democratic politics and capitalist economics.

New issues in international relations preoccupied the former Warsaw Pact members. Poland was mostly concerned with working out its disputed border with newly reunited Germany. Czechoslovakia was faced with dismantling its important arms export industry and would soon be embroiled in the process of splitting into two states. Hungary showed its concern for its large minorities abroad while spearheading various "Danubian" cooperative arrangements with its Central European neighbors. Between 1997 and 1999, all three of these countries were, with grudging Russian approval, admitted into the ranks of NATO.

John K. Cox

Soviet Union Disintegrates

During late 1991, very poor economic conditions, a falling standard of living, growing nationalist separatism, and a coup attempt destroyed confidence in the central Soviet government and led to the creation of the Commonwealth of Independent States.

What: Political reform; Social reform
When: August-December, 1991
Where: Commonwealth of Independent States (the former Soviet Union)
Who:
MIKHAIL GORBACHEV (1931-   ), first secretary of the Soviet Communist Party from 1985 to 1991
BORIS YELTSIN (1931-   ), president of the Russian Republic from 1991 to 1999
GENNADY I. YANAYEV (1937-   ), vice president of the Soviet Union from 1990 to 1991
LEONID KRAVCHUK (1934-   ), president of the Ukraine from 1991 to 1994
STANISLAV SHUSKEVICH (1934-   ), chairman of the Belorussian Supreme Soviet

Communism Breeds Discontent

The year after the Bolshevik Revolution in 1917, Vladimir Ilich Lenin established the Soviet Republic of Russia. All land, housing, and large factories were taken over and managed by the government, and the Communist Party and the government became one and the same. The Union of Soviet Socialist Republics (or Soviet Union) was founded in 1922.

Joseph Stalin took over in 1924, after Lenin's death. Under Stalin, the Soviet Union's central government created collective farms and began to develop factories. Because these farms and factories were not very productive, living standards fell and people began to protest. In the 1930's, Stalin arrested hundreds of thousands of people who stated their opposition; they were put into labor camps or executed.

Nikita Khrushchev, who rose to power after Stalin's death in 1953, brought a few political reforms but was unable to solve the Soviet Union's economic problems. When his farm program collapsed in 1964, he was forced to retire.

Leonid Brezhnev continued centralized economic planning and stifled political reform in the late 1960's and the 1970's; the Soviet Union's economic growth slowed even more. It became necessary to import food, and after 1979 there were major shortages of consumer goods.

The Union Is Undone

Mikhail Gorbachev became first secretary of the Soviet Communist Party in 1985, and the next year he began an economic reform program called *perestroika* (restructuring), along with a new *glasnost* (openness) in Soviet society. In 1987 Gorbachev stopped giving government money to industries; instead, he gave them more freedom to buy and sell. He also allowed individuals to farm private plots and sell their produce freely. To reduce the demand for consumer goods, he raised the government-assigned prices of various goods.

At first, many Soviet citizens were pleased with Gorbachev's reform plans, but by late 1989 most people blamed Gorbachev for the worsening economy. Workers worked longer and harder, but often their wages were lower. Prices went up and goods were more difficult to find. Support for Gorbachev decreased, and membership in the Communist Party declined.

In 1988 Gorbachev also began changing the Soviet government, so that more democratic elections were allowed. Meanwhile, the nations of Eastern Europe began throwing out their communist governments. In 1989, the first Soviet republic declared that it would no longer be ruled by the Soviet Union, and in 1990, the Party officially gave up control over the country.

In mid-1991, Gorbachev planned a new Union Treaty, to reshape the country into a "Union of Sovereign Soviet Republics." The treaty called for a new constitution that would allow national elections and weaken the Communist Party and the Soviet central government. The treaty was to be signed August 20, 1991.

Instead, on August 18, a coup began. The coup leaders, a group of hard-line communists, wanted to dismiss Gorbachev and return the Soviet Union to strong Party rule. On August 19, Vice President Gennady Yanayev became acting president, and an emergency committee took power.

In Moscow, Boris Yeltsin, president of the Russian Republic, called for nationwide resistance to the coup, and citizens gathered in Moscow to put up barricades to protect Yeltsin's government.

With such strong displays of protest, the coup began to fall apart on August 20. The next day, Soviet tanks and troops began returning to their bases across the Soviet Union, the Communist Party denounced the coup, and crowds celebrated in Moscow. By August 22, the coup was over, and on August 29, the Communist Party officially disbanded.

The coup also sparked independence movements in the Soviet republics. Latvia and Estonia declared their independence, and soon others joined them. On August 26, leaders of these breakaway republics declared central Soviet authority dead. On September 5, the Congress of People's Deputies gave its power to the republics.

As central power crumbled, republican leaders took control. On December 7, Yeltsin, Ukrainian president Leonid Kravchuk, and Belorussian chairman Stanislav Shuskevich decided that it

CAUCASUS AND FORMER SOVIET REPUBLICS OF CENTRAL ASIA

was impossible to improve the economy within the Soviet structure. These three leaders formed a "Commonwealth of Independent States" and proclaimed the Soviet Union over on December 8.

On December 17, Gorbachev agreed to dissolve the Soviet Union. Five days later, Russia, Ukraine, Armenia, Belorussia, Kazakhstan, Kyrgyzstan, Moldavia, Turkmenistan, Azerbaijan, Tajikistan, and Uzbekistan approved the Commonwealth agreement. Gorbachev formally resigned as president of the Soviet Union on December 25, and on December 31, 1991, the Soviet Union, the world's first communist state, ceased to exist.

Consequences

Yet the new commonwealth was in poor political and economic health, and nationalism surged in the republics. Fears grew that the Soviet military, now under a shared command, might cause political upheaval. People stockpiled food in preparation for famine.

In Russia, Yeltsin began turning property, agriculture, and industry over to private ownership. In January, 1992, he ended most price controls, hoping to stimulate production, match supply with demand, and reduce rising prices. The result was that many people lost their jobs, the ruble (Russia's currency) lost value, and prices rose even higher—yet there

were still shortages of goods. So it was that discontent grew quickly after the formation of the Commonwealth of Independent States. Many wondered how long the commonwealth would last.

Martha Ellen Webb

FORMER SOVIET EUROPEAN NATIONS

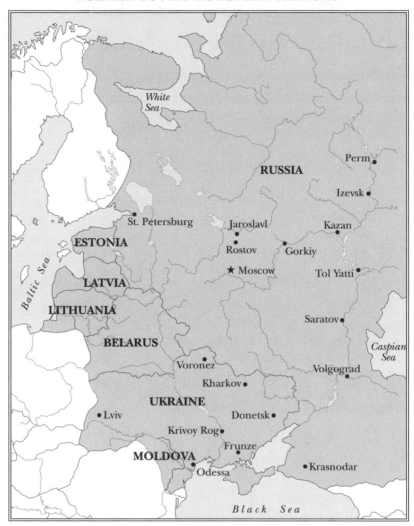

2375

President George Bush Announces Nuclear Arms Reduction

With the end of the Cold War, U.S. president Bush announced extensive unilateral cuts in stocks of nuclear arms.

What: International relations; Weapons technology
When: September 27, 1991
Where: Washington, D.C.
Who:
GEORGE HERBERT WALKER BUSH (1924-   ), president of the United States from 1989 to 1993
MIKHAIL GORBACHEV (1931-   ), president of the Soviet Union from 1988 to 1991
RICHARD (DICK) CHENEY (1941-   ), United States secretary of defense from 1989 to 1993

A Longtime Dream

Ever since the end of the nineteenth century, the world's nations have sought to limit arms. The introduction of mechanized and technologically advanced weapons spurred further resolve. Nations at the Hague Peace Conference of 1899, however, only succeeded in deciding what the rules of war should be, not how to disarm for peace.

The dangers of an arms buildup are evident from a study of World War I. An arms race is acknowledged to be one of the causes of that fighting. When the war ended, nations of the world held a number of multilateral talks aimed at reducing weapons. At first the governments arrived at partial solutions: Each nation agreed to keep a limited number of weapons dependent on its size and power. With the appearance of aggressive fascism in the 1930's, however, even these limitations were ignored, and a new arms race began. As with World War I, the arms race

contributed to the outbreak of World War II. The goal of universal disarmament remained an unfulfilled, longtime dream.

World War II brought the end of fascism and the emergence of the United States and the Soviet Union as superpowers. It also brought nuclear bombs, weapons by which the human species could destroy itself and the entire planet. Although the United States and the Soviet Union had been allies in World War II, they now faced each other in a "Cold War," one in which their soldiers did not meet in battle. Instead both countries threatened each other with an ever-enlarging nuclear arsenal. Beginning in the 1950's, the superpowers began to talk to each other about how to reduce armaments, both nuclear and conventional weapons. The situation was complicated because other countries, such as France, England, and China, soon had nuclear weapons as well. Furthermore, the superpowers did not trust each other. Each thought the other's proposals for disarmament were only schemes to gain superiority. While some agreements were reached, real disarmament did not occur and the weapons actually became more deadly as technology advanced. By 1980, there were enough nuclear weapons on the face of the earth to destroy the planet many times over.

In 1985, when Mikhail Gorbachev became leader of the Soviet Union, a cultural revolution occurred within the Soviet sphere of influence. The Soviet Union and its satellite countries became more like the United States—open societies where agreements could be made which Washington would trust. The two superpowers began a period of friendlier relations. Furthermore, because of dire economic needs Gorbachev was anxious that disarmament agreements succeed. He wanted to direct less money to the military and more to civilian needs.

In August, 1990, the Soviet people stood firm against an attempt by hardline communists to oust Gorbachev and reverse the democratic process. The United States government then realized more than ever that disarmament would work. In fact, it appeared that there was more danger from a nuclear accident caused by instability in the Soviet Union than from a direct attack, and a reduction in arms became a necessity.

Reducing Nuclear Arms

On September 27, 1991, President George Bush announced to the world that he would make unilateral cuts in the United States' nuclear arsenal. He also announced that the constant readiness of the United States Air Force's Strategic Air Command (SAC) would be modified. He invited Soviet president Mikhail Gorbachev to make similar reductions but stated that the American plans would go ahead in any case. This announcement heralded a major shift in

U.S. policy. The proposal projected a reduction of 2,400 American nuclear weapons, chiefly of the battlefield tactical type. The specific proposal included the removal from overseas bases of all long- and short-range nuclear weapons, both missiles and cannon shells.

Those weapons that the military did not destroy, Bush said, would be stockpiled in the United States. His orders would affect land-based weapons as well as those on submarines and surface vessels. He canceled the constant alert status of B-1 and B-52 SAC bombers. Their nuclear bombs would be placed in storage. Bush's plan ended the alert status of 450 Minuteman II Intercontinental Ballistic Missiles (ICBMs). These were to be phased out over the next years, and the president promised to hasten the reduction if arms talks with the Soviet Union were completed.

President Bush also scrapped some weapons programs already under construction, including the mobile MX and short-range nuclear attack missiles. He stated, however, that the Strategic

At the Pentagon, General Colin Powell explains President George Bush's plans to reduce nuclear arms.

Defense Initiative (SDI), popularly known as Star Wars, the Midgetman ICBM, and the B2 Stealth bomber programs would continue.

The president finished his announcement by inviting Gorbachev to make similar cuts in the Soviet arsenal.

Consequences

Some people argued that there should be even more reductions than those announced by Bush. However, the president and Secretary of Defense Richard (Dick) Cheney made it clear that cuts in addition to those proposed in Bush's speech and related measures would damage the country's defense. The proposals did have an immediate effect, because Gorbachev afterward praised the president's statement and announced similar reductions in the Soviet nuclear arsenal. The two countries made progress in other areas of arms control as well. Leaders around the world greeted the proposals with enthusiastic applause.

Frederick B. Chary

Senate Confirms Thomas's Appointment to Supreme Court

> *Despite contentious Senate Judiciary Committee hearings, which included accusations of illegal behavior, conservative African American nominee Clarence Thomas was successfully confirmed associate justice of the U.S. Supreme Court.*

What: National politics; Law
When: October 15, 1991
Where: Washington, D.C.
Who:

CLARENCE THOMAS (1948-　　), judge on U.S. District Court of Appeals for the District of Columbia

ANITA FAYE HILL (1956-　　), professor at the University of Oklahoma School of Law

GEORGE HERBERT WALKER BUSH (1924-　　), president of the United States from 1989 to 1993

The Nominee

When Justice Thurgood Marshall retired in the summer of 1991, the U.S. Supreme Court lost its first African American justice and an activist for civil rights as defined by liberal legislative and governmental policies and Court decisions since the mid-1960's. To replace him, President George Bush sought a nominee who was African American but preferred one who shared his generally conservative values. In March of 1991, he nominated and the Senate confirmed Clarence Thomas to the U.S. District Court of Appeals for the District of Columbia, and on July 1, 1991, he nominated him for Marshall's position.

Thomas was born in rural southern Georgia, raised by his maternal grandparents, and educated in Roman Catholic schools through university. He attended Yale Law School and briefly practiced law. In 1981, President Ronald Reagan appointed him assistant secretary for civil rights in the U.S. Department of Education, despite Thomas's having avoided all posts involving mi-

nority group issues in the past. In the following year, Reagan named him chairman of the Equal Employment Opportunity Commission (EEOC), in which post he served for eight years.

During this time, Thomas made a mark as a conservative who preferred to support individual rights rather than minority group rights and sought to undo some of the policies that had been fundamental to advancing the civil rights agenda in the United States. Affirmative action for African Americans had relied upon numerical quotas (required hiring of a certain percentage of minority employees) and timetables related to these, and Thomas undermined both. Prosecution for violation of such laws had often taken the form of class-action suits (representing a large number of people rather than one aggrieved party) that were based on statistical evidence that seemed to indicate racial discrimination rather than actual actions by a defendant, and Thomas backed the EEOC away from these. He also published and spoke regarding his belief in natural law, which opposed artificial governmental or legal remedies for inequality, and therefore drew criticism from those who supported such remedies. Although his appointment to the district court went smoothly, his nomination to the Supreme Court angered a great many liberals who were familiar with his record and beliefs.

The Judiciary Committee Hearings

From the time of his nomination, Thomas was attacked for his views and previous actions. The National Organization of Women opposed him for disagreeing with the Supreme Court decision in *Roe v. Wade* (1973), which provided federal protection to a woman's right to abort her fetus.

2379

AP/Wide World Photos

Justice Byron White (right) swears in Clarence Thomas (second from left) as associate justice of the Supreme Court.

African American organizations such as the Congressional Black Caucus and the National Association for the Advancement of Colored People opposed Thomas for his views on federally mandated racial preference and attacked Bush for nominating a conservative African American rather than a progressive one. They also accused Thomas of hypocrisy in opposing affirmative action, a policy that was responsible for his own admission to Yale Law School. Labor unions such as the American Federation of Labor-Congress of Industrial Organizations (AFL-CIO) raised fears that as a conservative, Thomas would vote against protections for workers' safety and welfare. The American Bar Association rated him a "qualified" candidate—a cautious endorsement. However, openly opposing an African American nominee made many, especially politicians, very nervous.

The Senate Judiciary Committee hearings concerning Thomas's appointment opened on September 10, 1991, and Thomas faced close and determined questioning by committee Democrats. Thomas insisted that as justice, he would make his decisions according to a strict constructionist principle of relying on the intent of the authors of the U.S. Constitution and precedents set by previous court rulings. Several senators, most pointedly Chairman Joseph Biden, questioned his ideas of natural law and the role it would play in his voting, and Thomas summed up his contention by saying, "I have no agenda, Senator." He rather defiantly refused to discuss his views concerning abortion. By September 27, the committee was deadlocked seven to seven, and confirmation by the full Senate seemed imminent. Then, in early October, a political bombshell exploded.

Accusations and Aftermath

On October 7, 1991, University of Oklahoma Law School professor Anita Hill conducted a press conference at which she leveled charges of sexual harassment at Thomas. She had served as his personal assistant at the Department of Education and the EEOC, a position she had left in 1983 to teach law at Oral Roberts University. Thomas categorically denied her claims the following day, and the Senate decided to delay its vote and reopen hearings. A full month earlier, Hill had been approached by Democratic staffers who had heard of her story, but she was reluctant to make her claims public.

On October 11, Hill began her testimony before the Senate and a huge television audience. Although no one could unequivocally support her claims, many came forward as character witnesses on her behalf. Republican senators such as Orrin Hatch of Utah and Arlen Specter of Pennsylvania defended Thomas's character and attacked Hill's and raised grave doubts about her motivations and credibility. Feminist organizations rushed to her defense, especially after

critics questioned why she had remained with Thomas so long and had moved with him to the EEOC if she were so uncomfortable with his behavior. The phrase "he said/she said" summed up the impasse: One party was lying, but the truth could not be confirmed independently. Thomas responded before the committee as not only one wrongfully accused but also as victim of a "high-tech lynching for uppity blacks who in any way deign to think for themselves."

The hearings adjourned on Sunday, October 13, and two days later, the full Senate voted 52-48 to confirm Thomas. He was sworn in on October 18. The entire episode continued to polarize opinion: Liberals considered Thomas unfit, immoral, and his views indefensible; conservatives vilified Hill, defended Thomas's character and philosophy, and attacked the process that so damaged his reputation, and, arguably, that of Hill and the committee itself.

Joseph P. Byrne

Middle East Peace Talks Are Held in Madrid

After decades of tension between Israel and the Palestinians, the Madrid Peace Talks brought a new attempt to resolve differences.

What: International relations
When: October 30-November 4, 1991
Where: Madrid, Spain
Who:

YITZHAK SHAMIR (1915-   ), prime minister of Israel from 1983 to 1984 and from 1986 to 1992

JAMES BAKER III (1930-   ), United States secretary of state from 1989 to 1992

HAFEZ AL-ASSAD (1930-2000), president of Syria from 1971 to 2000

YASIR ARAFAT (1929-   ), chair of the Palestine Liberation Organization (PLO)

Quest for the Promised Land

The Arab-Israeli conflict has roots in the Jews' long hope of returning to the "promised land" of biblical times. The Zionist movement of the late 1800's and early 1900's encouraged a number of Jews—especially those who had suffered persecution in Russia and Eastern Europe—to buy land and settle in Palestine. Great Britain's Balfour Declaration of 1917 supported the idea of a Jewish homeland in Palestine.

After World War I, Great Britain was given a mandate to govern Palestine, and the British tried to help the local Jews develop self-governing institutions. Arabs in Palestine were alarmed, and some of them began fighting the Jews. Although Jews and Arabs come from the same ethnic background (Semitic), they now saw each other as enemies struggling for the right to live in the same territory.

In 1947, after World War II, the United Nations proposed a plan to partition the land of Palestine. This did not satisfy anyone, and when the

British left Palestine on May 14, 1948, the Jewish National Council announced the establishment of the State of Israel. War broke out between the Jews and the Arabs—the first of a long series of major conflicts.

Because of the 1948 war, many Palestinians had to leave their homes. For many of them, camps set up in Israel and other countries became their homes. Very few Arab rulers were willing to welcome them permanently; besides, the Palestinians hoped eventually to return to their old homes in Palestine (now Israel).

Two more wars followed, the Suez War of 1956 and the Six-Day Arab-Israeli War of 1967. With its 1967 triumph, Israel occupied various territories that had been held by Arab states: Egypt's Sinai Peninsula, the West Bank (which had been claimed by Jordan), Syria's Golan Heights, and the old city of Jerusalem. In November, 1967, the United Nations Security Council passed Resolution 242, which urged Israel to withdraw from these territories, called for a solution to the "Palestinian problem," and recognized the right of all Middle Eastern states to live in peace. As time passed, however, some of the Palestinians turned to terrorism as a way of resisting the Israeli government.

Israel was attacked by Egypt and Syria in 1973, but this war, too, was brief. U.N. Security Council Resolution 338 quickly called for a cease-fire and again urged peaceful solutions for Middle Eastern problems. With the Camp David Agreement of 1978, Israel returned Sinai to Egypt. Egypt and Israel were now on better terms, but many other issues were left unresolved.

Waving the Olive Branch

Israel and a number of Arab states cooperated with the United States in the Gulf War of 1991. Now U.S. officials saw an opportunity to bring an end to the long tensions between Israel and its Arab neighbors. The first step was to bring the two sides together for face-to-face talks. The organizers of the talks found it difficult to get everyone to agree on a place and a time, the issues that would be discussed, and the groups that would be allowed to participate.

Israel wanted to meet in Jerusalem, but the others would not accept this idea, because it would have seemed that they were agreeing with Israel's claim to the city. Washington, D.C., was also unsuitable, because meeting there would make it look as if the peace talks were being managed by the United States. Madrid was chosen as a neutral spot.

As the host nation, Spain had the hard task of providing proper living and meeting places for all the participants, and of setting up security arrangements. Everyone knew that terrorists might try to break up the talks through acts of violenceg. There were also separatist groups in Spain that hoped to draw attention to their cause by demonstrating during the meetings.

Choosing the participants was also difficult. Israel refused to negotiate with Arabs they considered extremist, yet the Palestinians wanted strong-minded representatives who would be tough bargainers. For many years, the Palestinians had been demanding a homeland for themselves and Israel's withdrawal from the occupied territories. If the Palestinian representatives to the peace talks were not determined, they might be tempted to make compromises and give up these demands.

Consequences

In the end, the delegates who were selected had various opinions—some radical, some moderate. In the true spirit of the meeting, they arrived in Spain's capital waving olive branches, the traditional symbol of peace.

In its opening statement, Israel insisted that it

Israelis (left) meet with delegates from Palestine and Jordan in Madrid.

AP/Wide World Photos

would never give up an inch of its territory, while Syria demanded that every inch of the lands taken by Israel in 1967 be returned. The Lebanese asked Israel to withdraw from the "security zone" it had established in southern Lebanon. Meanwhile, the Palestinians showed that they were ready to look for a practical solution. They were willing to consider limited self-rule—something that they had rejected when it was mentioned in the Camp David Agreement of 1978.

The Palestinians saw self-rule as the first step toward an independent state, while the Israelis saw it as the biggest compromise they would consider.

In spite of all the differences of opinion, the meeting was a breakthrough. It showed that most leaders in the Middle East realized that they needed to depend on negotiation rather than force or violence.

Felicia Krishna-Hensel

European Economic Community Tries to Create Single Market

Member nations of the European Economic Community moved toward economic union during the early 1990's.

What: Economics; International relations
When: 1992
Where: Brussels, Belgium
Who:
ALTIERO SPINELLI (1907-1986), an Italian communist member of the European Parliament
JACQUES DELORS (1925-   ), president of the Commission of the European Communities
POUL SCHLÜTER (1929-   ), prime minister of Denmark from 1982 to 1993

Dream in the Making

The dream of a united Europe goes back to the seventeenth century. Yet plans for such a union usually failed because they gave unfair advantage to one or a few countries or because the union lacked the power to prevent its members from acting independently. After World War I, the victorious Allies established the League of Nations, but the rise of fascism brought about its demise. While League leadership rested with the European countries, the organization included nations from around the world.

A new organization established after World War II, the United Nations, included many more non-European countries than the old League, but its power was centered in the non-European world. Therefore, several West European countries, including the ancient enemies Germany and France, now looked for a regional European union that would provide some of the economic advantages of a single country.

France, Germany, Italy, Luxembourg, Belgium, and the Netherlands created a historic union in 1952: the European Coal and Steel Community (ECSC). These six countries agreed to eliminate tariffs and to share resources. Great Britain had been invited to join but declined.

Over the next two decades, the ECSC increased economic cooperation in all areas. In 1957, by signing the Treaty of Rome, the countries came closer together and renamed the union the European Economic Community (EEC), popularly called the Common Market. The Common Market broke down barriers to European cooperation and helped create a miraculous economic recovery on the war-torn Continent. It also cooperated in political, diplomatic, and military matters. Its members joined the North Atlantic Treaty Organization (NATO), which established military links with the United States, Great Britain, and several other countries that were not members of the EEC.

Outside the Common Market, Great Britain, Switzerland, Sweden, Austria, Denmark, Norway, and Portugal in 1957 formed a rival economic union, the European Free Trade Association (EFTA). Countries under the control of the Soviet Union formed an economic union of their own, the Council for Mutual Economic Assistance (COMECON).

In the 1960's and 1970's, the Common Market and the EFTA expanded. Iceland joined the EFTA, and Finland became an associate member. The Common Market, though, became the major economic association on the Continent, eventually absorbing members of the EFTA. Great Britain, Ireland, and Denmark joined in 1973, and Greece joined in 1981. In 1986, Spain and Portugal joined, after democracy had been established in those countries.

2385

Strengthening the Union

In the 1980's, the Common Market moved toward an even firmer union. The original Treaty of Rome of 1957 had included provisions for seeking a more unified federation. Over the years, moves toward unity—such as the introduction of a European currency and the creation of the European parliament—grew out of the original treaty. The EEC studied various proposals for closer cooperation, but national differences always impeded movement toward genuine union. By the early 1980's, however, Common Market leaders recognized that a true single market would be needed to ensure European growth.

In 1982, a new round of discussions on union began. Altiero Spinelli, an Italian member of the European parliament, drafted a plan to turn the Common Market into a genuine political union. While Common Market leaders received the draft coolly, they did agree, through their political arm, the Council of Europe, to consider union proposals. In 1986, the members of the EEC agreed to the Single European Act (commonly called Act 1992), which stated that, by the end of 1992, all members of the EEC would be part of a single economic market. The market would be "an area without frontiers in which the free movement of goods, persons, services and capital is ensured." A committee chaired by Jacques Delors reported on the financial issues necessary for the change.

Many objections, especially concerning the loss of national sovereignty and particular labor, economic, agricultural, and environmental issues in the various nations, were voiced by the members of the EEC. In general, however, the members supported the proposals. Act 1992 would give member nations great economic advantages, and nonmember nations such as Austria, afraid of losing commerce to the EEC, began to apply for membership. Others, including the United States, sought special trading relationships.

Gradually, Act 1992 was approved. In December, 1991, the heads of government and foreign ministers of the twelve nations signed the final accords at Maastricht, Belgium. The treaty, now known as the Maastricht Treaty, provided that each of the twelve countries would ratify the agreements the following year.

Originally, only nine of the twelve member nations had agreed to the proposals, but Italy and Greece later joined; Denmark, however, remained outside. From the time of its entry into the Common Market, Denmark, being a small nation, had expressed reservations about a possible loss of national sovereignty, and a vocal minority in the Danish parliament opposed European cooperation. In a referendum in June, 1992, Prime Minister Poul Schlüter linked Danish approval of the single market to the kingdom's continued membership in the EEC. Nevertheless, the Danes rejected ratification.

The treaty received a boost when the Irish public, in a referendum held on June 18, became the first to ratify participation in the new, closer union. Belgium's lower house of parliament approved the treaty in July and passed it on to the Senate; Luxembourg's parliament approved on July 2; Greece's parliament, on August 1; French voters, on September 20. Ratification was expected in Spain and Portugal, but doubts remained about Britain, Germany and the Netherlands. Denmark's rejection of the treaty, it was thought, could be overcome, but not before January, 1993, when the treaty was originally to have gone into effect.

Consequences

Until 1991, Western European cohesiveness was motivated by Cold War security issues. With the dissolution of the Soviet Union, however, this motivation began to dissolve as well. At the same time, the emergence of new nations and the ethnic battles in Eastern Europe divided many of the EEC members. Yugoslavia, for example, entered a civil war that resulted in new, smaller countries, and the members of the EEC disagreed over which of these nations to recognize. This political chaos placed the treaty in jeopardy.

Given the high economic, military, and political stakes involved, it was no surprise that the original vision of the treaty—movement toward ever closer European union—would take a long time and seemingly endless negotiation. Nevertheless, France's narrow approval of the treaty in September of 1992 held out the hope that some form of union would eventually result from nearly half a century of efforts in that direction.

Frederick B. Chary

Scientists Find Strong Evidence for "Big Bang"

Scientists found a pattern of radiation in the universe that provided strong evidence for the big bang theory, giving possible explanation for the origins of stars and galaxies.

What: Astronomy
When: 1992
Where: Berkeley, California
Who:

GEORGE F. SMOOT III (1945-   ), an American physicist who analyzed data from the Cosmic Background Explorer (COBE) satellite

JOHN C. MATHER (1946-   ), an American astronomer at Goddard Flight Center and project scientist for COBE

JEREMIAH P. OSTRIKER (1937-   ), an American physicist at Princeton University who studied background radiation

The "Big Bang"

Astronomers believe that the universe originated about fifteen billion years ago in a tremendous explosion known as the "big bang." Such an explosion would give off a great amount of heat radiation, in the same way that a fire heats up the area surrounding it. The radiation from the big bang filled the early universe, and since the radiation cannot get out of the universe, it is still here—a relic of an event that happened when the universe was very young. The radiation is known as the universal background radiation.

The universal background radiation from the big bang was first discovered in 1964 at Bell Laboratories in New Jersey by Robert Wilson and Arno Penzias. Since that time, astronomers have been studying the radiation very carefully, trying to understand the nature of the early universe.

The detailed pattern of the universal background radiation revealed that the early universe was perfectly smooth and had no irregularities of any kind. A serious question arose: How could the complex structure of stars, galaxies, clusters, and other elements of the universe have developed from an early stage of the universe that was uniform and had no structure?

Cosmic Background Explorer to the Rescue

The Cosmic Background Explorer (COBE) is a spacecraft that was launched by the National Aeronautics and Space Administration (NASA) in 1989 with the specific purpose of studying, for a five-year period, the universal background radiation from the big bang. Initial data became available early in 1992. The satellite's instruments were shut down at the end of 1993, after the project completed its scheduled mission.

The COBE studies radiation in the universe by measuring the intensity, or brightness, of that radiation. It measures the weak microwave radiation between the stars rather than the strong radiation from the stars themselves. Microwaves are an invisible form of radiation that is used, for example, in a microwave oven. Although microwaves are invisible, they can easily heat up an object. In some ways, the whole universe is like a gigantic microwave oven that is filled with weak microwave radiation left over from the big bang. It is this radiation that the COBE measures.

The radiation from the early universe comes from a time when there were no stars or galaxies anywhere in the universe, only atoms and molecules that had not yet been assembled by gravity into stars. (All objects in the universe, from tiny atoms to people to planets to huge stars, are attracted to each other by the force of gravity.) The laws of physics state that there should be less radiation coming from regions where matter was concentrated and more radiation coming from regions where matter was scarce. Astronomers are interested in finding places in the early universe where matter was more concentrated; such places could be the birthplaces of stars. The stronger gravity there would collect matter until there was enough to form a star.

When the background radiation was first studied, it seemed to be perfectly smooth and uniform. There was exactly the same amount coming from everywhere. This was a mystery. Why did the radiation not have bright and dim spots? Since the present universe is full of stars, there should have been many dim spots where those stars would have begun to form. Where did the stars come from? How did the present universe, with its complex structure, evolve from an earlier time when there was no structure?

The high-precision instruments on the COBE allowed it to measure the background radiation more accurately than had ever been done before. The measurements revealed that the radiation was not perfectly uniform. As the data collected by the COBE early in 1992 were analyzed, astronomers were thrilled to discover that they had at last found bright and dim spots in the background radiation. The COBE took a picture of the universe as it existed fifteen billion years ago. This picture revealed many small spots that were exactly what the scientists working on the big bang theory wanted to see. Cosmologist Stephen Hawking called this development "perhaps the greatest discovery of the century."

Scientists may never be able to use the COBE's data to find the exact locations where the various structures of the present universe originated, but the fact that they have demonstrated that the early universe had spots where such structures could have been born is an extremely important result for astronomy.

Consequences

Although almost all astronomers believe that the universe originated in some sort of a big bang, there have been various theories to explain the way in which the universe developed after the big bang. A universe that begins with a big bang does not necessarily have to produce stars and galaxies, any more than an exploding firecracker has to produce round pieces of paper.

The universe is full of interesting structures. There are stars with planets around them, the stars form galaxies, the galaxies form clusters, and the clusters form superclusters. The galaxies are also organized into vast sheets and walls, with huge empty spaces between them. These large structures are so vast that it takes light, which travels at 186,000 miles per second, hundreds of millions of years to travel from one side of a structure to the other.

As astronomers discovered such huge structures, they tried very hard to figure out how they could have originated. The most likely explanation was that the early universe had small concentrations of matter that gravity was able to turn into much larger objects as the universe expanded. If this explanation were true, however, the early universe should have had spots where gravity could go to work. The COBE has found these spots.

Karl W. Giberson

Astronomers Detect Planets Orbiting Another Star

> *Two American astronomers discovered the first planets to be found outside the solar system.*

What: Astronomy
When: 1992
Where: Arecibo, Puerto Rico
Who:
ALEXANDER WOLSZCZAN (1946-), an American astronomer
DALE A. FRAIL (1961-), an American astronomer

The Search for Distant Planets

For centuries, scientists and philosophers have attempted to determine whether life exists on other planets. Humans have explored as many of the other planets in the solar system as technology has allowed, and they continue to do so by any means available. Until the 1990's, however, the existence of planets outside the solar system seemed too remote to contemplate.

In 1991, however, the possibility of finding planets around other stars seemed real. Three British astronomers, Matthew Bailes, Andrew Lyne, and Setnam Shemar, announced that they had detected a planet orbiting a neutron star, the pulsar PSR1829-10. "PSR" stands for "pulsating source of radio," a reference to the fact that a pulsar is a spinning, very dense star that emits bursts of radio waves at extremely regular intervals, similar to the beam of light from a lighthouse; the numbers refer to the pulsar's coordinates in the sky. The team of astronomers found that there were periodic variations in the arrival times of the bursts of radio waves on Earth over a six-month span. Since pulsars are very regular in the pulses of energy they emit, the variation that the British team found seemed to be a significant indication of another force affecting the pulsar's beam. They concluded that this force must be the gravitational pull of a planet orbiting the pulsar, causing it to be pulled slightly nearer to or farther from Earth.

The scientific community was excited by—and skeptical of—this discovery. The reason for the excitement was clear. Proof of the existence of planets outside the solar system opened up a whole new branch of astrophysics and raised a number of new questions. The skepticism was founded in the existing theories for the origin of pulsars. Pulsars are believed to be the remains of medium-weight stars that, at the end of their lives, undergo a huge explosion called a supernova. As the elements that make up the star undergo a series of reactions that produce a large amount of energy, the core of the star gains mass. When the mass of the core exceeds a certain limit, it collapses inward as a result of its own gravity. The particles that make up the elements are forced together, producing a ball of densely packed neutrons known as a neutron star. As a neutron star spins, it produces pulses of radiation that sweep through the sky. These pulses can be measured when they hit Earth.

Yet if pulsars are left behind after a supernova, how can they have planets orbiting around them? The explosion that forms the pulsar should either destroy any existing planets or cause such a rapid change in gravity that a planet previously in orbit around a star would be thrown off into space. The British discovery raised more questions than it answered.

Retraction and Redemption

In January of 1992, the British team retracted their findings, announcing that they had made a mistake in their calculations. They concluded that they had not taken into consideration certain factors that had skewed their data. In the same month, however, two American astronomers declared that they had discovered not one but two planets orbiting a pulsar by using the same techniques as the British team.

Alexander Wolszczan and Dale Frail found slight irregularities in the arrival times of the pulses from a pulsar known as PSR1257+12. Their findings indicated the gravitational pulls of two planets orbiting the pulsar at approximately the same distance that Mercury is from the Sun. The planets, which have masses of at least 2.8 and 3.4 times the mass of Earth, orbit the pulsar approximately once every ninety-eight and sixty-seven days, respectively. There were two major differences between the American discovery and that announced by the British team. First, the existence of two planets made the patterns of variation in the rate at which pulses arrived at Earth much more complex than the variations shown by the British team. This implied that a simple miscalculation, or the disregarding of Earth's own orbital irregularities, was unlikely to be the cause of such a pattern, thus supporting the argument that the variations must be caused by other forces acting on the pulsar. Second, PSR1257+12 is a millisecond pulsar, a very fast-spinning pulsar that rotates every 6.2 milliseconds. This high speed of rotation is rare in older pulsars such as the one the British team observed; pulsars that form from supernovae spin fast at first but slow down over the years. Millisecond pulsars, however, have their rotation period increased dramatically by streams of gas that fall on them from a neighboring star. This gas, scientists theorize, could also provide material that then could coalesce to form planets.

Consequences

Because of these two major differences, the American discovery did not present the conceptual difficulties to the scientific community that the British one did. In the months following the original research publication, astronomers around the world wrote articles and letters proposing theories that would help to explain the previously undiscovered phenomenon of planets orbiting pulsars. They also recommended further study that would validate the data already found. In particular, it was suggested that the planets would exert gravitational pulls on each other, as well as on the pulsar, that would eventually be measurable, changing the paths of the orbits. If these measurements were made and the findings were consistent with the hypothesis, such measurements would constitute proof that the original discovery was correct. In 1994, Wolszczan and Frail provided this proof, which was virtually identical with that predicted by the theories.

The confirmation of the existence of planets outside the solar system has far-reaching implications for the study of the universe. The discovery of a second pulsar with planets was reported in 1994 by two Russian astronomers, and more discoveries were expected. The Russian discovery resurrects some of the questions of the British assertion; the pulsar in question is relatively young and has a much longer period, so any planets orbiting it must have existed before the supernova that formed the pulsar occurred. Regardless, the data gained from the study of these planetary systems should provide scientists with new information regarding the life cycles of stars and the formation of planets. These findings seem to indicate that planets may form more easily and in a greater variety of ways than previously supposed, and so may be found in what were thought to be unlikely places. If this is the case, the apparent possibilities for extraterrestrial life will have expanded exponentially.

Margaret Hawthorne

Yugoslav Federation Dissolves

Yugoslavia disintegrated into component countries that were soon recognized by the rest of the world.

What: Political independence; International relations
When: January 15, 1992
Where: Yugoslavia
Who:

TITO (JOSIP BROZ, 1892-1980), prime minister of Yugoslavia from 1945 until 1953, then president from 1953 until his death

SLOBODAN MILOŠEVIĆ (1941-   ), president of Serbia from 1990 to 1997

FRANJO TUDJMAN (1922-1999), president of Croatia from 1990

A Violent End to Federation

By the beginning of 1992, most European countries had concluded that the Socialist Federal Republic of Yugoslavia had ceased to exist. The first step in the federation's demise came in October, 1991, when Slovenia, Croatia, and Bosnia-Herzegovina, three of its six republics, formally declared their independence.

Germany recognized Slovenia and Croatia on December 23. The most significant international acknowledgment of the situation came on January 15, 1992. On that date, the other eleven members of the European Community (EC) recognized the new countries. The United States withheld recognition until April 7.

Another Yugoslav republic, Macedonia, repeatedly had claimed independence, although a dispute with Greece had slowed international acceptance. In January of 1992, Bulgaria recognized the new country. Turkey, Slovenia, and Croatia soon followed.

The two remaining republics, Serbia and its ally Montenegro, proclaimed the creation of another "new" country, the Federal Republic of Yugoslavia, on April 27. This proclamation marked tacit recognition by Serbian president Slobodan Milošević that the dream of an intact Yugoslavia dominated by Serbia had failed.

In the meantime, both Milošević and Croatian president Franjo Tudjman were asserting their countries' territorial claims on the battlefield. Serbia had already invaded those parts of Croatia inhabited by Serbian minorities. Similar minorities were resisting efforts at independence by Bosnia-Herzegovina; they declared their own "Serbian Republic of Bosnia-Herzegovina." Open violence broke out in early April in Sarajevo, the capital of Bosnia-Herzegovina, with firings on peace demonstrators. Further conflict followed in other parts of the republic, and Sarajevo came under direct and intensive attack by Serb forces on April 22.

The Croatian minority in Bosnia-Herzegovina soon declared the creation of yet another entity, "Herceg-Bosna," and Croatian forces subsequently took control of much of western and central Bosnia-Herzegovina. That republic was quickly becoming a bloody microcosm of the entire Yugoslav conflict.

A Young but Divided Country

Yugoslavia had existed as a country only since the end of World War I and had suffered from internal dissent for most of its history. In 1918, Serbia, which had freed itself from Turkish control in the nineteenth century, merged with the regions of Croatia, Slovenia, and Bosnia-Herzegovina, all of which had been under the rule of Austria-Hungary until that empire collapsed at the end of the war. Another part of the new country, Montenegro, had been independent for several centuries.

The country was known at first as the Kingdom of Serbs, Croats, and Slovenes. It was not renamed Yugoslavia until 1929. In addition to the peoples identified in the country's original name, Bosnian Muslims, Albanians, Macedonians, and Montenegrans inhabited the country. There were four major languages (Serbo-Croatian and the closely related Croato-Serbian as well as Slovenian and Macedonian), three religions (Muslim, Eastern Orthodox, and Catholic), and two alphabets (Latin and Cyrillic), with the linguistic and religious lines frequently cutting across internal boundaries. With all these internal divisions, it was not surprising that the kingdom proved hard to govern. Violence broke out between the Croats and the Serbs as early as 1928, leading the country's Serbian king, Alexander, to assume dictatorial power.

Alexander was assassinated in 1934 by Croat and Macedonian extremists. He was succeeded by his cousin Paul, who acted as regent and remained in control with the help of the army. By 1941, however, Prince Paul's German sympathies

led to his ouster and to the installation of Alexander's son Peter as king. Within weeks, Germany and Italy invaded Yugoslavia, and the royal family was driven into exile. A German puppet government was set up in Croatia, and massacres of Serbs and Jews followed. The Serbs themselves were responsible for the murder of Croats and Muslims. Even resistance to the invaders was divided—between royalists based in Serbia, and communist-dominated forces led by Tito and supported by Bosnia, Croatia, Montenegro, and Slovenia. The communists established their own government and eventually outmaneuvered the royalists.

The communist Federal Republic of Yugoslavia was proclaimed at the end of World War II, and a new constitution established six constituent republics: Bosnia-Herzegovina, Croatia, Mace-

PAST AND PRESENT YUGOSLAVIA

donia, Montenegro, Serbia, and Slovenia. Tito ruled the country until his death in 1980.

The internal differences, which Tito had managed to keep under control, soon reasserted themselves. In 1987, Serbian leader Slobodan Milošević began consolidating his support with a program of open hostility toward the many Albanians living in the south, indirectly upsetting ethnic and nationalist balances throughout all of Yugoslavia. In an effort to placate Serbia, Croatian president Franjo Tudjman suggested in 1991 dividing Bosnia-Herzegovina (peopled by a volatile mixture of Croats, Muslims, and Serbs) between Serbia and Croatia.

Slovenia and Croatia had declared their right to secede months before, but when they actually did so in June of 1991, Milošević invaded the two republics in the name of the federation. At a meeting held in July, Serbia agreed to withdraw its troops and Slovenia and Croatia agreed to suspend their declaration for three months. Despite the agreement, fighting intensified in Croatia. At the same time, clashes between Muslims and Serbs in Bosnia-Herzegovina grew common.

Consequences

By the end of 1992, reports of atrocities, prisoner-of-war camps, and "ethnic cleansing" (a term for the forced displacement of ethnic groups) were filtering out of the former federation, particularly from Bosnia-Herzegovina and southern Serbia. Thousands of soldiers and civilians had been killed, and millions had been driven from their homes. In May of 1992, the EC and the United Nations imposed economic sanctions against the new Federal Republic of Yugoslavia in a failed attempt to end the fighting. Efforts by the EC and North Atlantic Treaty Organization were equally unsuccessful. Despite mounting intervention from the United States and Western European nations, conflicts in the former Yugoslav republics continued unabated into the twenty-first century.

Grove Koger

El Salvador's Civil War Ends

> *The government of El Salvador and leftist rebels signed a pact to end the country's devastating twelve-year-old civil war.*

What: Civil war
When: January 16, 1992
Where: Mexico City
Who:
Alfredo Cristiani (1947-   ), president of El Salvador from 1989 to 1994
Javier Pérez de Cuéllar (1920-   ), secretary general of the United Nations from 1982 to 1991
Roberto D'Aubuisson (1943-1992), founder of the right-wing ARENA party

The Peace Accord Signed

On January 16, 1992, representatives of the Salvadoran government and the Farabundo Martí National Liberation Front (FMLN) met in Mexico City and signed an accord to end the nation's twelve-year civil war. Gathered in Chapultepec Castle, government negotiators and guerrilla leaders ratified the documents. Then Salvadoran president Alfredo Cristiani, who had earlier vowed that he would not affix his signature to the accord, relented and gave his personal approval. Emotions ran high among the peacemakers. Some participants were teary-eyed to think that their nation had finally turned away from more than a decade of unspeakable violence.

The agreement ratified in Mexico City initiated a process whereby the combatants laid down their weapons and resolved their disputes in the political arena rather than on the battlefield. According to its provisions, the Salvadoran National Assembly was to create a National Commission for the Consolidation of Peace (COPAZ), with representatives from the various factions, to implement the terms of the treaty. By February 1 the government and the FMLN were to provide observers from the United Nations with data on the size of their forces and the quantity of their weapons. A general cease-fire would then take effect on that date.

In simultaneous steps during the following months, the Salvadoran and FMLN armed forces were to congregate and demobilize. By March 2 the government was to release all political prisoners and appoint a national procurator for the defense of human rights. The pact envisioned a 50 percent reduction in the size of the Salvadoran military and the elimination by year's end of the counterinsurgency units trained by the U.S. security forces, including the National Guard and Treasury Police, who were responsible for much of the death-squad violence. They were to be disbanded and replaced by a new National Civil Police. Lands which the FMLN had redistributed in rebel-held territories were to remain in the hands of the peasants.

To help the transition from war to peace unfold, the United Nations agreed to provide more than a thousand observers, particularly to oversee the military and police forces and thereby protect the guerrillas who laid down their weapons. On November 30 the demobilization was to be completed and peace fully restored.

The War No One Could Win

In 1989 the tiny Central American nation of El Salvador reached the climactic stage of its civil war, which had been raging since 1980. On the domestic front, the presidential election of March 20 brought to power the candidate representing the right-wing Nationalist Republican Alliance (ARENA), Alfredo Cristiani, a wealthy coffee planter with ties to the traditional oligarchy. Divisions among the Christian Democrats and the Left's refusal to participate in the election enabled Cristiani to win an outright majority.

AP/Wide World Photos

Alfredo Christiani (center), shown at the 1996 annual convention of the National Republican Alliance.

Critics of ARENA immediately predicted that Cristiani's victory would lead to an upsurge in right-wing violence. The founder of ARENA, Roberto D'Aubuisson, was widely linked to the activities of death squads that had murdered opponents of the oligarchy. On taking office in June, however, Cristiani surprised ARENA critics by announcing that he favored peace talks with the FMLN. Later that year, in November, the FMLN launched its largest offensive of the war but failed to generate much popular support. During the offensive, the Salvadoran military murdered six Jesuit priests at the Centroamerican University, leading to world outrage.

By 1989, the Cold War had also started to wind down, changing international conditions which had prolonged the conflict. Mikhail Gorbachev's reforms in the Soviet Union and the collapse of Soviet power in Eastern Europe ended military and economic aid to Cuba and Nicaragua, the FMLN's chief backers. In the United States, Ronald Reagan's presidency ended, and the new administration of George Bush showed less concern about the communist threat in El Salvador.

Thus, international conditions no longer fueled the war, and neither the Salvadoran Left nor the Right achieved a total victory over the other. In fact, both sides were reaching the conclusion that the Salvadoran people considered continuation of the war unacceptable. This presented an opening for third parties to mediate between the warring factions. Presidents of other Central American nations persuaded U.N. secretary general Javier Pérez de Cuéllar to press for the resumption of peace negotiations between the government and the FMLN. On April 4, 1990, representatives of the two sides agreed to resume talks. Mexico, Spain, Colombia, and Venezuela also helped press for a settlement. In May, 1990, the Cristiani government and the guerrillas agreed on guiding principles for their negotiations, including not only military and political issues but also those related to human rights and the country's social and economic predicament. Evidence of the changing atmosphere was the conviction in September of two military officers for their role in the assassination of the Jesuit priests. The U.S. State Department called

2395

the verdict "an historic achievement for El Salvador's judicial system and the cause of human rights."

Fighting continued, but so did the negotiations. The two sides hammered out a statement on human rights, which the legislature included in the constitution. Under Pérez de Cuéllar's prodding, they also agreed to create a new police force to replace those security forces that had been so heavily involved in death-squad activity. Meeting at the United Nations in December, 1991, representatives of the government and FMLN finally accepted the general provisions of a peace settlement to be ratified in Mexico City on January 16, 1992. David Escobar Galindo, a government representative at the negotiations, judged that Pérez de Cuéllar had "unraveled the Gordian knot" and predicted that "there are now all the conditions for an end not only to the armed confrontation but to the war itself." Pérez de Cuéllar also established a Truth Commission to investigate the worst outrages committed during the war, although it had no power to prosecute the perpetrators.

Consequences

On signing the January 16 accord, President Cristiani stated: "We understand that what be-gins to happen now in El Salvador is not the reestablishment of a peace that existed before, but the beginning of a real peace founded on a social consensus." The challenge of building a political and economic system to deal peacefully with the nation's ills was perhaps even greater than that of bringing the fighting to an end. Joaquín Villalobos, the FMLN strategist during the war, declared on February 1, when the cease-fire officially took effect, "Today we stop being enemies and become political adversaries." He also, however, warned that the former guerrillas would not tolerate any treaty violations by the government.

A fragile peace had come to the exhausted nation, although seventy-five thousand people had died and Cristiani believed that his country needed $1.8 billion in aid to overcome the economic devastation. The peace accords were a great achievement for the war-weary people. Even D'Aubuisson, who in recognition of the war's growing futility had begun to moderate his views in the late 1980's, found cause for optimism despite the throat cancer that took his life on February 20. Still, it remained to be seen whether the nation could construct the "real peace" that Cristiani had so eloquently described.

Kendall W. Brown

Kenyans Demonstrate Against Moi's Government

In the first major test of an amendment to Kenya's constitution legalizing a multiparty political system, 100,000 Kenyans staged a protest against President Daniel arap Moi's government.

What: Civil rights and liberties; National politics
When: January 18, 1992
Where: Nairobi, Kenya
Who:
DANIEL ARAP MOI (1924-   ), president of Kenya from 1978
OGINGA ODINGA (1912-1994), candidate for the presidency of Kenya in 1992

The Courage to Protest

On January 18, 1992, 100,000 Kenyan citizens staged a demonstration against their president, Daniel arap Moi, and their government. For years, Kenyans had been dissatisfied with government officials, who had stolen millions of dollars from the very poor country. It was illegal in Kenya to criticize the president or challenge the government, and many people had been thrown in jail and even tortured for speaking out. This important demonstration was the first time so many Kenyans had dared to say publicly that they wanted a change.

The demonstration was organized by members of a new political party in Kenya, FORD (Forum for the Restoration of Democracy). FORD had tried to stage a rally several weeks earlier, but President Moi had forbidden it, threatening to arrest anyone who showed up. Shortly before the rally was to take place, the organizers called it off but made it clear that they would try again. They soon announced that the next attempt would come on January 18.

Over the next weeks, people talked about problems of the government and the need for change. As more people entered the discussion, they became more confident, even though the government issued warnings that anyone participating in a demonstration would be jailed. It soon appeared that so many Kenyans were planning to attend the demonstration that the police would not be able to stop it.

On the day of the demonstration, approximately 100,000 Kenyans turned out in Nairobi, the capital city. They listened to speeches given by people who wanted to be part of a new government in Kenya. One of these was Oginga Odinga, a candidate for president. Many members of the audience waved branches, the sign of FORD, or made hand gestures to show their support for DP, the Democratic Party. Demonstrators did not all agree on how the new government should be formed, but they all agreed that it was time to act.

The demonstration lasted all day, and Kenyans went home to their villages with new ideas for social change. The next day, the government-run newspapers claimed that only five thousand Kenyans had attended the demonstration and that those few people were traitors and opportunists who did not represent the will of the people. Most Kenyans ignored these newspaper accounts. They knew that something important had occurred.

Independence Without Democracy

Kenya was ruled by the British beginning in the nineteenth century. The best farming land in the country had been taken from native Kenyans and given to white settlers from Great Britain. Even after the Africans fought with the British in two world wars, they were not given a fair share of the wealth gained from their own country.

On December 12, 1963, Kenya became an independent nation, with Jomo Kenyatta as its president. To help unify the country, the ruling KANU party (Kenya African National Union) ab-

Daniel arap Moi.

sorbed all opposing parties. Health care and educational programs were created, and many coffee and tea plantations were transferred to African control. The average standard of living improved substantially, but the distribution of wealth was far from equal. Most people believed that members of Kenyatta's tribe, the Kikuyu, were receiving better land, more money, and most of the powerful military and government positions. Kenyatta himself gathered a huge personal fortune.

Oginga Odinga, vice president of KANU, broke away from the party in 1966 to form the KPU, the Kenya People's Union. Shortly thereafter, the KPU was banned and Odinga was detained without trial. For the next twenty-five years, KANU would be the only political party in Kenya. This meant, for example, that in a presidential election there would be only one candidate.

When Kenyatta died in 1978, he was replaced by another member of KANU, Daniel arap Moi. For a time Moi took action against corrupt gov-

ernment workers, and he released all political prisoners, including Odinga. Odinga was not permitted to run for president in 1979, and in 1982 the constitution was changed to officially ban other political parties. Over the next ten years, Moi's government showed itself to be greedy and corrupt. International groups including Amnesty International accused Moi and his highest officials of human rights abuses and of redirecting international monetary aid to their own bank accounts.

Within Kenya, protest slowly grew, but most people were afraid. People who spoke out against government corruption were arrested and tortured. When journalists printed stories demonstrating corruption, they also were arrested, and issues of their publications were seized and destroyed.

As the government clamped down harder, international donor countries pushed Moi to accept democracy or lose aid. Kenyans increasingly demanded the right to vote in truly free elections. Finally, on December 3, 1991, Moi announced that the constitution of Kenya would be changed to allow multiparty elections. Moi retained the right to say when elections would be held. With the freedom to form political parties and present candidates for government positions, however, Kenyans would have more say in how their country was run.

Consequences

For several months after the demonstration, the country experienced a massive wave of violence and intimidation. Large groups of Kenyans attacked and killed members of other tribes. Moi claimed that the violence was a natural result of the multiparty state and that Kenyans were not ready for democracy. Opponents inside and outside the country demonstrated, however, that the violence was led by the government itself.

Meanwhile, the leaders of the new political parties were unable to settle their own differences and create a strong opposition. Their fighting among themselves caused each party to split into smaller, weaker parties. When Moi finally allowed the country's first free multiparty election in December, 1992, he was reelected to the presidency.

As Moi's new term approached its end in 1997, continued political corruption finally led the International Monetary Fund (IMF) to suspend payments on loans that were helping pay for Kenya's development, claiming that high-level Kenya officials were stealing the money. Political parties again began nominating candidates to oppose Moi, and again violence spread throughout Kenya. Nevertheless, Moi was reelected for another five-year term in December, 1997, the same month he established the Anti-Corruption Authority. One of Moi's first acts in the new term was to fire thirty thousand medical workers who had opposed him in the election.

Cynthia A. Bily

Muslim Refugees Flee Persecution in Myanmar

Reacting to widespread human rights abuses, Muslim Rohingya refugees fled in large numbers from Myanmar into Bangladesh.

What: Human rights; International relations; Religion

When: February-April, 1992

Where: Myanmar and Bangladesh

Who:

BEGUM KHALEDA ZIA (1945-), prime minister of Bangladesh from 1991 to 1996

AUNG SAN SUU KYI (1945-), jailed political dissident and Nobel Peace Prize winner

U NE WIN (1911-), leader of Burma, 1962-1988

SAW MAUNG (1928-1997), general and head of the military government until April, 1992

THAN SHWE, general who replaced Saw Maung as head of the military government

Persecution and Flight

Between February and April, 1992, more than 220,000 Muslim refugees, known as Rohingyas, fled from Arakan State in Myanmar to neighboring Bangladesh. They joined some 30,000 fellow Rohingyas who fled from Myanmar in 1991. Coupled with some 20,000 non-Rohingya, ethnic Burmese who fled into Thailand to escape from military repression, these refugee flows marked Myanmar as Southeast Asia's largest producer of refugees in 1992.

Stimulating this outflow of refugees were the human rights abuses of the State Law and Order Restoration Council (SLORC), composed of the military leaders who seized power in 1988. Contributing to refugee flows were repressive policies of the SLORC against opposition groups, the democracy movement headed by Aung San Suu Kyi (1991 Nobel Peace Prize winner), and minority ethnic groups. The SLORC opened its offensive against the Muslim Rohingyas in early 1991.

Arakan State, where the Rohingyas live, had long been a seat of secessionist movements. Burmese governments had engaged in more than a dozen military campaigns since 1948 to root out Arakanese separatist groups. The SLORC's campaign in 1991-1992 was particularly brutal and repressive. Rohingya refugees reported that the government engaged in widespread use of forced labor, torture, beatings, and rapes of the Muslim population. Destruction of mosques and Muslim cemeteries was also reported, together with large-scale arrests of Muslims. Killings of Muslims forced into the service of the military as porters were reported, as were killings of those attempting to flee. Some refugees stated that they were forcibly uprooted by the military so that their lands could be settled by Buddhists. The SLORC steadfastly denied such charges in March, 1992.

By April of 1992, the outflow of Rohingyas had slowed considerably. By the end of the same month, General Than Shwe replaced ailing General Saw Maung as head of the SLORC. On April 28, the SLORC signed an agreement with the government of Begum Khaleda Zia in Bangladesh to allow the repatriation of all Rohingya refugees who could prove former residence in Myanmar. The SLORC also promised to suspend military operations against Arakanese separatists and to seek a negotiated settlement to the long-lived insurgency. By the end of 1992, only six thousand Muslim Rohingyas had agreed to repatriate, although nearly fifty thousand did so in the following year, after the situation in Arakan State improved. Concern about the safety of the returning refugees continued, given the inability of the United Nations High Commissioner for Refugees to monitor the repatriation and the possibility that Bangladesh had fallen short of guaranteeing that all the refugees were repatriating voluntarily.

Failed Democracy and Authoritarian Control

The flight of Muslim refugees from Myanmar in 1992 was only the latest in a long series of disputes and battles in Arakan State. Upon independence in 1948, the democratic government of Burma (designated Myanmar in 1989 by the military regime) promised autonomy for ethnic minorities in various regions. Little progress was made, and insurgent groups arose to oppose the central government. The government in response launched offensives against insurgent groups in Arakan State. Insurgents, in turn, have sought to secede from Myanmar. No government in Myanmar had been willing to permit this, but the SLORC was particularly brutal in seeking to eradicate the insurgency in 1991-1992, as it had been in stamping out the prodemocracy movement in the years since 1988.

The military had been in power in the country since 1962, when General U Ne Win overthrew the democratic government, established a one-party state, and embarked on a plan of socialist isolation. Ne Win resigned as the government's leader in July, 1988, in the midst of widespread popular opposition. The governments that followed failed to establish order, and within three months direct military rule was proclaimed by the SLORC.

Under pressure from ongoing demonstrations for democratic reforms, the SLORC permitted elections in 1990. These led to an overwhelming victory for Aung San Suu Kyi's National League for Democracy, which won 80 percent of the legislative seats against the SLORC's party. Aung San Suu Kyi had been under house arrest since July, 1989, and was not permitted to take office. The SLORC nullified the results of the election. Aung San Suu Kyi won the Nobel Peace Prize in 1991 based on her efforts on behalf of human rights and democracy in Myanmar, and she was released from house arrest in July, 1995.

The SLORC's heavy hand has been felt both by ethnic minorities and by those who have had the courage to oppose the military regime and seek democratic reforms. Repressive policies met with increasing international opposition and pressure. The U.S., Canadian, and Australian governments severely criticized the SLORC's abusive policies in July, 1992. United Nations bodies and private human rights groups such as Amnesty International also condemned SLORC policies.

Aung San Suu Kyi.

AP/Wide World Photos

Consequences

The year 1992 saw the culmination of SLORC repressive policies. The Muslim refugee flows inflamed international opposition, and the SLORC began to pursue a policy of improving its international image. This began with the April, 1992, agreement to repatriate refugees and negotiate political settlements with insurgent groups. The slackening of repression saw some fifty thousand refugees return to Myanmar in 1993. The SLORC also released some two thousand political prisoners in 1992 and 1993, although numerous opposition leaders, including Aung San Suu Kyi, remained in detention. (She was not released until 1996.) After many years of socialist isolation, the SLORC began to implement economic reforms and gradually opened its doors to foreign aid and capitalist economic development. These trends offered hope that Myanmar would gradually move toward democratic reforms, but events over the next decade failed to fulfill that hope.

Robert F. Gorman

Russia and United States End Cold War

After four decades of competition and confrontation between the two superpowers, President George Bush and President Boris Yeltsin issued a formal statement of principles declaring the end of the Cold War and the beginning of a new relationship.

What: International relations
When: February 1, 1992
Where: Camp David, Maryland
Who:

GEORGE HERBERT WALKER BUSH
(1924-   ), president of the United
States from 1989 to 1993

BORIS YELTSIN (1931-   ), president of
the Russian Republic from 1991 to
1999

MIKHAIL GORBACHEV (1931-   ),
Communist Party general secretary
from 1985 to 1991, and president of
the Soviet Union from 1990 to 1991

RONALD REAGAN (1911-   ), president
of the United States from 1981 to
1989

The Camp David Declaration

U.S. president George Bush and Russian president Boris Yeltsin conferred at the presidential retreat at Camp David, Maryland, on February 1, 1992. Following their businesslike discussions, the two leaders approved and signed a declaration of principles outlining the future relations of the two nations. The leaders agreed that their countries would commit in the future to a cooperative and friendly relationship, with the objective of reducing and eventually eliminating the numerous causes of disagreements that fueled the dangerous Cold War following the end of World War II. Both presidents announced this statement of principles at a Camp David press conference the same day.

The six-point Camp David Declaration included many issues and objectives. The two nations promised no longer to view each other as potential adversaries. As a second objective, both governments promised to eliminate any remnants of past hostility. A third goal emphasized concern for the well-being of both societies. Freedom of economic development and international trade was included as the fourth principle. A fifth objective promised common support for the development and strengthening of democratic values and institutions. The sixth goal supported joint efforts to prevent the spread of nuclear weapons, work for peaceful resolution of regional disputes, promote high environmental standards, cooperate in efforts to halt the trade in illegal drugs, and prevent instances of world terrorism.

Nothing in the joint statement required legislative approval, as the declaration did not constitute a formal and binding international treaty. The document lacked specifics on military budgets, weapons, and personnel, and it did not contain precise details related to economic matters such as trade, investments, or loans. The major significance of the Camp David Declaration was that it committed both governments to work more cooperatively in the future, as the Cold War became less hostile and dangerous.

The Road to Improved Relations

The February, 1992, Russo-American statement of principles logically evolved from improved relations between the superpowers beginning in the mid-1980's. Under the prior leadership of President Ronald Reagan and Communist Party general secretary Mikhail Gorbachev, the two nations between 1985 and 1989 negotiated the resolution or management of many issues dividing them. Most discussions dealt with military matters such as comparative military strength and budgets, conventional and nuclear weapons, and strategic policies shaping the foreign policies of both nations. Although

2403

the dangers related to possible nuclear war still existed, the atmosphere for discussing and occasionally resolving these longtime differences gave evidence of significant change. The suspicion and confrontation of previous decades seemed increasingly irrelevant as the 1980's came to an end.

The 1989-1990 period saw dramatic shifts in the Cold War. The collapse of communist governments in Europe included the tearing down of the infamous wall dividing East and West Berlin. In 1990, East and West Germany united as one democratic country. These rapid changes led to further steps to reduce Cold War suspicions and military systems.

President Bush, upon taking office in January, 1989, continued the Reagan administration's conversations and negotiations with the Soviet government. An important agreement signed in late 1990 reduced the levels of military forces and weapons systems of NATO (North Atlantic Treaty Organization) and the Soviet-led Warsaw

Pact, the two major alliance systems opposing each other in Europe. This represented a major step in reducing the likelihood of war. Another significant step in improving relations was the Strategic Arms Reduction Talks (START I), which resulted in a treaty signed in July, 1991. It provided for substantial reductions in Soviet and American long-range strategic nuclear missiles. Other events creating opportunities for better relations included abolition of the Warsaw Pact of communist states in mid-1991.

Gorbachev's authority as Soviet leader and Communist Party head ended by late 1991 with the disintegration of the Soviet Union as a nation as a result of internal disputes and economic crises. The collapse of the Soviet Union and the Gorbachev regime made Boris Yeltsin, elected president of the Russian Federation in 1991, the most significant Russian leader. Future American negotiations and agreements on arms control (and all other issues) henceforth would be made primarily with Yeltsin and the new Russian

AP/Wide World Photos

Russian president Boris Yeltsin (left) and U.S. president George Bush, during Yeltsin's June visit to the United States.

government. Yeltsin's visit to the United States in early 1992 provided the first opportunity for high-level discussions between Bush and Yeltsin. President Bush, on January 28, announced additional American military reductions, and President Yeltsin made similar proposals at the same time. The moment was right for the two leaders to meet face to face in a common effort to agree on the future relationship between the two superpowers. These conversations took place on February 1 (Yeltsin's birthday) at Camp David.

Consequences

The positive Russian-American relationship continued throughout the year. The American government increased its economic support for Russia. President Yeltsin visited the United States in June, agreeing to a faster elimination of strategic missiles. The U.S. Senate and Russian Parliament ratified START I in the fall. President Bush signed START II with the Russians in Moscow in January, 1993. This new arms agreement further reduced nuclear armaments, to their lowest levels since the 1960's. Although substantial problems remained to be resolved in the future, the Cold War appeared to have ended, and Russia and the United States maintained cordial relations through the ensuing decade.

Taylor Stults

United States Resumes Repatriating Haitian Boat People

The United States Supreme Court lifted a lower court injunction, thereby allowing the government to resume repatriation of Haitian boat people.

What: International relations; Law

When: February 1, 1992

Where: Guantánamo Naval Base, Cuba

Who:

WILLIAM P. BARR (1950-), United States attorney general from 1991 to 1993

GEORGE HERBERT WALKER BUSH (1924-), president of the United States from 1989 to 1993

RAOUL CÉDRAS, leader of Haiti's military government from 1991 to 1994

JEAN-BERTRAND ARISTIDE (1953-), president of Haiti from 1990

Involuntary Repatriation Resumes

On January 31, 1992, the United States Supreme Court by a vote of six to three, with Justices Clarence Thomas, Harry A. Blackmun, and John Paul Stevens dissenting, granted the government's motion to stay an injunction prohibiting the forcible return of Haitians who had fled their country. Attorney General William Barr stated that the government now had clear authority to return the more than nine thousand Haitian refugees held at Guantánamo Naval Base in Cuba. That authority would remain in force until a decision on repatriation could be made by the United States Court of Appeals of the Eleventh Circuit in Atlanta, and that decision appealed to the Supreme Court.

In the weeks preceding the lifting of the injunction, the number of refugees fleeing Haiti had increased dramatically. When the government could no longer accommodate the refugees aboard the Coast Guard cutters patrolling the waters between Haiti and the United States,

the naval base at Guantánamo was prepared for them. The base quickly became overcrowded as more refugees arrived daily. The United States State Department estimated that only one-third of the refugees qualified for political asylum. The remaining refugees were fleeing economic conditions.

Hours after the Supreme Court lifted the injunction, the Coast Guard cutter *Steadfast*, with 150 Haitians on board, sailed from Guantánamo on the ten- to twelve-hour voyage to Port-au-Prince, the Haitian capital. Additional Haitians were sent back in the next two days. Previously, about four hundred Haitians had returned voluntarily through the United Nations High Commissioner for Refugees.

President George Bush had strongly condemned the growing political violence in Haiti. On January 25, 1992, he recalled Ambassador Alvin P. Adams, Jr., following an attack by police officers in civilian clothes on a political meeting called by Communist Party leader René Theodore, who had been nominated prime minister in a compromise concluded several weeks before the meeting. President Bush believed that chances for a political settlement in Haiti were decreasing and was concerned that the increasing number of Haitian refugees would be too large for the United States to handle. Opposition to the administration's repatriation policy was raised by civil rights groups and others who charged that the policy was racist. Their attempt to stop the program by injunction had delayed the return of the boat people but did not stop it.

The Refugee Problem

In February of 1986, the military seized power in Haiti after a thirty-year rule by the Duvalier family. François ruled from 1956 to 1971 and his son Jean-Claude from 1971 to 1986. After 1986, the military continued the extremely repressive and brutal policies of the Duvaliers. In 1990, popular demonstrations and a general strike brought about the downfall of the military, then open elections in December. Jean-Bertrand Aristide, a Roman Catholic priest, was elected president. He advocated far-reaching political reforms and social changes. He was able to remain in office not quite a year before being overthrown by the military.

Amnesty International reported that under military rule the Haitian people were living in a climate of fear and repression and that hundreds of people were extrajudicially executed or were detained without warrant and tortured. Many were brutally beaten in the streets. Freedom of the press was severely curtailed, and property was destroyed by members of the military and the police.

The nations of the Organization of American States, including the United States, imposed an embargo to pressure Haiti for the return of the duly elected president to power. The embargo added to the problems of the Haitian people, who were the poorest in the Western Hemisphere. Shortages of fuel, food, medicines, and potable water developed, and unemployment increased because factories had to close when they ran out of fuel. More than 140,000 Haitians fled after the coup. During the month of January, 1992, 6,653 were picked up at sea. The Bush administration felt compelled to return those Haitians it regarded as economic, rather than political, refugees.

The United States government had begun returning Haitians in November, 1991, but at the end of the year had been restrained by the injunction obtained by civil rights advocates in Miami, Florida. The Bush administration feared that at least an additional twenty thousand Haitians would flee the island. The lifting of the injunction in January, 1992, gave the administration a victory. The return of refugees resumed with the voyage of the *Steadfast*.

Consequences

The Supreme Court again refused, on February 11, 1992, to end the return of the refugees, and the repatriation program continued. When opponents failed to stop the program by court order, they turned to Congress. Representative Charles E. Summer, a member of the House Subcommittee on International Law, Immigration, and Refugees, introduced legislation granting Haitians protected status that would permit them to stay in the United States until Haiti's military government was overthrown. The proposed legislation also granted a quota of two thousand refugees from Haiti. The legislation was not passed.

The repatriation problem was solved in September, 1994, when President Bill Clinton sent former president Jimmy Carter, General Colin L. Powell, and Senator Sam Nunn to negotiate an agreement with General Raoul Cédras, the ruler of Haiti, for the return of Aristide to serve the remainder of his term. United States forces temporarily occupied Haiti to ensure an orderly transition of political power and the development of democratic government. The United States obtained a foreign policy victory in 1994 with the return of Aristide to the presidency in Haiti. At the same time, the problem of Haitians fleeing to the United States was solved. Civil rights groups in the United States and abroad, however, criticized United States policy on the basis of humanitarian and civil rights concerns. Others charged the United States with racism. Overall, the gamble by President Clinton when he sent troops into Haiti redounded to his credit and gave a boost to his foreign policy.

Robert D. Talbott

International Agencies Pledge Funds for Ethiopia

Hoping that the country would find respite from its protracted agonies from war and famine, the Ethiopian government and international donors formulated political and economic strategies for rehabilitation and reconstruction.

What: Social reform; Economics; International relations

When: February 6, 1992

Where: Ethiopia

Who:

ABDULEMEJID HUSSEIN, Ethiopian minister of foreign economic cooperation

MELES ZENAWI (1955-   ), interim president of Ethiopia from 1991 to 1995

MICHAEL CARTER, chief of country operations for the World Bank's Africa Division

Rehabilitating Ethiopia

On February 6, 1992, the World Bank, the International Monetary Fund (IMF), and other organizations pledged $672 million in loans to Ethiopia. The money, to be disbursed over the course of thirty months, was to be used to build roads, bridges, and schools as well as to satisfy other basic needs. The first $7 million was designated as emergency medical aid.

The Ethiopian People's Revolutionary Democratic Front (EPRDF) had taken control of the country in May, 1991. It was determined to rehabilitate the country after thirty years of civil war, famine, and drought. The government, however, had few resources. According to World Bank figures, Ethiopia was one of the four least developed countries in the world. Its infrastructure was in ruins. The agricultural sector needed rehabilitation, and farmers were faced with a lack of fertilizer, equipment, and capital. Close to a million demobilized soldiers of the former

Marxist government were without jobs.

In order to secure economic assistance from the World Bank, Ethiopia had to agree to specific conditions stipulated by the bank. First, the business principles of the bank were to be upheld. As long as Ethiopia was interested in securing aid from the bank, some form of financial partnership had to be established to protect the bank's investment. The World Bank and IMF commonly stipulate, as a condition of loans, that recipients follow specified budget plans. Ethiopia also had to protect its creditworthiness by agreeing to repay its loan plus interest on schedule. As early as September, 1991, the Addis Ababa Chamber of Commerce was commissioned by the government to draft an economic recovery plan to assist the government in its negotiations with the World Bank and other potential lenders.

Privatization of the economy was an essential aspect of conditions for the loans set by the lending agencies. This meant allowing unrestricted market activity in the production and distribution of goods and services. The government agreed to reduce inflation, control interest rates, reduce government expenditure, and focus its development efforts on agriculture, reconstruction, infrastructure development, and enhancing capital formation by promoting export-oriented industries. Ethiopia's willingness to work with the World Bank to rehabilitate the country's economy produced a favorable response from lenders. The World Bank and the IMF pledged additional loans in the following years.

Civil War and Famine

Ethiopia has survived many traumatic events in its history. In 1974, a military regime known as

2408

the Derg overthrew Emperor Haile Selassie after nearly fifty years of reign. The Derg ruled the country for seventeen years, perpetrating widespread violations of citizens' rights. Between 100,000 and 200,000 civilians were killed during the 1977-1979 Red Terror. The Derg endorsed radical Marxism and structured Ethiopia on Stalinist lines. It nationalized land, industry, commercial banks, and insurance companies as well as introducing programs of agricultural collectivization that contributed to famines in 1984 and 1987. Its villagization program uprooted people from their traditional homes in the north and settled them in the southern part of the country, where many died from exposure to unfamiliar climatic conditions. About a million and a half Ethiopians are estimated to have perished as a result of the famines, dislocation, and war that accompanied the Derg rule.

The Derg inherited a protracted civil war in Eritrea. Eritrea, a former Italian colony, was annexed by Haile Selassie in 1962. The Eritrean parliament was dissolved. Haile Selassie muzzled the Eritrean press and curtailed labor associations. The Eritreans resorted to armed struggle, which caused the overthrow of Haile Selassie in a 1974 revolution. The Derg sought a military solution to the Eritrean crisis and squandered the country's meager resources on armaments. In 1975, other guerrilla movements, which in subsequent years formed the EPRDF, arose and joined the Eritreans in overthrowing the Derg. Eritrea became an independent state, and the EPRDF took control of Ethiopia's government. The new leaders soon discovered that the Derg had spent more than $10 billion fighting the war.

The government took immediate steps toward recovery of the devastated economy. The New Economic Policy (NEP) was drafted in June of 1991. The goals of the plan were to set the country on a rational economic course and to express the country's commitment to economic liberal-

Many Ethiopians and Eritreans attribute their countries' problems to mistakes made by Haile Selassie during his long reign as emperor.

Library of Congress

ization. Key objectives of the plan included encouraging private investment, limiting the state's role in the economy, securing external support, promoting disciplined macroeconomic policies, and reforming and deregulating the agricultural sector.

Once the new economic plan was issued, the government actively sought international aid, particularly from the World Bank. The bank expressed optimism in the country, noting that the government had introduced improvements in governance, accountability, and the democratic process. The bank's loans to Ethiopia were expected to contribute to infrastructure development, manpower training, and capital development.

Consequences

The involvement of the World Bank in Ethiopia's economy led to several achievements. First, the government shifted its priorities from defense spending to reconstruction and rebuilding. Second, the bank recommended lowering the value of the birr, the Ethiopian currency. This devaluation attracted foreign investment in Ethiopia. Third, Ethiopian export earnings from such goods as coffee, hides, oilseeds, leather, and leather products increased. Such earnings would help in paying back the loans. Fourth, the rate of inflation was reduced from 30 percent in 1990 to 9 percent in 1994. Ethiopia became one of the few African countries with a strong currency supported by a credible free market. The World Bank announced additional commitments for 1994 and 1995 in support of the country's economic reform program.

Tseggai Isaac

Israelis Assassinate Hezbollah Leader

> *Using helicopter gunships, the Israeli army fired rockets at the car in which Sheikh Abbas al-Mussawi was riding in southern Lebanon, killing him, his wife, his young son, and several bodyguards.*

What: Assassination
When: February 16, 1992
Where: Lebanon
Who:
ABBAS AL-MUSSAWI (1952-1992), leader of Hezbollah
AYATOLLAH RUHOLLAH KHOMEINI (1902-1989), Islamic leader of Iran from 1979 to 1989

The Assassination

In the 1980's and 1990's, Hezbollah (Party of God) militants mounted a number of attacks on Israel and on Israeli-controlled sections of southern Lebanon. Israel's military saw Sheikh Abbas al-Mussawi, the leader of Hezbollah, as the chief planner of those actions. Israel believed that the assassination of the sheikh would leave the Hezbollah organization without leadership.

Using sources in southern Lebanon, the Israeli military learned about a route that Mussawi would take in his travels on February 16, 1992. Helicopter gunships were dispatched. They fired rockets that destroyed the automobile in which he, his wife, and his son were riding. Several of the sheikh's bodyguards also died in the attack.

Born in Lebanon but educated at the sacred city of Najaf in Iraq, Mussawi was a religious scholar of the Imami (Twelver) Shiite sect. Attracted by the teachings of Ayatollah Ruhollah Khomeini and an admirer of the Iranian revolution of 1978-1979, Mussawi was one of the founders of the Hezbollah movement in Lebanon.

Mussawi was named secretary general of Hezbollah in May of 1991. Although it was a purely ceremonial office, Israeli intelligence officials became convinced that Mussawi was a religious radical involved in planning and inciting violence against Israel. The Israeli government decided that he should be killed. The automobile trip in 1992 provided the opportunity for an attack.

The Shiites of Lebanon

Lebanon's mountains and valleys have for many centuries provided a refuge for religious dissidents. Although Sunni Muslims are the largest single group in the Arab Middle East, Shiites of both the Ismaili (Sevener) and Imami (Twelver) persuasions constitute significant minorities. In addition, Christians, especially those of the Maronite church, as well as sects such as the Druze and Alavites that broke away from the Ismailis, found a haven in southern Lebanon.

The Ottoman Empire never attempted to exercise direct control over most of Lebanon. The Ottomans occasionally sent punitive expeditions to the area with the goal of extracting tribute, but on the whole the Ottomans avoided involvement in Lebanon. Each mountain valley retained its independence, and villagers obeyed their local chiefs.

At the end of World War I, France received a mandate from the League of Nations over Syria and Lebanon. Believing that religious minorities would be more loyal to them than would members of the majority, the French actively recruited Maronites, Druze, and Alavites into the colonial police and army. France wrote Lebanon's constitution in 1940. That document sought a compromise among all the local groups. A Maronite Catholic was supposed to be the country's president, and its prime minister had to be a Sunni Muslim. That political compromise included only the groups most powerful in 1940. Imami Shiites were excluded from the agreement. Imamis were politically deprived and, on average, poorer than members of other religious groups. By the 1970's, the Imami Shiite population had grown. The group began demanding a real voice in Lebanon's affairs.

As a young man, Mussawi experienced the political and economic frustrations of his fellow Imamis. He decided to become a religious scholar and traveled to the city of Najaf in Iraq. Najaf is the site of some of Shiism's most holy shrines. A number of religious academies are also located there. Ayatollah Ruhollah Khomeini, having been exiled from his native Iran, was teaching in Najaf. Mussawi was attracted to Khomeini's style of Islamic thought.

Khomeini stressed that Shiite Islam was the religion of the oppressed of the earth. The oppressors included the great powers of the modern world—the United States, the Soviet Union, Great Britain, and France—together with their clients, such as Israel and Syria. Islam was, from Khomeini's perspective, a force actively opposing the evil of the superpowers. Strict adherence to Islamic teaching appealed to many who believed they sought justice for the world's impoverished majority. In 1978 and 1979, a revolution partly inspired by those teachings occurred in Iran, with Khomeini as its leader.

When Mussawi returned to Lebanon in 1978, southern Lebanon's Imami Shiites were beginning to mobilize to protest their inferior political and economic status. Mussawi was one of the founding members of an Imami Shiite political party called Hezbollah. The use of the title, meaning "Party of God," appealed to the broadest possible following. It implied that Hezbollah was above any petty considerations of private gain.

The founding of Hezbollah coincided with the disintegration of Lebanon's fragile constitutional compromise. Warlords of the various religious communities began a civil war, and international intervention fueled domestic warfare. Still in the grip of the Cold War, the United States and the Soviet Union provided weapons to their Syrian and Israeli clients, who in turn passed the guns on to the warlords' militias. Iran became the most important backer of southern Lebanon's Shiites.

Palestinians who arrived in Lebanon after 1948 incited more internal violence. The Palestinians wanted southern Lebanon as a staging point for attacks on Israel. When they launched short-range rockets in that direction, the Israelis counterattacked. Finally, in 1982, Israel invaded Lebanon and besieged the city of Beirut. Israel attempted to drive the Palestinian armed forces out of Lebanon. The invasion of Lebanon proved to be Israel's costliest war in terms of both human casualties and financial expenditures.

Israel also incited a reaction among groups such as Hezbollah. Hezbollah became a frontline group willing to use suicide bombings and other guerrilla tactics to oppose Israel.

Consequences

Hezbollah remained one of Israel's main opponents in the Middle East. Bombings and guerrilla raids sponsored by the group continued in both Lebanon and Israel. Israel signed a peace accord with the Palestinian Liberation Organization in September, 1993, that Hezbollah opposed, leading to speculations about further violence that were later realized.

Gregory C. Kozlowski

Civil War Intensifies in Sudan

Historic ethnic and religious rivalries between the Muslim north and Christian and animist south erupted into full-scale civil war.

What: Civil war
When: March 17, 1992
Where: The Sudan
Who:

OMAR HASSAN AHMAD AL-BASHIR (1944-   ), Sudan prime minister and minister of defense from 1989

JOHN GARANG (1943-   ), founder and leader of the Sudanese People's Liberation Movement (SPLM), 1983-1991, and leader of the Torit faction of the Sudanese People's Liberation Army (SPLA) from 1991

HASSAN AL-TURABI (1932-   ), founder of the Sudan chapter of the Muslim Brotherhood, later known as the National Islamic Front

The Fundamentalist Revival

During 1992, Hassan al-Turabi, considered the real power in Sudan, intensified the government's drive to impose Islamic law (*sharia*) and culture on the southern Sudan, where most people practice Christianity or local religions. On February 21, the government drove 400,000 squatters out of Khartoum into the desert at gunpoint. Strengthened by fundamentalist recruits and foreign assistance, on March 17 the government launched a major offensive against the rebel Sudanese People's Liberation Army (SPLA). By late May the government forces had seized more than ten rebel strongholds, including Bor, Kongor, Yirol, and Kapoeta. By the end of the year, the government had lifted the siege on the southern capital of Juba and captured SPLA leader John Garang's home base of Torit near the Uganda border.

Libyan military advisers and jet fighters were instrumental in the push, as were heavy artillery and road-building equipment from China. Islamic fundamentalist militia from the paramilitary Popular Defense Force, who supplemented regular Sudanese troops, were sent to the front as a part of a holy war (*jihad*) against the SPLA. The Popular Defense Force was organized with Iranian advice along the lines of the Iranian Revolutionary Guard.

In 1992, the government began using food as a major weapon in the civil war. Since 1991, Operation Lifeline Sudan, a joint relief effort including the World Food Program and the United Nations Children's Fund, had fed 7.1 million people, including 1.5 million in the south. In the wake of the new government offensive, however, on April 13 the United Nations temporarily suspended relief operations. In August, the government executed two Sudanese employees of the Agency for International Development for "collaborating," and in November, a Norwegian journalist and three relief workers were murdered. Warring factions of the SPLA blamed one another for the murders. These killings led most international relief organizations to evacuate their personnel. By early 1993, a small number of relief workers had come back to the south, but they were still hampered by the politics of civil war.

The North/South Divide

British rule in the Sudan established a number of obstacles to eventual unification. From the time of occupation in 1899, British development efforts were almost exclusively devoted to the fertile farming regions of the north. Legislation effectively prevented economic, political, and cultural ties between north and south. British determination to work through the Muslim elite ensured the dissatisfaction of southern groups, who mutinied against the British government in 1955.

Since independence in 1956, the Sudanese government has alternated between democracy

John Garang.

and military rule. A coup in 1958, following a series of impotent coalition governments, brought General Ibrahim Abboud to power. His brutal policy of forced Arabization and Islamization in the south further divided the country. In May, 1969, Gaafar Nimeiri broke the political stranglehold of parties affiliated with the northern Khatmiyya and Ansar sects, assuming power in a bloodless coup. His initial popularity enabled the government to sign the Addis Ababa accords in 1972. These accords created an autonomous southern regional government, guaranteeing the south a distinct political and cultural identity. Although they provided a measure of peaceful north/south relations, the accords led to tension in the south as various ethnic representatives vied for the limited number of government positions.

As the Arab oil-producing states grew rich in the wake of rising oil prices in 1973, Islamic fundamentalists sought to develop the Sudan as the breadbasket of the Arab world. Poor planning and unequal distribution of resources nevertheless weakened the Sudan. During the 1970's, more than 80 percent of all government expen-

ditures went to the area around Khartoum and almost none to the south. Between 1978 and 1982, the foreign debt rose more than 70 percent. The destruction of the formal economy led to a huge black market that quickly came to be monopolized by Islamic fundamentalists.

The weakened economy led Nimeiri to seek the support of fundamentalist bankers. After the discovery of oil in the southern town of Bentiu, Nimeiri abrogated the Addis Ababa agreement and in January, 1984, imposed Islamic law, including flogging and amputation of limbs for crimes such as theft and consumption of alcohol. Six months earlier, Garang had formed the Sudanese People's Liberation Movement (SPLM), calling for a "national revolution" that would correct social and regional inequalities. Early military successes and an attempt to implement an International Monetary Fund austerity program led to Nimeiri's ouster on April 5, 1985. Five different coalition governments followed. The last of these was overthrown by a military coup in June, 1989, that brought General Omar Hassan Ahmad al-Bashir to power.

In 1989, an ethnic factor was introduced into the conflict as the SPLM began to fracture. In August, the Nasir faction under commanders Riek Mashar and Lam Akol broke away from the SPLA, principally over control of relief supplies. They drew most of their support from the Nuer ethnic group. The mainstream Torit faction of the SPLA commanded the allegiance of Garang's own Dinka tribe.

In 1990, al-Bashir declared the Sudan to be an Islamic state, and in the following year he backed Iraq in the Gulf War. By early 1991, the SPLA held most of the south and was fighting the government as far north as Darfur province. After the May, 1991, overthrow of one of the SPLA's biggest backers, Ethiopian president Haile Mengistu Mariam, the rebels began to lose ground.

2414

Consequences

Civil war and violence in the Sudan worsened following the withdrawal of international aid agencies. Sudanese officials were accused of harboring terrorists at the bidding of al-Turabi, including radical Islamic Egyptian rebels linked to the June, 1995, attempt to assassinate Egyptian president Hosni Mubarak.

In February, 1994, al-Bashir launched an offensive against southern rebels, halting relief aid to some 2 million people and driving 100,000 refugees into Zaire (later renamed Democratic Republic of Congo) and Uganda. By the end of 1994, almost 400,000 Sudanese were refugees in other countries, with 2.5 million being displaced within the country. Talks between two rebel delegations and government negotiators in Nairobi in March, 1994, proved inconclusive. Between 1983 and 1995, there were more than 1 million war-related deaths, and the fighting continued into the next century.

John Powell

South Africans Vote to End Apartheid

> *South African whites formally approved, through a referendum, government moves to create a democratic, nonracial system.*

What: National politics; Civil rights and liberties
When: March 17, 1992
Where: South Africa
Who:
NELSON MANDELA (1918-   ), leader of the African National Congress and president of South Africa from 1994 to 1999
FREDERIK WILLEM DE KLERK (1936-   ), president of South Africa, 1989-1994

The Referendum

On March 17, 1992, South African voters were asked the question, "Do you support the continuation of the reform process which the State president began on February 2, 1990, and which is aimed at a new constitution though negotiations?" This referendum was the point of no return in a political revolution spanning several decades. It was but one step in a series of political developments that moved South Africa from being a pariah among nations because of its apartheid policies to taking a place as a potential major player on the African continent.

In February, 1990, following decades of pressure, both internal and international, the government of President Frederik Willem de Klerk had unbanned political parties and released Nelson Mandela, who had served twenty-seven years of imprisonment on charges of political violence. Talks began between the government and the dominant black political party, the African National Congress (ANC), led by Mandela. These talks eventually resulted in the referendum, one step along the path to total political transformation.

De Klerk sought the referendum as confirmation of the voters' desire to continue the process of ending apartheid, which he announced in his address to the opening of Parliament in February, 1991. The vote was variously described as a huge gamble and a brilliant stratagem. In an election with 86 percent voter turnout, the referendum was approved by more than two-thirds of voters, all white. It strengthened de Klerk's hand as he moved to complete the revolution he had undertaken in conjunction with Mandela.

This event marked one of the rare occasions when a ruling regime voluntarily agrees to relinquish power to an opposing force. As President de Klerk observed, the nation had taken the opportunity to rise above itself. In this case, political wisdom prevailed over intractability, even in the face of staunch opposition from white conservatives on de Klerk's political right and black radicals on the political left.

The Process of Change

Negotiations had been taking place between the government and nineteen political parties across the political spectrum. Primary negotiators were the ruling National Party and Mandela's ANC. CODESA (Conference for a Democratic South Africa) was working toward an interim political arrangement that would facilitate transition from the existing regime to a new, democratic system as well as developing a model for the new system that would ultimately emerge. These negotiations led to the democratic constitution, which includes a bill of rights that conforms to the general values accepted in the democratic West. What emerged from the political evolution in South Africa was the opportunity for whites to affirm that they would yield power sensibly in a process that would result in a liberal constitution, rather than fighting a futile battle with uncertain consequences.

South Africa had long felt the pressures of sanctions, international condemnation, internal

2416

religious objections, and a crumbling social fabric. The process of change was furthered by an international wave of democracy that followed the end of the Cold War as well as elections in several neighboring states. De Klerk took the bold step of moving away from the core political value of his National Party, that of a white government for the country. The party had held that value since its founding in 1948. The vote was not presented as a timid step in a new direction; it was confirmation of bold—and unquestionably irreversible—steps that de Klerk had already undertaken. Had he lost the referendum, his government would certainly have collapsed, to be replaced by a more conservative and nonaccommodationist white regime. That would likely have led to even more rigid segregation, and that, in turn, would have likely produced an escalation in black violence and increased internal pressure and world condemnation. The fact that the transition was

supported by those whose backing was essential to its success indicated a great political victory.

These steps ended fifty years of legal segregation and hundreds more of social and economic dominance by the white population. For a generation, black political movements had engaged in sporadic guerrilla attacks. In the midst of this low-intensity domestic conflict, the political leadership decided to form a new government and eventually a new country.

Consequences

This bold move affirmed the desires of the great majority of the white population to move beyond the rigidity of apartheid and reach out to the majority black population of South Africa. Subsequently, South Africa was readmitted to the United Nations, sanctions by most of the world's countries were lifted, the much-desired reentry into the world in international sport (especially rugby and soccer) occurred, and there was a wave of tourism and investment. The referendum was seen as a vote of confidence in the leadership of Mandela and de Klerk. Not least among the results was South Africa's formal readmission into the Olympic movement in time for participation in the 1992 Barcelona games.

In December, 1993, Mandela and de Klerk received the Nobel Peace Prize. Earlier, they had received the Houphouet-Boigny Peace Prize awarded by the United Nations Educational, Scientific, and Cultural Organization. CODESA negotiations proceeded to establish the outline of a new political system. The next dramatic step was the nonracial election in April, 1994. The tricameral parliament that represented the three nonblack groups was replaced by a nonracial, bicameral legislature, an elected president, and two vice presidents. Mandela was chosen as president, and de Klerk became one of the vice presidents. After Mandela completed his term in June, 1999, he retired from politics and was succeeded as president by Thabo Mbeki.

Progress was not without problems. Although the country's black population had been granted a meaningful voice in politics, racial tensions remained. Mandela and his successor faced the challenge of balancing political forces during an era of transition.

AP/Wide World Photos

Frederik Willem de Klerk.

Richard A. Fredland

Kurds Battle Turkey's Government

In a continuation of a long-standing conflict between the Kurds and the government of Turkey, the Turkish military responded to Kurdish guerrilla actions by attacking Kurdish bases.

What: Military conflict; Ethnic conflict; International relations; Political independence
When: March 25, 1992
Where: Eastern Turkey and northern Iraq
Who:
Tansu Ciller (1946-   ), prime minister of Turkey from 1993 to 1996; the country's first female leader
Mehdi Zana (1941-   ), Kurdish activist in Turkey
Jalal al-Talabani (1933-   ), leader of a rival Kurdish movement in Iraq

Attacks on Kurdish Bases

After the Persian Gulf War of 1991, Kurds had attempted to gain independence from Iraq but quickly discovered that they were no match for the Iraqi army. They fled to Turkey and Iran, and United Nations forces came to their assistance. The United Nations established a safe zone for the Kurds in northern Iraq. Intended to protect the Kurds from attacks by the Iraqi army, the safe zone quickly filled with bases from which Kurdish rebels attacked military and civilian targets in Turkey. In 1992 and again in 1995, the Turkish government responded to those covert attacks by launching air strikes and sending troops over the border to suppress the Kurdish bases in the Iraqi safe zone. In an attack on March 25, 1992, at least seventy persons were killed.

The Kurdish population of Turkey numbered about twelve million in 1994. This gave Turkey the world's largest population of Kurds. About five million Kurds lived in Iraq, six million in Iran, and another two million in Azerbaijan, Syria, and elsewhere around the world. The position of the Kurds in Turkey was unique in that the Turkish government refused to recognize Kurds as a distinct ethnic group. The Turkish government's official vocabulary refers to Kurds as "Mountain Turks."

Several political movements have appeared among the Kurds since 1924. Some have moderate ambitions such as having the government of Turkey recognize them as a separate cultural group. Other Kurdish groups—notably the Kurdish Workers Party (PKK)—have radical agendas that include establishing an independent Kurdish nation of Kurdistan. Operating in Istanbul and Ankara, the major cities of Turkey, but particularly in eastern Turkey, groups aligned with the PKK have conducted a campaign of political assassination and bombing. Those attacks increased after the defeat of Iraq's Saddam Hussein and the creation of a protective cordon around the Kurds of northern Iraq. In the lawless atmosphere following the war, weapons were readily available. Iraqi Kurds passed those across the border to Kurds living in Turkey.

Under Prime Minister Tansu Ciller, Turkey's government responded to PKK attacks within its borders by mounting air raids on Kurdish strongholds in Iraq. Strikes by jet fighters and helicopter gunships were not effective in the mountainous terrain of that area. Eventually infantry, supported by tanks and heavy artillery, were used to destroy Kurdish bases. In March of 1995, Turkey committed some fifty thousand ground troops to the effort. Turkish soldiers seized some stocks of weapons, but most of the Kurdish fighters retreated into the hills. After a month, most of the Turkish military withdrew. Sporadic warfare was expected to continue until some political compromise was reached between the Kurds and the Turkish government.

A Stateless People

People now identified as Kurds have lived in the mountainous areas west of the Caspian Sea for at least two thousand years. Kurdish belongs to the Iranian language family. It is related to, but not mutually intelligible with, the Persian language. The Kurds accepted the Islamic religion in the eighth through the tenth centuries C.E. Most Kurds belong to the Sunni branch of Islam, but a significant minority adheres to the Imami Shiite sect dominant in Iran. Despite their religious connection, Kurds have remained culturally distinct from the Arab, Iranian, and Turkish Muslims surrounding them.

The majority of Kurds live in villages scattered in the mountainous regions of modern Turkey, Iraq, and Iran. They usually combine pastoralism with small-scale agriculture. In the past, Kurdish warriors sometimes hired themselves out to different kings in the Middle East. One Kurdish leader, Saladin, founded the Ayyubid dynasty that briefly ruled Egypt and Syria. For the most part, however, the Kurds have lived in relative obscurity. In political terms, they were organized by personal allegiance to one or another family of chiefs. The rugged terrain of their homeland prevented any sort of broader government.

For almost five centuries the Ottoman Empire dominated the Middle East. Kurdistan was recognized as a province of the Ottoman state. Generally, the Ottomans were content to extract tribute from the Kurds while leaving them to handle their own affairs. The situation changed dramatically in 1918 when the Ottoman Empire, an ally of Germany and Austria-Hungary in World War I, was demolished. The Arabic-speaking portions of the Ottoman state became independent countries including Syria, Jordan, Saudi Arabia, and Iraq. The cosmopolitan empire of the Ottomans was reduced to the Turkish-speaking regions of Anatolia.

A military revolution led by Kemal Atatürk in the 1920's based itself on an assertive Turkish nationalism. The Turkish language was the only

Cemil Bayik, the commander of the military wing of the Kurdish Workers Party, and his fighters take a break in a mountain hideout in northern Iraq.

2419

one recognized for use by schools and the press. This left the Kurds with the choice of abandoning their own culture or preserving it but losing educational and economic opportunities. Although some Kurds accepted the demands of Turkey's nationalism by adopting Turkish family names, the majority tried to ignore the government. As the powers of the Turkish state increased, the Kurds began to clash with it. The conflict escalated in the 1970's and seemed destined to be carried on by violent means for the foreseeable future.

Consequences

The United Nations established a safe haven for the Kurds of northern Iraq with the hope of protecting them from attacks by the Iraqi government. An unintended consequence gave the Kurds of Turkey a refuge from which they could launch a more extensive guerrilla campaign in-side Turkey. The Turkish government responded to those developments by attacking Kurdish bases from the air and by sending ground troops to demolish Kurdish strongholds in Iraq. Those tactics proved effective in the short run but could not prevent an ongoing conflict. Because the Kurds harbor their own nationalist ambitions, they will inevitably confront Turkish nationalism. Only a political compromise seemed likely to bring a permanent halt to the Kurdish-Turkish conflict.

A breakthrough appeared to come in February, 2000, when leaders of the outlawed PKK, which had been at the head of the Kurdish struggle against the Turkish government, renounced the use of violence. They promised to give up their war and work for peaceful change within the Turkish political system.

Gregory C. Kozlowski

Commonwealth of Independent States Faces Economic Woes

> *Following the dissolution of the Soviet Union, the fragile Commonwealth of Independent States teetered on the brink of economic disaster.*

What: Government; Politics
When: Spring, 1992
Where: Commonwealth of Independent States
Who:
BORIS YELTSIN (1931-   ), president of Russia from 1991 to 1999
MIKHAIL GORBACHEV (1931-   ), president of the Soviet Union from 1988 to 1991
LEONID KRAVCHUK (1934-   ), president of the Ukraine from 1991 to 1994

Dire Straits

The spring of 1992 witnessed troubled times for the recently formed Commonwealth of Independent States (CIS), the confederation of countries which had made up the old Soviet Union. The CIS, consisting of only eleven of the republics of the defunct Soviet Union, had to organize new relations with the unaffiliated republics as well as the rest of the world on an individual and collective basis. While Boris Yeltsin, the president of Russia, the largest republic, wielded the most power in the confederation, he also had to shoulder much of the blame for its failures. Discontent threatened to unseat him just as it had his predecessor, Mikhail Gorbachev, the former president of the now disbanded Soviet Union.

Many of the CIS's problems stemmed from its slow change from a controlled economy to a free market. The value of the Commonwealth's currency based on the Russian ruble fluctuated widely. This led to a lack of confidence by the international community. The dissolution of the former economic institutions, the sale of Soviet and Communist Party property, and the full im-plementation of a market system led to major disruptions in the economy. Massive shortages, particularly in the smaller cities, continued to exist. Where products were available prices were exorbitant for the average citizen. Furthermore, ignorance of modern economics, such as pricing mechanisms, made the situation worse. Some world leaders feared that the Commonwealth's dire straits might lead its scientists to sell their expertise abroad and contribute to the proliferation of nuclear weapons.

The Commonwealth was plagued by various ethnic and national rivalries: Russians and Romanians fought each other in Moldovia, Armenians and Azerbaijanis fought each other in the Caucasus, and Russia and Ukraine quarreled over the distribution of the former Black Sea fleet. Several of the minor nationalities that were in autonomous republics within the independent states demanded their complete freedom. Cooperation between the former Soviet Republics, economically dependent on one another, deteriorated.

Yeltsin found himself in an untenable position. He was attacked by both those who wanted to return to the former socialist system and those who claimed he was not producing results quickly enough. By mid-1992 the breakup of the Commonwealth was a real possibility. Those who supported the CIS hoped that the new economic plan Yeltsin shepherded through the Russian parliament on April 15 and the promised fifty-five billion dollars of aid from the West would, in the long run, save the Commonwealth.

A New Leader, a New Commonwealth

In 1985 Mikhail Gorbachev became the general secretary of the Communist Party of the Soviet Union, ushering in a new era. Gorbachev attempted to save the moribund economic and political systems of the Soviet Union by introducing *glasnost* (freedom of expression) and *perestroika* (economic reform). However, the Soviet Union's economy proved to be beyond repair, and the new freedoms led a majority of the Soviet people to demand fundamental changes modeled on the democratic and market systems of the West. Gorbachev introduced some changes, but these only led to increased demands. Some of the non-Russian republics of the Union demanded more autonomy or independence. Gorbachev, while gaining international renown as a peacemaker, lost prestige at home because of his economic failures.

In August, 1991, hard-line communist conservatives tried to oust Gorbachev and restore the old system. The dramatic intervention of Moscow's citizens, led by Yeltsin, restored Gorbachev, whose prestige was now further damaged. Yeltsin, on the other hand, now became the most powerful figure in the Soviet Union.

Under Yeltsin the move for total and fundamental reform was speeded up. In December, 1991, the Soviet Union was dissolved and replaced with a confederation of independent countries, the CIS. These countries were eleven of the former republics of the Soviet Union: Russia, Ukraine, Belarus, Moldovia, Armenia, Azerbaijan, Kyrgystan (Kirghizia), Uzbekistan, Turkmenia, Tajikistan, and Kazakhstan. Four republics refused to join: The three Baltic states of Estonia, Latvia, and Lithuania, which had been the most vociferous in demanding complete independence under Gorbachev, preferred to stay out of the new confederation, as did Georgia in the Caucasus.

In order to deemphasize the role of Moscow, the confederation leaders chose Minsk, the capital of Belarus, as the new capital. The CIS was a genuine confederation, and each country was truly independent. This led to a whole series of problems, including economic relations between the states, status of minorities, borders, and the division of property belonging to the former Soviet Union. One particular problem was the stationing of nuclear weapons. Under the Soviet government nuclear weapons had been kept in Russia, Belarus, Kazakhstan, and Ukraine. All

leaders decided to place them in Russia, but some of the countries proved slow to implement the agreement. The CIS also agreed to maintain the international treaties of the Soviet Union, with Russia again taking the lead. Moscow, for example, assumed the permanent seat of the Soviet Union on the United Nations Security Council.

The second largest country in the CIS was Ukraine, which had never really had an independent existence and whose nationals have felt oppressed by Russian domination for centuries. At first it appeared that Ukraine might join the Baltic states and Georgia in remaining outside of the CIS. However, Leonid Kravchuk, the president of Ukraine, who won elections on a pronational ticket, convinced his country to join the confederation. Nevertheless, problems between the two large Slavic states continued. There were disagreements over the rights of the large number of Russians living in Ukraine, the dividing of the Soviet property, particularly military and naval property, borders in the Crimean region, the status of an independent Ukrainian military, and economic issues.

In Russia the most serious problems for Yeltsin remained economic. As under Gorbachev, the Russian people were anxious for immediate results. Despite the new situation, the shortages and inefficiencies remained. The winter of 1991-1992 was particularly hard. When spring came, the economic situation in Russia put the whole CIS in a precarious position.

Consequences

The conversion of the ruble to an international currency, the promise of help from the international community, and the resiliency of the Russian people in coping with hard times all served to give new hope for the confederation's recovery and survival. The West provided financial aid and worked with the CIS in creating programs to employ Soviet scientists in order to lessen the dangerous possibility of the proliferation of nuclear arms. Though ethnic rivalries and economic problems challenged the confederation, the democratic governments and political freedoms, although imperfect, seemed to be surviving.

Frederick B. Chary

Nepalese Police Fire on Kathmandu Protesters

> *Police in Kathmandu, Nepal, killed seven people when they fired on crowds of protesters demonstrating against economic conditions.*

What: Civil strife; Social reform; Economics

When: April 6, 1992

Where: Kathmandu, Nepal

Who:

BIRENDRA BIR BIKRAM SHAH DEVA (1945-2001), king of Nepal from 1972 to 2001

GIRIJA PRASAD KOIRALA (1924-   ), prime minister of Nepal from 1991 to 1994

MAHENDRA BIR BIKRAM SHAH DEVA (1920-1972), king of Nepal from 1955 to 1972

JAWAHARLAL NEHRU (1889-1964), India's first prime minister from 1947 to 1964

Protesters Killed by Police

On April 6, 1992, during a strike called by Nepal's leftist opposition party, the Nepal Communist Party-Unity Center, riot police fired into a crowd of protesting youth armed with stones, bricks, and soda bottles. One man, shot in the head, died on the scene. Of the dozens of wounded people taken to Kathmandu's Bir Hospital, four soon died; two more succumbed to their wounds within days. In a related disturbance, another Nepalese civilian was reported killed by the police in nearby Patan.

The dispute that triggered the police action stemmed from leftist discontent with the Nepali Congress Party. The party came into power in 1991, in Nepal's first multiparty elections since 1960, when King Mahendra Bir Bikram had fired government officials and suspended Parliament because he considered the parliamentary system unsuited to current conditions in the country. The protesters demonstrated because of Nepal's staggering inflation, which was not accompanied by a corresponding increase in income, and because of the Congress Party's practice of nepotism in its administration and in appointments to the country's educational institutions.

The communists demanded that business activity cease during the strike The demonstrations occurred after party supporters attacked workers in offices and businesses that refused to observe the mandate. Initially, police fired tear gas into the crowd and bullets into the air above it. However, the scattered protesters reassembled, attacking police officers with rocks and bottles. Some of the police then fired into the crowd, resulting in the reported injuries and deaths. Estimates placed the number of injured people at more than fifty.

Nepal, one of the world's poorest countries with an estimated gross domestic product of $186 per capita, spends more than twice what its domestic revenues of $388 million provide annually. The country is landlocked and isolated by its mountain barriers. Development of its resources has been hampered by seemingly insuperable transportation problems. It imports goods worth about $600 million a year while exporting goods worth about $150 million. Its economic problems are intense.

Attempts at Representative Government

The contest between monarchy and representative government in Nepal surfaced noticeably in 1950 when the country's prime minister, Maharaja Mohan Shumsher Rana, deposed King

2423

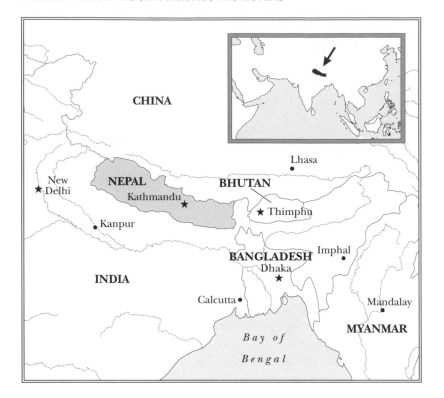

his reign by exercising strong control over the government. He discouraged reform efforts that were under way. By the late 1970's, many of the increasingly oppressive king's leftist subjects had participated in riots that severely threatened his regime.

To buttress his sagging hold on political power, Birendra held a referendum in 1980 in which the citizens voted on the form of government they preferred. The majority vote was in favor of a nonparty panchayat (monarchically guided) system, much like what had existed, but demanded some modifications from the existing system. Elections were held in 1981 and again in 1986. By the early 1990's, a new prodemocratic wave imposed itself on Nepal's complex political scene, causing the king to end the ban on political parties.

Bir Bikram Tribhuvana. The king, who traditionally served a largely ceremonial function, evoked Rana's ire when he intervened in a domestic political dispute fueled by Nepalese dissidents living in India.

At the urging of India's prime minister, Jawaharlal Nehru, Tribhuvana was restored to the throne by February, 1951. Members of the moderate Congress Party were appointed to the cabinet. When Tribhuvana died in 1955, however, his son, Mahendra Bir Bikram, assumed the kingship. In 1958, he issued Nepal's first modern constitution, which instituted a democratic system. In 1959, for the first time in Nepal's history, the country held an election for a two-party parliament.

By 1960, however, King Mahendra had disbanded the democratically elected government and parliament. He issued a new constitution in 1962 that promised land reform and eliminated caste discrimination. In a country where 80 percent of the population was composed of subsistence farmers, land reform, with all of its socialist and communist overtones, was an important consideration.

Upon Mahendra's death in 1972, his son Birendra Bir Bikram took the throne, beginning

In a country oppressed by poverty, removed from the mainstream of world politics by its geographical isolation, and populated by peasant farmers, the notion of land reform that would divide the large property holdings of affluent proprietors, often absentee owners, is extremely appealing. For subsistence farmers, such land reform is the only vestige of hope.

When the Nepali Congress Party was unable to deliver on its pre-election promises, it was understandable that the more extreme elements of the electorate viewed demonstrations and their accompanying violence as their sole means of accomplishing social and economic ends. The 1992 uprising in Kathmandu foreshadowed future conflicts.

Consequences

Despite attempts to institute a representative system of government in Nepal, the country's leaders have not succeeded in arriving at a mode of leadership that the electorate trusts and supports. As in the past, when the situation becomes too volatile, an authoritarian form of government is likely to emerge.

On July 4, 1993, a general strike was called that paralyzed commerce in Nepal for nearly a month. This strike was brought about by the suspicious deaths of two Communist leaders in a highway incident on May 16. Violent protests related to the general strike caused the police to fire into crowds on July 4, 1993, killing thirteen people.

The Communist Party of Nepal prevailed in the 1994 parliamentary elections, although it fell short, by 17, of the 103 members required for a majority. The lack of a majority forced formation of a coalition government that included the pro-monarchy National Democratic Party. Prime Minister Girija Prasad Koirala received a no-confidence vote three years into his five-year term and was forced to resign on July 10, 1994. Man Mohan Adhikari took the office in November.

R. Baird Shuman

Keating Is Sentenced for Securities Fraud

Charles H. Keating, Jr., received state and federal prison sentences for securities fraud and racketeering connected with the failure of Lincoln Savings and Loan Association.

What: Law; Business
When: April 10, 1992
Where: Los Angeles, California
Who:
CHARLES H. KEATING, JR. (1923-), financier
M. DANNY WALL (1939-), chair of the Federal Home Loan Bank Board, 1987-1989, and director of the Office of Thrift Supervision, 1989-1990
ALAN CRANSTON (1914-2000), senator from California from 1969 to 1993

The Sentencing

On April 10, 1992, Charles H. Keating, Jr., was sentenced to ten years in the California state penitentiary. He had been convicted on seventeen counts of securities fraud. On July 8, 1993, he was sentenced in federal court to twelve years and seven months for his conviction on seventy-three counts of fraud and racketeering. The federal and state sentences were to run concurrently.

Early in 1989, Lincoln Savings and Loan Association of Irvine, California, failed. The ultimate cost of the Lincoln Savings and Loan failure to U.S. taxpayers was estimated to be $2.6 billion, making it one of the most costly of such failures in U.S. history. The failure involved regulatory problems, personal greed, corruption, and political influence, all part of the larger savings and loan crisis of the period.

Keating, the head of American Continental, a construction company planning a huge real estate development in Arizona, was allowed to acquire Lincoln Savings and Loan in 1984. This was despite the fact that he had been accused of fraud by the Securities and Exchange Commission (SEC) in 1979. The acquisition of Lincoln at that time, in a lax regulatory climate, literally created a "money machine" for Keating to finance his other activities.

Lincoln failed after Keating had led it to engage in numerous questionable practices. As a means of obtaining funds, American Continental, the parent company of Lincoln, sold junk bonds (securities with low safety ratings) in branches of Lincoln. These junk bonds were misrepresented as insured deposits and sold primarily to older, unsophisticated investors who were attracted by the high interest rate offered. Sales of these bonds resulted in many of the charges against Keating.

Deregulation and Lax Regulation

In 1982, the Garn-St. Germain Act was passed in response to an immediate crisis that faced thrift institutions. Money market funds and mutual funds were paying high interest rates, and thrift institutions, which by law could not compete with these rates, were losing deposits to these funds.

The Garn-St. Germain Act authorized creation of money market deposit accounts by thrift institutions. These were insured deposits not subject to either reserve requirements or Regulation Q ceilings on the interest rate they could pay. Both conditions made these accounts different from standard checking and savings accounts. The legislation also broadened the lending power of thrifts, allowing them to make commercial loans of up to 10 percent, and consumer loans of up to 30 percent, of their assets. The act also permitted interstate and interindustry mergers of failing institutions. In addition, for the first time in fifty years, commercial banks could acquire failing savings and loan associations in other states.

The effect of the passage of the Garn-St. Germain Act and the Depository Institutions Deregulation and Monetary Control Act two years earlier, in 1980, was to increase competition and to blur the distinctions among traditional depository institutions. The climate of deregulation created by these acts encouraged savings and loans to pursue high returns that would match the high interest rates paid to depositors. This often involved making new types of loans that involved higher risks.

When Keating took over Lincoln, he removed all of Lincoln's conservative loan officers and internal auditors within days. This was despite the fact that he had promised the regulatory authorities that he would not drastically alter the institution's lending policies. Keating used the funds provided by insured depositors to speculate and to make questionable and inappropriate investments. Lincoln acquired vast tracts of Arizona desert land, interests in hotels, common stocks, junk bonds, and even foreign currency futures.

At the time, there was a critical shortage of savings and loan examiners, allowing Lincoln to escape any serious regulatory supervision until 1986. The San Francisco office of the Home Loan Bank Board finally caught up with Lincoln and recommended that the federal government seize

the institution. The examiners initially discovered that Lincoln had exceeded the 10 percent limit for equity investments by $600 million, and they found evidence that Lincoln was deliberately attempting to mislead savings and loan examiners.

Keating fought the attempt to seize Lincoln, accusing the bank examiners of being biased. He sued the federal government over the 10 percent limit on equity investments. He bragged about hiring seventy-seven law firms and spending $50 million to fight federal regulators. Keating also attempted to use political influence to ward off regulators. He gave a total of $1.5 million to the election campaigns of five senators—Dennis De Concini and John McCain of Arizona, Alan Cranston of California, John Glenn of Ohio, and Donald Riegle of Michigan. These senators were later to become known pejoratively as the "Keating Five."

In April of 1987, these five senators met with Edwin Gray, the chairman of the Federal Home Loan Bank Board. They later met with the top four regulators from the San Francisco office to complain that regulators were being too tough on Lincoln. Gray took no action. Shortly after this meeting, he was replaced by M. Danny Wall. In September of 1987, Wall, in an unprecedented action, transferred the investigation of Lincoln to the board in Washington, D.C., removing the case from the San Francisco jurisdiction. No action was taken for ten months. One of the San Francisco examiners later characterized the situation as Lincoln having fallen into a "regulatory black hole." Wall went on to become director of the Office of Thrift Supervision.

Consequences

Wall was forced to resign from the Office of Thrift Supervision for his failure to act. After months of hearings, the U.S. Senate Ethics Committee reprimanded Cranston, finding "substantial credible evidence" of misconduct. The investigation of the case against Cranston was left open, a procedure that could have led to a vote by the full U.S. Senate to censure him. He suffered from prostate cancer and did not run for reelection in 1992. De Concini, McCain, Glenn, and Riegle all were given mild reprimands. The investigations against them were closed.

Daniel C. Falkowski

Charles H. Keating, Jr.

United Nations Allows Sanctions Against Libya

The United Nations Security Council allowed sanctions to be imposed against Libya for failing to hand over to the United States two suspects in the bombing of Pan Am flight 103.

What: International relations; Terrorism
When: April 15, 1992
Where: United Nations, New York City
Who:

MUAMMAR AL-QADDAFI (1942-   ), president of Libya beginning in 1977

BOUTROS BOUTROS-GHALI (1922-   ), secretary general of the United Nations

Sanctions Against Libya

Pan Am flight 103, going from London to New York, blew up in midair on December 21, 1988. Debris from the plane and human bodies began to fall on the small Scottish town of Lockerbie shortly after 7:00 P.M. All 259 passengers and flight crew, in addition to 11 persons on the ground, were killed in the incident.

Investigators began the grim task of piecing together the evidence, scattered over an area of 80 square miles. They concluded that the explosion was caused by a device hidden in a suitcase that was placed in the forward luggage compartment. When it exploded, it tore off the cockpit.

Various groups and individuals called media organizations to claim responsibility or to blame others for the attack. The international team of Scottish, British, German, and American investigators advanced the theory that the explosion was the work of the Syrian-based Popular Front for the Liberation of Palestine-General Command (PFLP-GC). They said that members of the PFLP-GC, working on behalf of Iran, blew up the plane in retaliation for the earlier shooting down of an Iran Air passenger plane (Iran Air flight 655) by the USS *Vincennes* over the Persian Gulf on July 3, 1988. The plausibility of this theory was strengthened by strained U.S.-Iran relations and by the fact that German security officials had earlier arrested alleged members of the PFLP-GC and found them to be in possession of materials used to make bombs.

The PFLP-GC, Syria, and Iran denied any responsibility for the incident. Despite the repeated denials, investigators held on to their hypothesis until evidence surfaced that linked the bomb-containing suitcase to Malta. Investigators theorized that the suitcase came to Frankfurt on an Air Malta flight. It was then put on a smaller Pan Am flight to London, then loaded on the jumbo jet of flight 103.

One crucial piece of evidence was a tiny remnant of a timing device, found embedded in the luggage container. The timing device was manufactured in Switzerland and was sold to Libya in 1985.

Nearly nine months after the Lockerbie incident, on September 19, 1989, another plane was blown out of the sky. Flight 772 of the French company UTA was on the second leg of its Brazzaville-N'Djamena-Paris journey. The plane exploded over the Tener desert region of the Sahara near Bilma, Niger, about four hundred miles northwest of N'Djamena, Chad. All 170 passengers and crew members were killed. Investigators of this incident determined that an explosive device hidden in a suitcase and placed in the forward luggage compartment caused the tragic crash. Traces of the explosive Pentharite were found. The French government accused Libya of masterminding this incident, but Libya denied any responsibility.

2428

Libyan Terrorism

On September 1, 1969, Colonel Muammar al-Qaddafi led a successful coup against the conservative pro-Western Sanusi monarchy of Libya. Since then, Libya has steered a course of radical international and domestic policies. Libya nationalized its oil industry and played an important role in the Organization of Petroleum Exporting Countries (OPEC). Under Qaddafi's leadership, Libya extended money and arms to various radical and revolutionary groups and organizations operating in various countries and regions of the world.

Libya claimed possession of a uranium-rich section of northern Chad and invaded that country. Libyan troops remained in Chad between 1975 and 1987. The relationship between the United States and Libya took a severe downturn during the presidency of Ronald Reagan. On March 6, 1981, Reagan ordered the closure of the Libyan embassy in Washington, D.C. A more direct confrontation with the United States took place when Libya claimed the Gulf of Serte as its territorial waters. The claim resulted in a confrontation between the Libyan air force and U.S. fighter planes from the Sixth Fleet, stationed in the Mediterranean. A dogfight on August 19, 1981, resulted in the downing of two Libyan fighter jets. Matters between the two countries became worse after April 5, 1986, when Libyan agents were accused of planting a bomb in a West Berlin nightclub. Ten days later, U.S. jets bombed the Libyan cities of Tripoli and Benghazi, causing many casualties. The relationship remained tense, and tensions increased after the Lockerbie incident.

By November, 1991, the U.S. Justice Department, in continuation of its investigation of the Lockerbie incident, had indicted two alleged Libyan intelligence agents. They were accused of masterminding the plot to blow up Pan Am flight 103. The United States sought the extradition of the two suspects, but Libya refused to hand them over. Diplomatic wrangling and negotiations between the two sides through intermediaries produced no result. Libya offered to hand over the suspects to the World Court, but this compromise was unacceptable to the United States. The continued impasse prompted the United Nations Security Council to convene. On March 31, 1992, it passed a resolution that banned air flights and arms sales to Libya. On April 15, the United Nations allowed the imposition of sanctions on Libya as long as Libya refused to comply with the extradition request.

Consequences

The U.N. Security Council resolution succeeded in isolating Libya from the international community. The economic sanctions hurt Libya's economy. Despite occasional reports of a breakthrough in the impasse over extradition, the two suspects were not handed over.

Libya, however, appeared to be moderating its political posture in regional and international politics. Following the end of the Cold War and the breakup of the Soviet Union, and in response to the changed international balance of power, Libya appeared to be withdrawing its support of revolutionary groups that it had supported in the past. For example, during the 1991 Persian Gulf War a military coalition led by the United States ousted Iraqi forces from Kuwait. Libya did not side with Iraq. Nevertheless, the United States continued to ostracize the North African nation for its covert support of terrorism. In July, 2001, the U.S. Senate voted to extend economic sanctions against Libya for an additional five years.

Mahmood Ibrahim

Acquittals of Police Who Beat King Trigger Riots

In the wake of the acquittal of four police officers charged with criminal assault in the arrest of Rodney King, South Central Los Angeles erupted into riots resulting in fifty-eight deaths and nearly $1 billion in property damage.

What: Civil strife; Law
When: April 29, 1992
Where: Los Angeles, California
Who:
RODNEY KING (1965-   ), felony suspect beaten by Los Angeles police officers
STACEY C. KOON (1951-   ), Los Angeles police officer in charge
LAURENCE M. POWELL (1963-   ), Los Angeles police officer charged with beating King

Los Angeles Erupts in Riots

On April 29, 1992, four white Los Angeles Police Department (LAPD) officers were acquitted of state criminal assault charges stemming from the arrest of African American felony suspect Rodney King. The verdict delivered by a predominantly white jury, seen as unjust by many in the black community, sparked rioting and looting in the South Central area of Los Angeles. Although the riots and looting occurred primarily in the South Central district, criminal activity spread throughout the Los Angeles area and as far north as San Francisco and as far east as Atlanta, Georgia. In each of those two cities, approximately five hundred people were arrested for looting and other crimes, and curfews were ordered.

The 1992 riots in Los Angeles were one of the most violent and bloody occurrences in twentieth century America. Fifty-eight people were killed during three days of rioting, and approximately twenty-four hundred people suffered riot-related injuries. Property loss from looting, arson, and vandalism was estimated at $1 billion or more in the Los Angeles area alone.

Those who have studied urban riots, particularly those occurring in the late twentieth century, have been somewhat cautious in defining this event as a "civil insurrection" or a 1960's-style "people's rebellion." Although the announcement of the "Rodney King Trial" verdict by the media may have touched off the initial disturbance, evidence suggests that the vast majority of criminal activity reported had little to do with the verdict.

Most of the property damage caused by the looting and arson was not directed at institutions of the "white power structure" but instead primarily at businesses owned by Korean Americans and others of Asian background. Black and Latino businesses were looted and burned as well. Police were unable to respond in sufficient numbers in the early hours of the riots, creating a climate favorable to rampant criminal activity.

Police arrest records indicate that a major motivation of those involved in the rioting was to loot businesses of electronic gear, liquor, auto parts, and other merchandise. According to a study done by the RAND Corporation's Peter Morrison, looting was the primary motivation for most rioters who were caught, and few were consciously political. The *Los Angeles Times* reported that most of those convicted of felonies during the riots did not cite anger or other social grievances to explain their actions. According to the newspaper, police reported only one riot-related arrest in which the suspect mentioned or alluded to the King verdict. Many of those arrested in the first hours of rioting did not know what the verdict was or even that a verdict had been rendered.

2431

AP/Wide World Photos

A fire rages near Vermont Street in Los Angeles during the riots caused by the acquittals of four police officers in the beating of African American Rodney King.

The Videotaped Arrest

On the evening of March 3, 1991, King, who had been released recently from prison after serving time for armed robbery, became involved in a high-speed felony chase. After an eight-mile pursuit involving speeds in excess of 115 miles per hour, King was forced to stop.

There were three people in King's car: King, who was driving, and two black passengers. All three were ordered out of the car. The two passengers exited the car as ordered and put up no resistance. They were handcuffed and taken into custody without incident. King, however, decided to resist the arrest. He refused to obey the officers' verbal commands to submit and exhibited aggressive, threatening verbal and physical behavior toward the officers. The officers initiated escalating levels of nondeadly force to effect the arrest.

The first two levels, police presence and verbal commands, were ignored by King. Overpowering King by "swarming" him with four LAPD officers also proved to be ineffective because of King's large physical stature and violent reactions.

Sergeant Stacey Koon, the officer in charge at the scene, decided to use a stun gun in an attempt to immobilize King without reverting to deadly force. Two jolts of fifty thousand volts each from the stun gun did not render King immobile, and he continued to display violent, aggressive behavior.

After this attempt to subdue King, Koon ordered his men to implement the last level of nondeadly force available, the use of police batons. Officers administered several strikes to King's backside and legs. Each round of strikes was followed by a pause and back-off by the police to allow King the opportunity to submit. After receiving numerous blows, King eventually gave up and was arrested.

King was taken into custody and to a nearby hospital, in accordance with standard police procedure. Emergency room physician Antonio Marcia requested that King be placed in restraints because of his violent nature. Marcia diagnosed King as having overdosed on phencyclidine (PCP), a psychedelic drug, and as having superficial facial lacerations. King was released into police custody as an outpatient the same night.

The entire arrest episode lasted only a few minutes. Amateur photographer George Holliday, who lived across the street from the arrest scene, captured the arrest on videotape. The scene had been illuminated by multiple police squad cars from three different agencies and by an LAPD helicopter. Holliday's videotape was approximately 81 seconds in length and showed not only the police action toward King but also King's prior resistance and the actions of the two other passengers in King's car, who had been peacefully taken into custody.

The jury in the police officers' trial viewed the entire tape and heard testimony explaining the audio portion of the tape. Much of America, however, never saw the entire tape but only its last 30 seconds, which showed King being hit by police officers. Television stations showed this portion repeatedly, often in slow motion, distorting the amount of time actually depicted. Public perceptions, based primarily on the videotape as presented by the media, were that King was a victim of an unprovoked beating by police officers. Acquittal of the LAPD officers came as a shock to people with that opinion and prompted accusations of injustice and racism.

Consequences

The U.S. Justice Department charged the four LAPD officers with violating King's constitutional rights and tried them in 1993. Although technically not double jeopardy, or a second accusation of a given crime, the federal charges against the officers were essentially the same as those for which the state jury had already failed to convict.

In the federal trial, officer Laurence Powell was convicted of using unreasonable force and Sergeant Koon was convicted of permitting the violations to occur. Both were sentenced to two and one-half years in federal prison. The two other LAPD officers acquitted in the state trial were acquitted in the federal trial as well.

Frank Andritzky

Muslim Fundamentalists Attack Egyptian Coptic Christians

> *Muslim radicals attacked Coptic Christians at Manshiet Nasser during an escalation of religious violence in Egypt accompanying the growth of Muslim fundamentalism.*

What: Civil strife; Religion
When: May 4, 1992
Where: Manshiet Nasser, Asyut
   Governorate, Egypt
Who:

HOSNI MUBARAK (1928-   ), president
   of Egypt from 1981, chairman of
   the National Democratic Party from
   1982
ANBA SHENOUDA III (1923-   ),
   religious leader in the Coptic
   Orthodox church
SHEIKH JAD AL-HAQ ALI JAD AL-HAQ
   (1917-   ), grand sheikh of al-Azhar
   University beginning in 1982; mufti of
   Egypt, 1978-1982

Deadly Religious Clash

On May 4, 1992, Muslim fundamentalists attacked Coptic Christians in the farming village of Manshiet Nasser, about two hundred miles south of Cairo in central Upper Egypt. The origin of this clash, the deadliest in a decade, was a feud between a Christian homeowner and a Muslim family, which had insisted that the Copt sell his house to a Muslim. Other versions of the clash mentioned the refusal of some Copts to pay "protection money" to local Muslim militants as well as revenge for the killing of a Muslim fundamentalist leader in a gun battle between Muslims and Copts two months earlier.

In the attack, whatever its cause, Muslim fundamentalists surprised Coptic farmers at dawn. They killed several farmers immediately upon arriving at their farms. Another group of militants stabbed a Christian doctor to death in his home.

A third group erupted into a classroom, killing the teacher and wounding five schoolchildren. The casualties were fourteen deaths and the five wounded children. Two of the fatalities were Muslims trying to defend their Coptic neighbors. Some of the attackers were apprehended immediately.

This incident was one of several since 1981 in which Muslim militants attacked Coptic Christians, their property, and their churches. Former Egyptian president Anwar el-Sadat had accused the Muslim radicals of fanaticism and the Coptic church leaders of sectarian sedition, removing Pope Anba Shenouda III in September of 1981. He was restored to office by President Hosni Mubarak in January of 1985. The attack at Manshiet Nasser was attributed to the Gamaa al-Islamiya (Islamic League), a local splinter group of the Muslim fundamentalist *jihad* (holy war), one of several such groups in the growing movement. The more moderate and older Muslim Brotherhood, Egypt's largest and politically most powerful Islamic organization, condemned the killings as repulsive to Islam, as did the grand sheikh of Cairo's al-Azhar religious university, Sheikh Jad al-Haq Ali Jad al-Haq. President Mubarak, on March 30, 1992, already had criticized "religious fanaticism" when he alluded to earlier Muslim militant attacks against Copts.

A History of Conflict

The native Egyptian Christians, 80 percent of whom follow the Orthodox rite, represent about 10 percent of the country's total population, the overwhelming majority of which is Sunni Muslim. These Christians trace their history to Saint Mark in the first century C.E. Their predecessors did not convert to Islam after the Arab Muslim invasion of Egypt in the seventh century C.E., thus conserving the principal characteristics of their ancient Egyptian ancestors.

During fourteen centuries of living side by side, the two communities, Muslim and Coptic, tolerated each other for the most part, with occasional outbursts resulting from religious tensions. At times, Copts have achieved prominent positions in Egyptian private and public life, particularly in finance and foreign affairs. The Copts tended to be better educated, more prosperous, and more Westernized than Muslims. In part because of this, the fundamentalist movement in Egypt, which saw a resurgence after Ayatollah Ruhollah Khomeini's successful Islamic revolution in Iran in 1979, intermittently but overtly focused on the Copts as the targets of religious violence. One of the purposes of this conservative fundamentalist movement was to purify Muslim society from the distortions and perversities that allegedly had crept into it and to go back to a strict Islamic state with traditional interpretations of the Koran, the professed revelations by Allah to the prophet Muhammad.

Even though Egyptian Christians are a minority of Egypt's population—they claim about six million, but the government's estimate is half of that—Copts make up about 20 percent of the population of Upper Egypt south of Cairo. There is an especially heavy concentration in Asyut Governorate, where the village of Manshiet Nasser is located.

The reactions of the Egyptian government to the conflicts were intermittent crackdowns on Muslim radicals beginning in September, 1981, a few days before President Sadat's assassination by members of another fundamentalist group that also advocated the restoration of a pure Islamic

society in Egypt. Fundamentalist militancy continued, not only against Copts but also, after 1992, against vulnerable points of the Egyptian economy such as tourism and the secular government itself. The government responded even more harshly, executing twenty-nine Muslim militants in 1993 alone, the largest number to be executed in any year until then.

In the meantime, the government had come under the scrutiny of the United Nations. Its special rapporteur on religious intolerance, in his 1992 report, noted the alleged actions against Copts, including the imprisonment and torture of individuals who had converted from Islam to Christianity, as well as actions taken against Coptic churches and organizations. On their part, Coptic leaders charged that at the very least the government had been lax in protecting their community from the Muslim fundamentalist onslaught.

Consequences

This cycle of violence, repression, and denial of human rights led to a rising chorus of criticism by Western governments, nongovernmental organizations such as Amnesty International and Human Rights Watch, and the United Nations. Partly because of worldwide attention and partly because the Copts had started to fight back, the incident at Manshiet Nasser placed the subject of religious violence at the forefront of Egypt's public policy agenda in the 1990's. Even though, after the May 4 event, the government cracked down relentlessly on members of various Muslim militant groups, the final outcome remained tied to the future of Islamic fundamentalism in Egypt and to the improvement of economic conditions there.

Peter B. Heller

Lebanon's Prime Minister Karami Resigns

> *Collapse of the government left Lebanon firmly under Syrian control, exposed to continued civil strife, wracked by a failed economy, and battered by the Arab-Israeli conflict.*

What: National politics
When: May 6, 1992
Where: Beirut, Lebanon
Who:

OMAR KARAMI (1921-   ), prime minister
of Lebanon from 1990 to 1992

HAFEZ AL-ASSAD (1930-2000), president
of Syria from 1971 to 2000

ELIAS HRAWI (1926-   ), president of
Lebanon from 1989 to 1998

RASHID SOHL (1926-   ), Karami's
successor in 1992

Karami Resigns

Reappointed as Lebanon's prime minister by pro-Syrian president Elias Hrawi on December 24, 1990, Omar Karami, a Sunni Muslim, confronted almost the same intractable problems he had faced upon his previous appointment to that office in 1984. During Karami's new term in office, Lebanon continued to be torn by civil strife in Beirut and elsewhere in the country between Christians and Druze and Muslim militias. Fighting also raged in southern Lebanon between the Palestine Liberation Organization (PLO) and Hezbollah guerrillas on one side and Israeli armed forces on the other. Warlords further splintered the country.

Beirut, Lebanon's capital, lay in ruins. The Lebanese government remained officially and uneasily divided. There were strong pro-Syrian, pro-Christian, pro-Sunni Muslim, and pro-Shia Muslim parties and factions, as well as remnants of the proindependence forces of General Michel Aoun, which had sought the ouster of all Syrian influence in Lebanon. Furthermore, the nation's economy had been disastrously affected by fifteen years of civil warfare; by Syrian, Israeli, American, French, Italian, and United Nations interventions; and by terrorist hostage-taking.

Prime Minister Karami's downfall on May 6, 1992, was perhaps more the result of popular reactions to Lebanon's dying economy than of any other factor. Karami had lengthy experience in Lebanese ministries and, accordingly, much was expected of him. A graduate of Egypt's Fuad University, he had served variously as minister of economic and social affairs in the 1950's, minister of the interior as well as prime minister in the same decade, and minister of finance in the early 1960's. His leadership of the Parliamentary Democratic Front afforded him a strong political base.

Syria's military presence in Lebanon and its dominating influence over Lebanese governments prevented the Karami government from giving priority to the country's economic plight. Tourism, which had always brought in substantial revenues, came to a halt with the rising incidence of terrorism and hostage-taking during the 1980's. Inflation soared to 50 percent per year, and the value of Lebanese currency plummeted by more than 60 percent during three months of 1991-1992. In addition, foreign aid had been halted until the Lebanese government gained control of guerrilla forces in southern Lebanon and until its purportedly neutral government joined in multilateral talks with Israel that were part of the evolving Arab-Israeli peace process. Amid riots in Beirut, the burning of public buildings, and attacks upon officials and their property, the Karami ministry fell.

The Buffer State

Lebanon had gained a partial independence in 1920 and formal independence in 1943, but it always remained a buffer state. First it was part of the Middle Eastern aspirations of Great Britain and France. Later, it figured in the ambitions—political, territorial, and ideological—of neigh-

2436

boring Syria, Jordan, Israel, Iraq, and Iran. Eventually the PLO and other anti-Israeli organizations used Lebanon as a battleground and strategic factor. The Cold War rivalry of the pro-Israeli United States and the seemingly pro-Arab Soviet Union further complicated tiny Lebanon's domestic politics and foreign policy.

Internally, Lebanon suffered religious divisions between Christians and Muslims, with subdivisions within these groups. Nevertheless, until the mid-1970's, Lebanon prospered. Beirut, which some perceived as the Paris of the Middle East, became an impressive center for international trade and finance. The prosperity soon ended.

Escalating factional warfare, beginning in the early 1970's, erupted into full-scale civil war that raged through 1975 and 1976, pitting the PLO and left-wing Muslim groups against Christian organizations. Various Arab states, viewing the conflict in the context of their hatred of Israel, subsidized Muslims, while Israel financed and lent military support to Christians. To restore order, Syrian president Hafez al-Assad intervened with his troops—eventually sending forty thousand of them—in 1976. Israeli forces, seeking to crush the PLO, invaded Lebanon in 1981. The Israelis withdrew in 1985.

After a pan-Arab meeting in Taif, Saudi Arabia, in 1989, Syria agreed to redeploy its troops to Lebanon's Bekka Valley. By 1991, it had formally recognized Lebanon as a genuinely independent state. Syria, like Israel, was beginning to respond to resumed Middle East peace talks encouraged by the United States, other Western countries, and the United Nations. During Karami's first term as prime minister, battles between Muslim militias and Christians continued along with bombings, terrorist attacks, gang warfare, hostage-taking, and hijackings. Karami inherited similar circumstances for his second term. Unable to resolve Lebanon's problems, he resigned.

Consequences

Rashid Sohl, a Sunni Muslim lawyer as well as a former prime minister, succeeded Karami. Sohl's appointment was seen as a Syrian ploy to counter attempts by Lebanese Christians to weaken Syria's hold over their country. Assad presumed that a Sohl government would be unlikely to press him to remove Syrian troops.

Sohl's domestic priorities initially focused on organizing the first parliamentary elections in twenty years, this time under a new constitution that provided a more equitable sharing of power between Lebanese Christians and Muslims. By November, 1992, the Sohl ministry was supplanted by still another prime minister, Saudi-Lebanese billionaire Rafic Hariri. Hariri grandiosely planned the rehabilitation of Beirut and sought special powers to control the economy and public administration. He discovered that he had to be shielded by intensive security measures against assassination and car bombings. Syria remained the dominant force in Lebanese politics.

Clifton K. Yearley

Omar Karami, shown voting in the 1996 elections.

Thailand Declares National Emergency

> *Massive street demonstrations against Thailand's military junta, an unelected prime minister, and pervasive corruption resulted in a declaration of national emergency.*

What: National politics; Civil strife
When: May 18, 1992
Where: Bangkok, Thailand
Who:

BHUMIBOL ADULYADEJ, RAMA IX OF THAILAND (1927-), Thailand's constitutional monarch from 1946

SUCHINDA KRAPRAYOON (1933-), former general and coup leader who became unelected prime minister in 1992

ARTHIT KAMLANG-EK (1925?-), deputy prime minister

CHAMLONG SRIMUANG (1935-), ascetic anticorruptionist leader of the Palang Dharma Party who opposed Suchinda

NARONG WONGWAN (1925-), leader of the Justice Unity Party

Opposition to the Junta

Commencing in early May, 1992, antigovernment protesters packed the streets of Bangkok, Thailand. The government in power consisted of a military junta that had assumed authority after staging a bloodless coup in February, 1991. The immediate objects of Bangkok's predominantly middle-class protesters, who gathered in crowds as large as 100,000, were the former general and unelected prime minister, Suchinda Kraprayoon, and Deputy Prime Minister Arthit Kamlang-Ek. Suchinda previously had disavowed political ambitions and did not stand for office in the March, 1992, elections. Dissatisfied with the results of the elections, however, he allowed himself to be nominated for premier by a coterie of generals, disregarding the election returns.

By April, 1992, as Suchinda, functioning as unofficial premier, announced the new government's program, thousands of protesters gathered in Bangkok and continued to do so through the following month. Protesters rallied around Chamlong Srimuang. Chamlong was a former general turned ascetic. The anticorruptionist former Bangkok mayor led the Palang Dharma Party. In early May, Chamlong, in company with Chalard Vorachat and other opposition leaders, whose combined parties had registered 45 percent of the vote in the Thai parliament's lower house, began hunger strikes.

On May 17, a crowd of 200,000 opponents of the Suchinda regime, composed largely of members of the middle classes, including university students, gathered in Bangkok. According to the government, police were unable to control the demonstrators. On May 18, Suchinda, claiming that Thai democracy was endangered by communists, declared a national emergency and ordered the army to end the demonstrations. In doing so, the predominantly peasant army opened fire. At least fifty people were killed, six hundred were wounded, and hundreds more were arrested and imprisoned. Two days later, several generals appeared on Thai television. While minimizing the number of casualties, they all but acknowledged the moral victory of the opposition. Amid national shock, embarrassment, and fears of civil war, Thailand's relatively powerless but revered constitutional monarch, Bhumibol Adulyadej, intervened in the crisis. With Suchinda and Chamlong on their knees before him, as tradition required, the king urged them to cooperate in restoring order.

Generals, Classes, and Masses

Thailand, an independent nation of sixty million people, has groped its way toward democratization since the 1970's. It experienced rapid economic growth in the range of 7 percent to 10

Bhumibol Adulyadej.

(AP/Wide World Photos)

riched, business enterprise controlling important elements of the nation's infrastructure including the airlines, port facilities, communications, and a wide range of contracting.

The military appointed all members of the parliament's senate. Joined by many of Thailand's wealthy civilian business leaders and politicians, generals participated in the flagrant, widespread corruption that became a hallmark of Thai government. Corruption was magnified by vast concentrations of wealth and productivity in the Bangkok area. One-third of the rural population lived in poverty and gladly sold votes to politicians. Only Bangkok's growing middle classes and a few renegade generals provided impetus toward reform and democracy.

Ostensibly to end this situation and help satisfy middle-class demands, Suchinda and other generals engineered the February, 1991, coup, ousting allegedly corrupt Premier Chatichai Choonhavan and abolishing the 1978 constitution. The king endorsed the coup. Selecting the unimpeachably honest Anand Panyarachun as premier in March, 1991, while simultaneously promising a new constitution and new, free, elections, Suchinda briefly appeared intent on reform. Corruption, however, continued to flourish. Anand was replaced by wealthy businessman Narong Wongwan, whom the middle classes distrusted and the United States labeled as a drug trafficker. The United States halted military aid to the Thai government.

percent a year. Its pro-Western foreign policy constantly confronted its governments with serious security problems. For years it was the recipient of massive U.S. military aid. Prodemocracy movements, the country's rapid economic growth, and its governments' national security problems created a series of political crises and coups that helped explain the political tumult of the early 1990's and the events of May, 1992.

Thailand's pro-Western foreign policy had, by the 1970's, vastly increased the authority of the armed forces. Deeply involved in aiding the United States during the Vietnam War, the Thai military as a consequence faced Vietnamese-backed guerrilla and other insurgent attacks at home. By 1976, Thailand's generals either ran or significantly influenced Thai governments. Thereafter, relative peace and prosperity soon converted the armed forces into a vast, rapidly en-

Consequences

The bloody crisis of May, 1992, ruined Suchinda's political future, although he was given amnesty for his role in the massacre. In June, 1992, the generals appointed Air Marshal Somboon Rahong as prime minister. King Bhumibol refused to confirm him. Instead, the king invited

Anand Panyarachun, the former diplomat and caretaker premier, to resume office as prime minister. Allegations of corruption made against Anand in 1992, though unproven, swiftly brought Chuan Leekpai into office as a replacement. Afterward, it appeared that Thailand's cycle of coups—seventeen of them, nine successful, since 1932—might be broken with the election of Chuan. Still holding office in 1995, three years later, Chuan had acquired the distinction of being Thailand's longest-serving prime minister. Under his regime, economic growth resumed to about 8 percent per year and ambitious land reform and educational programs were under way. Furthermore, despite national concerns about the king's health, it appeared that Chuan had placed democratic practices and constitutional procedures for transfers of power on a solid, if not yet indestructible, foundation.

Clifton K. Yearley

China Detonates Powerful Nuclear Device

China revived the Cold War threat of international nuclear war with the explosion of one of the most powerful nuclear devices ever tested.

What: Military technology; International relations

When: May 21, 1992

Where: Xinjiang Province, People's Republic of China

Who:

DENG XIAOPING (1904-1997), leader of the People's Republic of China from 1978 to 1997

GEORGE HERBERT WALKER BUSH (1924-   ), president of the United States from 1989 to 1993

MIKHAIL GORBACHEV (1931-   ), president of the Soviet Union from 1985 to 1991

RICHARD A. BOUCHER (1951-   ), United States State Department spokesman from 1989 to 1992

The Explosion

On May 21, 1992, China detonated a one-megaton nuclear device at its underground testing site in the far western deserts of Xinjiang Province. The blast, equal in force to one million tons of TNT, was seventy times as powerful as that from the bomb dropped by the United States on Hiroshima in 1945. Shock waves from the massive explosion were detected eighteen hundred miles away in Hong Kong, where the Royal Observatory mistakenly reported the blast as a major earthquake.

The explosion also sent political shock waves throughout the world. A warhead of this size, mounted on China's advanced intercontinental missiles, could destroy any major city in the world, including Taipei, Washington, D.C., Tokyo, or Moscow.

At a time when Russia and the United States had reached agreement on reducing their nu-clear arsenals, China's massive nuclear test threatened a return to the Cold War. In the United States, President George Bush's administration immediately criticized the Chinese action. State Department spokesman Richard Boucher announced the U.S. government's regret that the Chinese had conducted the test and that the Chinese were not demonstrating the same restraint shown by the United States, Russia, and other nuclear nations.

Beginning in the late 1980's, relations between the United States and the Soviet Union had improved under Soviet president Mikhail Gorbachev's policies of *perestroika* (restructuring) and *glasnost* (openness). In 1990, the two countries agreed to the Threshold Test Ban Treaty, limiting nuclear tests to relatively small blasts of 150 kilotons (equivalent to 150,000 tons of TNT). In 1991, Bush and Gorbachev signed the Strategic Arms Reduction Treaty (START), an agreement to reduce the number of nuclear weapons in the two countries. These agreements eased the threat of nuclear war between the two countries and contributed to ending the four-decade Cold War.

The Chinese blast endangered this new global peace. It signaled to the world that China must now be taken seriously as a world leader with international nuclear capabilities.

Global Change

A number of domestic and international events occurring since the late 1980's influenced the Chinese decision to develop and detonate a powerful nuclear device. Under Deng Xiaoping's reform policies, China's socialist economic system was rapidly becoming more capitalist. Chinese were able to run their own businesses, sell goods in private markets, and even buy their own cars and houses. This greater economic freedom influenced many Chinese to seek greater political freedom. In the summer of 1989, hundreds of thousands of students and workers marching for democracy filled Tiananmen Square in Beijing; they were dispersed by tanks.

The Chinese government, however, was unwilling to institute political reform. Government leaders had witnessed the political, economic, and territorial collapse of the Soviet Union after Gorbachev had introduced economic and political reforms. They were determined not to allow the same sort of collapse in China. In June, 1994, the government broke up the democracy movement, ordering the use of troops and tanks against demonstrators. World leaders condemned the ruthless action of the Chinese government, enacting sanctions that would isolate China from the world community.

From this position, China watched the world order change and its allies disappear. Democratic legislatures replaced Communist governments in eastern Europe. The Berlin Wall fell, signifying the reunification of East and West Germany. The various states of the former Soviet Union sought aid and improved relations with the United States. China rapidly was becoming the last major Communist country left in the world. In order to strengthen its domestic and international authority, Chinese leaders needed to modernize its military.

The swift victory of United Nations forces during the Persian Gulf War in 1991 influenced Chinese military leaders. Modern radar systems, satellite surveillance, Patriot missiles, cruise missiles, and "smart" bombs enabled the United Nations forces easily to rout the Iraqi troops, which—like the Chinese army—were well trained but not well equipped. The lesson was not lost on the Chinese leadership. An editorial in *Jiefangjun Bao* (Liberation Army newspaper) reported the change in modern warfare.

After the Gulf War, China's military leaders began an extensive campaign to modernize the military by improving the quality—not merely increasing the quantity—of the country's armed forces and weaponry. They trimmed 10 percent of China's 3.2-million-member military, the largest standing army in the world. They purchased modern weapons from the former Soviet Union, including SU-27 fighter jets, one of the most advanced aircraft in the world. They also increased funding for research on missile development and nuclear technology.

The one-megaton nuclear explosion of May 21, 1992, emphatically indicated the success of China's military modernization. It raised China to the status of a major international nuclear power with the military strength to defend or dominate all of Asia.

Consequences

China's detonation of a massive nuclear device, as well as the strengthening of its military forces, upset the fragile balance of power in Asia. In particular, the Republic of China on Taiwan became vulnerable to a preemptive military strike from the mainland. To restore Taiwanese security, President Bush reversed U.S. policy limiting arms sales to Taiwan and agreed to the sale of advanced F-16 fighter jets. This policy strained an already tense relationship between China and the United States, clouding the prospects for reunification of Taiwan and the mainland.

Daniel J. Meissner

United Nations Holds Earth Summit in Brazil

A milestone meeting of many nations resulted in international agreements to protect the environment.

What: International relations; Social change

When: June 3-14, 1992

Where: Rio de Janeiro, Brazil

Who:

GEORGE HERBERT WALKER BUSH (1924-   ), president of the United States from 1989 to 1993

WILLIAM KANE REILLY (1940-   ), director of the Environmental Protection Agency and head of the U.S. delegation to the conference

MAURICE FREDERICK STRONG (1929-   ), Canadian hydropower company executive and secretary general of the conference

The Earth Summit

The United Nations (U.N.) Conference on Environment and Development, often referred to as the Earth Summit, was held in Rio de Janeiro, June 3-14, 1992. The meeting's purpose, as stated by U.N. organizer Maurice F. Strong, was to prevent further degradation of the environment. Goals were to produce a set of principles for environmental protection and responsible development, a program of environmental action for the twenty-first century titled *Agenda 21*, and treaties to curb climatic change and limit losses of plant and animal species.

Two years of preparation and negotiations preceded the conference. An unparalleled 118 heads of state attended the conference. About thirty-five thousand representatives came from 178 countries, and about eight thousand journalists attended. The United States' delegation was headed by William Reilly, head of the Environmental Protection Agency. At the same time as the U.N. conference, almost eight thousand nongovernmental organizations from 167 countries held their own conference nearby.

Idealism was dampened by the realities of strong special interests. The Vatican and some Muslim countries fought to exclude references promoting birth control in *Agenda 21*. The oil-producing states argued against statements that urged reducing the world's dependence on fossil fuels.

U.S. president George Bush was openly reluctant to attend the Earth Summit because of his concerns about the effects of various proposals on U.S. industry. The treaty on greenhouse gases was watered down to placate the United States. Furthermore, because of fears for its biotechnology industry, the United States alone refused to sign the biodiversity treaty aimed at protecting plants and animals. Reilly bravely attempted a compromise on this issue but failed when his confidential memo to Bush's domestic policy adviser was leaked to the press. World reaction was merciless. When Bush arrived, he told the press that he was resisting global pressure to force huge U.S. spending on ecological concerns. Presidential candidate Bill Clinton responded that the conference should have capped a year of energetic leadership rather than grudging participation and general denial of ecological problems.

Bush did make strong efforts to protect world forests. Some developing nations argued, however, that forests were their resources, not the world's. The summit concluded by allowing continuing development of forest resources.

A global warming pact was signed that would reduce industrial emissions of carbon dioxide and other "greenhouse gases" to 1990 levels. The treaty, however, omitted deadlines because of objections by some countries. Twelve European countries had already agreed to try to reach this goal by 2000 and reaffirmed that pledge in Rio.

2443

Toward Sustainable Development

Awareness of the world's environmental problems has been encouraged by events. In 1986, for example, the nuclear power plant in Chernobyl in the Soviet Union released large amounts of radioactivity after a meltdown. In 1987, a garbage barge became a symbol of mounting trash problems in the United States as it traveled up and down the coast in an unsuccessful attempt to dump its load. In 1989, the oil tanker *Exxon Valdez* lost 11 million gallons of oil when it ran aground, causing the largest spill ever in U.S. waters. In 1990, about 200 million people participated in events on the twentieth annual Earth Day. In addition, the Gulf War in 1991 showed massive environmental damage resulting from the fighting and also revealed that U.S. foreign policy was tied to oil.

Increasing population and development have driven the cutting of old-growth and tropical forests. This threatens entire ecosystems. Organisms in those systems may be becoming extinct at a rate as high as fifty thousand species per year.

Mounting evidence shows two major risks to the earth's atmosphere. The use of particular industrial and household chemicals called chlorofluorocarbons may be causing chemical reactions in the upper atmosphere of the planet that result in the loss of ozone gas. Ozone absorbs ultraviolet radiation from the sun. Loss of ozone would allow increased ultraviolet radiation to reach the surface of the planet, increasing the likelihood of people getting skin cancer and cataracts (a clouding of the eye lens causing impaired vision). Crops also could be damaged.

Second, greenhouse gases such as carbon dioxide are increasing in the atmosphere and are believed to be causing a rise in average global temperature, with uncertain consequences for life and agriculture. These gases come both from burning of fossil fuels in industry and the slash-and-burn clearing of forests in developing countries.

Growing urban areas and increasing industrialization have also brought acid rain, the loss of farmland, overfishing, and lack of wood for fuel. All organisms modify their environment, but many people believe that unless something is done quickly about these particular changes, hunger, poverty, sickness, and death will increase.

Many people believe that the solutions to these problems can be summarized in the phrase "sustainable growth." The concept in-

One of the problems the Earth Summit addressed was protecting the earth's oceans from pollution caused by sinking oil tankers.

PhotoDisc

volves promoting and allowing activities that can be pursued without irreparable damage to the environment. Economic systems need to be designed that take into account the effects of any activity on the whole planet and all of its people, now and in the future. To do this effectively, nations must cooperate, possibly through organizations such as the United Nations.

Consequences

Although the conference did not meet everyone's hopes, it did not entirely fail. Principles were established. More people began to see that development and the environment are linked and that the world must continue to work toward sustainable activities. Some nations retreated from their Rio promises, but others showed their intent to honor their commitments. A year after the conference, the United States signed the biodiversity treaty with a rewording that guarded U.S. biotechnology interests and patents.

Clearly, problems remained. Nevertheless, at the close of the Rio meeting, U.N. secretary general Boutros Boutros-Ghali noted that the conference had marked a pathway for future action.

Paul R. Boehlke

Navy Secretary Resigns in Wake of Tailhook Scandal

> *After an incident in which women were sexually assaulted, the secretary of the U.S. Navy resigned and the Navy was forced to adjust its policies regarding women in the military.*

What: Social reform; Economics
When: June 26, 1992
Where: United States
Who:

PAULA A. COUGHLIN (1962-   ), Navy lieutenant who complained about sexual harassment at the Tailhook convention

FRANK B. KELSO II (1933-   ), U.S. Navy admiral and chief of naval operations

H. LAWRENCE GARRETT III (1939-   ), secretary of the Navy who resigned as a result of the scandal

Military Personnel Gather to Celebrate

During the Labor Day weekend of 1991, approximately five thousand persons, including U.S. Navy and Marine Corps aviators, many of them fighter pilots, converged on Las Vegas, Nevada, for the annual convention of the Tailhook Association, an organization named after the device that drops from the tail of a jet during carrier landings. Tailhook conventions were noted for alcohol consumption and rowdiness, in particular the destruction of property and sexual excess. Tailhook 1991 marked the nadir of the tradition and set into motion events that brought a reexamination of behavior in the United States military.

On Saturday night, Navy lieutenant Paula Coughlin, a helicopter pilot and admiral's aide, arrived at the third floor of the Hilton hotel. As she moved through the corridor, she was touched, fondled, and caressed by dozens of men. They reached under her skirt and inside her blouse. When she finally broke free, she learned that she had experienced what was known as "the gauntlet."

Coughlin spoke with her superiors repeatedly about the incident, but it was not until October 2 that a letter went to the Navy high command. On October 11, 1991, the Naval Investigative Service (NIS) began an investigation. On April 30, 1992, the NIS report was released, stating that twenty-six women had been sexually assaulted in varying degrees but that only two suspects could be identified, and those not sufficiently to warrant disciplinary action.

The news media, which had been following the story, sensed a cover-up and pressed for more details. Stories appeared that high Navy officers, including Admiral Frank Kelso II, had been on a veranda off the third floor the night of the incident and so should have known about—and, according to naval regulations, halted—the abuses. Secretary of the Navy H. Lawrence Garrett III ordered the investigation reopened. Shortly afterward, a supplemental report placed Garrett himself at the Hilton during the events, and the Department of Defense ordered its own investigation. On June 24, Coughlin revealed that she had filed the initial complaint and detailed what had happened. Two days later, Garrett resigned.

In April, 1993, the Department of Defense released its Tailhook report, naming 140 officers as either suspects in or being involved in incidents of sexual abuse. These included senior officers who were present but took no action to prevent unacceptable conduct from taking place. Among them were thirty-five Navy admirals and Marine Corps generals, including Kelso, who was by then chief of naval operations, the highest-ranking position in the United States Navy.

The disciplinary actions that followed were anticlimactic. More than half of the cases were not pursued because of insufficient evidence. Forty-three cases went to "admiral's mast," a military judicial proceeding a step below court-martial, resulting in "letters of admonition" or "letters of caution" placed in personnel files. Only six cases ever faced court-martial, and of those, three were dismissed outright and the final three were dismissed on technical grounds.

The Boys' Club

Historically, the United States Navy has been the most conservative of the nation's armed forces, and its self-proclaimed elite units, especially its fighter pilots, have resisted accepting women and members of minority groups. During the 1970's and 1980's, this began to change. Increasing numbers of women qualified as pilots, only to find themselves denied the opportunity to fly fighter jets. Officially, a 1948 law prohibited women from serving in combat situations, but attitudes within the Navy also contributed to the denial of opportunity.

Women who had chosen the military as a career chafed against this restriction, which reduced their chances for promotion and recognition. Gradually, some restrictions were lifted, and by the time of the Persian Gulf War of 1991, women had clearly demonstrated that they could serve in combat. Several women were captured during that conflict, and many performed in exemplary fashion under fire. Their question was, When would the remaining barriers come down?

As the Tailhook Association and its convention clearly demonstrated, the Navy as an institution did not want the barriers removed. The "fighter jocks" and their supporters saw naval aviation as an exclusively male preserve that could not function with the presence of women. Women

pilots were restricted to support roles in helicopters or transports. The Tailhook convention was a celebration and reinforcement of an intense "male bonding ritual" that denied women a place in the upper reaches of naval aerial warfare.

Senior officers, such as Kelso, who urged further integration of women into the Navy were met with obstruction and defiance, either open or covert. Advances made during the presidency of Jimmy Carter were undermined during the administration of Ronald Reagan. Under President George Bush, attempts were made to reverse the modest advances women had made and ban them from all opportunities for real advancement and promotion.

The Tailhook scandal, along with the election of Bill Clinton as president, brought an opportunity for sweeping changes in the United States military establishment. The publicity resulting from the scandal, combined with intense pressure from career military women, led to profound changes in how the American military is organized and how it will fight future conflicts.

Consequences

In April, 1994, the United States Senate debated whether Kelso should be allowed to retire as a full four-star admiral, as was customary, or as only a two-star admiral as penalty for his possible role in the Tailhook scandal. By a narrow vote of 54-43, Kelso was accorded the higher rank, but a symbolic victory had been won for women in the military.

The year before, during the first hundred days of President Clinton's term in office, the new administration had ordered that women be allowed to compete for assignments in fighter aircraft, including those involved in combat missions. Women had become more nearly equal partners in the United States military.

Michael Witkoski

Group of Seven Holds Summit

Leaders of seven of the world's most highly industrialized nations met in Munich, Germany, in an attempt to resolve international economic and political problems.

What: Economics; International relations
When: July 6-8, 1992
Where: Munich, Germany
Who:
GEORGE HERBERT WALKER BUSH (1924-   ), president of the United States from 1989 to 1993
FRANÇOIS MITTERRAND (1916-1996), president of France from 1981 to 1995
BORIS YELTSIN (1931-   ), president of Russia from 1991 to 1999

Leaders Meet in World Summit

On July 6, 1992, the eighteenth annual meeting of the Group of Seven (G-7) nations convened in Munich, Germany. Dominant issues of the summit included worldwide economic stagnation and the increasing probability of violence in Yugoslavia. There was hope of a new global trade agreement through the General Agreement on Tariffs and Trade (GATT). Participants expected financial support to Russia to help modernize its economy and anticipated that funds would be approved to increase nuclear reactor safety in Eastern Europe.

Little was formally agreed upon during the first day's session. The largest roadblock to progress on GATT came about through a conflict between the United States and France over agricultural subsidies by the European Community.

The second day featured an early arrival of guest Boris Yeltsin of Russia. American president George Bush encouraged a more active role for Russia in future G-7 summits. In the end, none of the other leaders was ready to expand to a Group of Eight. Despite this setback, Russia benefited from an agreement to extend $24 billion in financial credits to the former Soviet republics through the International Monetary Fund.

On other issues, there was less consensus. The G-7 leaders declared support for the United Nations in providing relief to civilians suffering from the civil war in Yugoslavia but provided few details. The participating nations failed to agree on ways to ensure that deterioration of nuclear power plants would not result in a regional disaster.

The talks ended on July 8, 1992, with the nations having narrowed their differences slightly on world trade, but they were far from agreement on GATT. They pledged to solidify agreement by the end of 1992. A final joint communiqué was issued to summarize the results of the Munich meeting. In a sense, the final communiqué was little more than a list of good intentions.

A New World Order

The G-7 meetings were originally designed to be informal gatherings of leaders of the seven richest democratic nations in the world. These meetings began in 1975, in an era in which the Soviet Union was considered a major military threat. It directly controlled much of Eastern Europe and indirectly controlled the governments in Poland and Yugoslavia. For several years there was relative continuity with respect to the seats of world power and the leadership of major nations. Continuity began to disintegrate in 1989. A number of the major industrialized nations experienced changes in leadership. More important, the communist influence and threat diminished. In 1991, the Soviet Union collapsed.

With the demise of communism, competition arose among the different varieties of capitalist systems. No single economy was strong enough to dominate worldwide markets. There was a growing fear that countries might become convinced that the time had come to put up barriers to trade to protect domestic interests. By 1992, regional trading blocs appeared to be the future

2448

direction for trade in the Americas, Europe, and Asia.

The Munich summit took place at a time when each of the member nations was experiencing major internal problems. The United States was stuck in a recession. American exports had dropped off, and jobs were being lost to overseas competitors. Budget deficits thwarted attempts to keep American interest rates down. The United States saw its economic clout erode internationally, but no other nation was ready to take the leadership role.

Germany experienced federal budget deficits even greater than those of the United States. The country was borrowing heavily to support the process of reunification. In order to attract foreign lenders, Germany had to offer high interest rates. This caused a large spread between U.S. and German interest rates that helped to drive down the value of the U.S. dollar. Japan enjoyed a trade surplus, but its economy was growing slowly and faced increasing prices and interest rates. If economic growth was slow in these three countries, it was even slower in the other G-7 nations. To make matters worse, G-7 leaders were collectively unpopular at home, and several had to face elections within the next eighteen months.

Stimulation of trade appeared to be a partial answer to worldwide economic stagnation. Recent summits had ended in pledges that GATT would soon be approved. These pledges all had proved empty. By July, 1992, GATT remained stalled over a relatively narrow dispute over farm aid. France was Europe's most prominent agricultural power. French farmers were already angry because their leaders had agreed to elimination of some price supports under a European Community trade agreement. Now the United States was demanding, as a condition of signing GATT, that the European Community cut its grain subsidies even further. President François Mitterrand of France believed he had no choice but to decline. President Bush, facing his own constituency of angry voters, had no choice but to stand pat.

Consequences

Immediately after the Munich summit, Bush publicly declared a number of accomplishments. European officials presented a similar positive picture. A few days later, Bush administration officials admitted that the president had displayed unwarranted optimism. Those who had initially painted a positive picture of the meeting were perhaps looking at changing attitudes. Those who saw no progress were looking at the details. Indeed, no agreements had been reached on GATT or nuclear plant safety. Both issues would take years to resolve. Whether groundwork was laid by the Munich talks is debatable.

The G-7 meetings had evolved gradually into media opportunities for world leaders to display their statesmanship for constituents at home. Those who expected concrete agreements became disappointed. The Munich meeting may have been one of the most disappointing summits of all.

Victor J. LaPorte, Jr.

AIDS Scientists and Activists Meet in Amsterdam

More than ten thousand scientists and AIDS activists meeting in Amsterdam revealed the possibility of a new AIDS-like virus.

What: Medicine; Health
When: July 19-24, 1992
Where: Amsterdam, The Netherlands
Who:

Jonathan Mann (1947-   ), chair of the eighth international AIDS conference

Luc Montagnier (1932-   ), French AIDS researcher and head of the team at the Pasteur Institute in Paris that isolated HIV in 1983

Robert Gallo (1937-   ), AIDS researcher and physician at the U.S. National Cancer Institute

Anthony Fauci (1940-   ), head of AIDS research at the U.S. National Institutes of Allergy and Infectious Disease

Jonas Salk (1914-1995), AIDS researcher and discoverer of a polio vaccine in 1955

AIDS Conference Convenes

At the opening of the eighth international conference on AIDS (acquired immune deficiency syndrome) in Amsterdam, conference chair Dr. Jonathan Mann advocated greater government involvement in AIDS research and prevention as well as the creation of political parties devoted to AIDS issues. The conference, an annual gathering of AIDS scientists, researchers, and activists, was originally scheduled to meet in Boston but was moved to Amsterdam to protest U.S. regulations, which prohibit persons infected with HIV (human immunodeficiency virus) from entering the country without special permission and bar their immigration.

The goal of the conference was to present new results of basic, clinical, and epidemiological research to help strengthen understanding of the social dimensions of the disease. Conference delegates somberly discussed research into the deadliest and costliest pandemic in history. They projected that by the year 2000 between 38 million and 110 million adults will be infected with AIDS. That estimate far exceeded the World Health Organization's (WHO) projection of 30 million. It was noted that although AIDS spread rapidly in Africa, Asia, and South America, government efforts to stem the epidemic had been inadequate. AIDS remains one of the most puzzling diseases facing medical science.

The biggest surprise of the Amsterdam conference was the shocking revelation of a possible new AIDS-like mystery virus. Between twenty-four and thirty patients in the United States and Europe had developed a new illness, termed "CD4 positive T-lymphocyte depletion in persons without evident HIV infection." Their blood showed no signs of HIV, the virus that causes AIDS. News of the new disease dominated the scientific sessions and captured the attention of the international press.

Various theories were asserted as to the cause of this disease. AIDS experts and virologists attempted to rule out known viruses or mutated, rare animal viruses not yet identified. Some scientists believed the disease was a combination of several different diseases. Dr. Luc Montagnier believed that the presence of a mycoplasma, a primitive bacterium-like organism, played a prominent role in the disease. Health officials attempted to gather information, put findings in context, and calm public fears. They promised that the blood supply was not threatened.

Highlights of the conference were public policy issues involving human rights, women and AIDS, the role of nongovernmental organizations in the developing world's struggle against AIDS, and the emerging roles of organizations of persons with AIDS. Women's issues received attention for the first time, and it was revealed that the so-called second generation of AIDS reflects a profound shift in the characteristics of the pandemic regarding the sex composition of HIV and AIDS cases. As a consequence of heterosexual transmission, women worldwide are becoming infected with the virus as often as men.

Inclusion of activists in the program eliminated the disruption evident at past conferences. Most of the protests were aimed at the shortcomings of the American government's complacent leadership in HIV/AIDS issues. Among the activists was actress Elizabeth Taylor, who condemned President George Bush for not doing enough to further AIDS research. Critics of the Amsterdam conference stressed the fact that 85 percent of the participants came from industrialized countries and that, therefore, the conference was not truly representative of all nations threatened by the pandemic.

The Pandemic Emerges

AIDS, discovered in 1981, has been identified as a disease of specific groups in American society, initially gay men, then hemophiliacs and intravenous drug users, and finally African American females and their children. Scientists at first questioned whether AIDS and HIV would become established among white, middle-class heterosexuals and eventually the general population. Within a short time after their discovery, AIDS and HIV evolved into a deadly worldwide epidemic of immense magnitude.

AIDS is an infectious disease with no known cure and complicated symptoms, but with a single cause—HIV. The definitive diagnosis of AIDS is based on laboratory analysis of blood. In developing countries where laboratory facilities are unavailable or too expensive, diagnosis depends on clinical symptoms using definitions devised by WHO. Because of the time lapse between an individual becoming infected with HIV and the onset of AIDS and the fact that HIV is clinically invisible, the disease can be transmitted to others. Information on its spread must be obtained through surveys that analyze blood tests of representative samples of the population. Those samples include target groups such as prostitutes, blood donors, women visiting clinics, and hospital patients.

WHO created and sponsored international forums on AIDS through annual conferences at which the dimensions of the pandemic, its features, and its cultural, economic, social, medical, and scientific implications were explored. Representatives from numerous academic disciplines as well as the international press attended. The annual conferences, which began in 1985 in Atlanta, Georgia, continued in Paris, France; Washington, D.C.; Stockholm, Sweden; Montreal, Canada; San Francisco, California; and Florence, Italy, prior to the Amsterdam conference.

Consequences

AIDS conferences took place in Berlin, Germany, in 1993 and in Yokohama, Japan, in 1994. Little progress was reported toward a cure for the deadly disease, and pessimism prevailed regarding anti-HIV drugs. Participants stressed that education and prevention are vital to the effort to conquer AIDS and HIV. Biennial conferences were planned beginning in 1995. By 2002, the fourteenth conference was being planned.

AIDS awareness has increased. As one indication and prompter of awareness, many celebrities wear symbolic looped red ribbons at public events. President Bill Clinton's administration allocated increased resources to AIDS research, and the Food and Drug Administration has taken steps to hasten approval and marketing of experimental AIDS drugs.

Marcia J. Weiss

2451

Americans with Disabilities Act Goes into Effect

The Americans with Disabilities Act of 1990 was a comprehensive mandate to end discrimination against individuals with disabilities.

What: Social reform; Civil rights and liberties
When: July 26, 1992
Where: Washington, D.C.
Who:
TONY COELHO (1942-   ), United States congressman from 1979 to 1989
AUGUSTUS F. HAWKINS (1907-   ), United States congressman from 1962 to 1990 and chairman of the Committee on Education and Labor
JOHN D. DINGELL, JR. (1926-   ), United States congressman from 1955 and chairman of the Committee of Energy and Commerce
DICK THORNBURGH (1932-   ), United States attorney general from 1988 to 1991
JACK BROOKS (1922-   ), United States congressman from 1953 to 1994 and chairman of the Committee on the Judiciary

A Mandate for Nondiscrimination

On July 26, 1992, many of the provisions of the Americans with Disabilities Act of 1990 went into effect. This historic act had been passed to provide a means of ending discrimination against disabled persons, making it possible for these persons to enter the economic and social mainstream of American life. The Americans with Disabilities Act addressed four areas in particular: employment, public services, public accommodations and services, and telecommunications. A fifth section made miscellaneous provisions.

The first section prohibited employers from discriminating against persons with disabilities who were otherwise qualified for jobs. Title II, Public Services, prohibited discrimination in the area of designated public transportation. Buses or trains that ran on fixed schedules and routes had to be made accessible to disabled persons, including individuals who use wheelchairs. In the case of trains operating between cities, the act required at least one passenger car per train to be made accessible to the disabled.

Title III identified public accommodations subject to the law. The list included such public places as hotels; restaurants; places for public entertainment; food, clothing, and hardware stores; dry cleaners; banks; beauty shops; travel agencies; repair shops; funeral parlors; gasoline stations; providers of health-related services; public transportation terminals; museums; libraries; parks; zoos; educational institutions; and day care, senior citizen, and social service establishments.

Title IV, Telecommunications, provided for protection of people with hearing or speech impairments by requiring the use of a Telecommunication Device for the Deaf (TDD), a machine that transmits coded signals through a wire or radio communication system, and telecommunication relay services, which use telephone transmission to enable a hearing- or speech-impaired person to engage in communication with a hearing individual. The final section of the act dealt with miscellaneous matters and stated that if any part of the act should be found to be unconstitutional, it could be removed without affecting the remaining provisions. Implementation of various parts of the act on July 26, 1992, had far-reaching significance for more than forty-three million Americans with disabilities. No previous law associated with the civil rights of Americans had been as comprehensive in its coverage as was the Americans with Disabilities Act.

Loopholes for Discrimination?

Events leading to the enactment of the Americans with Disabilities Act of 1990 reach back to the early twentieth century. Congress showed some interest in the rehabilitation needs of disabled veterans returning from service in World War I. Bills introduced for this purpose were extended in 1917 and 1918 to include the "industrially disabled." Amendments between the 1940's and the 1960's also focused on rehabilitation rather than discrimination. Section 705 of the Civil Rights Act of 1964 established the Equal Employment Opportunity Commission. "Equal opportunity," however, was based only on the grounds of "race, color, religion, or national origin." Thus, the issue of disability as a basis for discrimination was not yet considered.

The Rehabilitation Act of 1973 provided additional protection for disabled persons. A declared goal was to "promote and expand employment opportunities in the public and private sector for handicapped persons and to place such individuals in employment." Title V of the act prohibited recipients of federal assistance from discriminating against persons with disabilities. This did not close loopholes that could allow discrimination to take place without breaking this law. Programs not funded with federal money could discriminate against the disabled if there were no state law prohibiting such discrimination.

It became clear that such loopholes must be closed. The United States Congress worked on various versions of bills to address the problem. After a Senate version of the bill passed, it was referred to the Committee on Public Works and Transportation, the Committee on Education and Labor, the Committee on the Judiciary, and the Committee of Energy and Commerce for

President George Bush signs the Americans with Disabilities Act on the South Lawn of the White House on July 26, 1990.

AP/Wide World Photos

2453

study. Public hearings were conducted to obtain input from various groups such as the National Council on Disability. Joint hearings were conducted before the Subcommittee on Select Education and the Senate.

Senate Bill 933 officially became the Americans with Disabilities Act and made discrimination against disabled persons illegal. When President George Bush signed the Americans with Disabilities Act into law on July 26, 1990, he noted that the act "signals the end to the unjustified segregation and exclusion of persons with disabilities from the mainstream of American life."

Consequences

Since the Americans with Disabilities Act went into effect, reactions have varied. Expressions of support have countered allegations that the "bill has no teeth." Detractors cite inadequate funding and vague wording in the bill. Other expressions of frustration target issues such as who the disabled are, who is liable, how to ensure compliance, and how to handle potentially dangerous employees or the mentally disabled.

The cost of compliance also has caused contention. Individual businesses reported varying costs of compliance. These were offset in part by tax deductions. Increases in employee benefit costs were predicted. Persistent questions remained concerning whether, despite the law, the disabled still experienced reduced rates of employment. Numerous lawsuits were filed after the act became effective. The issues addressed by these lawsuits include protection of AIDS patients, insulin-dependent diabetics wishing to drive, and nonsmokers.

Victoria Price

Violence Against Immigrants Accelerates in Germany

> *The German government acted to stem violence by right-wing groups against foreign immigrants and asylum seekers.*

What: Civil rights and liberties; Law; Ethnic conflict

When: August, 1992-May, 1993

Where: Germany

Who:

HELMUT KOHL (1930-), chancellor of Germany from 1982 to 1998

RUDOLF SEITERS (1937-), interior minister of Germany until his resignation in July, 1993

ALEXANDER VON STAHL (1938-), chief prosecutor of Germany, dismissed in July, 1993

ECKART WERTHEBACH (1940-), head of the Force for the Protection of the Constitution

Violence Against Immigrants and Asylum Seekers

Although violence against immigrants and asylum seekers had been increasing since the reunification of Germany in 1990 and the fall of communism in 1991, few were prepared for the violence that erupted between August 23 and August 26, 1992, in the Baltic Sea port of Rostock, located in the former German Democratic Republic (East Germany). Once a bustling port with a major shipbuilding industry, the city had fallen on hard times since reunification because its industries lacked the technology to compete in a world market A housing shortage was made worse by the presence of thousands of immigrants and asylum seekers. The unemployment rate approached 70 percent. Hardest hit by unemployment were the young, because they lacked necessary job skills. Many despaired of ever finding employment and were ready to re-sort to violence. Immigrants were blamed widely for taking jobs.

On the evening of August 23, a crowd estimated at one thousand youths poured into the working-class section of Rostock, called Lichtenhagen, set up street barricades of burning garbage, and then firebombed hostels housing Vietnamese "guest workers" and Romanian Gypsies. The next night, the rioters turned their fury on the massed police forces, resulting in many arrests.

Later the same year, in another part of Germany, violence took a tragic turn. On November 23, a building in Mölln housing Turks was firebombed, this time resulting in the deaths of a fifty-one-year-old grandmother who had lived in Germany for many years, her nine-year-old granddaughter, and a ten-year-old niece. There were attacks on nearly two thousand foreigners during 1992, including the beating death of an African black by five youths.

In 1993, the violence seemed to accelerate. On May 29, 1993, in the steel city of Solingen, twenty miles north of Cologne, a crowd estimated at seven hundred youths, many of whom could be identified as neo-Nazi extremists, firebombed a hostel in which Turkish families lived. Two young women and three girls died. A dozen others escaped with light injuries, mostly from smoke inhalation.

Unemployment, Frustration, and Neo-Nazi Ideology

Because of Germany's unhappy Nazi past, it would be easy to blame the violence on neo-Nazis inspired by the legacy of the Nazis. Nazism died with Germany's defeat in World War II, making its ideology simply a part of history for most of Germany's youth.

2455

By 1990, the year of reunification, Germany had become one of the most technologically advanced countries in the world, with jobs increasingly only for the highly educated and skilled. Strong labor unions protected the jobs of those who were working. An economic recession beginning in 1991 increased unemployment among the young to nearly 20 percent in what had been West Germany and to 60 percent in what had been East Germany.

Tension built over the presence, especially in the working-class sections of German cities, of hundreds of thousands of "guest workers," immigrants, and asylum seekers. Many of them were easily identified by their skin color, customs, or dress. Many were fed, clothed, and housed at government expense.

In the post-World War II years, thousands of foreign workers flocked to West Germany to become part of its "economic miracle." Often taking jobs Germans did not want and usually not eligible for German citizenship, they kept their language, dress, and customs. Particularly noticeable were the 1.7 million Turks living in Germany.

The "foreign problem" became especially acute after 1990, when Germany was reunified. Reunification led to increasing unemployment in eastern Germany as inefficient factories were forced to close in the face of world competition. In 1991, with the fall of communism, hundreds of thousands of people were now free to move. Many chose Germany not as job seekers but for what was called "asylum." In 1949, when West Germany wrote its constitution, it included one of the most liberal asylum laws in the world. Anyone entering Germany could simply say he or she was seeking asylum from persecution. Those persons would be fed and housed at public expense until their case went to court, which could take years. In 1972, 7,000 people sought asylum. This rose to 256,000 in 1991 and to 500,000 in 1992. In the first months of 1993, 1,000 persons a day streamed into Germany, overwhelming the re-

sources of local governments, unsettling voters, and becoming increasingly a source of resentment among Germans, particularly unemployed people who also depended on governmental support.

Another result of the influx of foreigners was the growth of radical, right-wing organizations that were racist, anti-immigrant, anti-Semitic, and violent. They adopted many of the slogans and even the uniforms of the outlawed Nazis. Police Chief Eckart Werthebach estimated the number of persons belonging to these organizations at forty thousand. Only about five thousand of them were violent, but those few caused tremendous problems.

Helmut Kohl, chancellor of Germany, condemned the violence and called the situation a national emergency. He promised to take action.

Consequences

Pressed by criticism both at home and abroad, the German government finally adopted two policies. Under Alexander von Stahl, the country's chief prosecutor, and Rudolf Seiters, the interior minister, basic civil rights were restricted for citizens judged violently hostile to democracy. Werthebach was authorized to increase the number of agents assigned to counter the rightist movements. Some of the more violent groups were banned.

The second step was to amend the constitution to modify the liberal asylum law. This required a two-thirds vote in the German parliament. The liberal Social Democratic Party was finally forced by public pressure to vote for the change, which was approved on May 26, 1993.

On October 13, 1995, four men were sentenced in connection with the fire on May 29, 1993. Felix Koehnen, Christian Reder, and Christian Buchholz, all tried as juveniles, received the maximum allowable sentence of ten years. Markus Gartmann was sentenced to fifteen years.

Nis Petersen

South Korea Establishes Diplomatic Ties with China

In a controversial move, the Republic of Korea (South Korea) established diplomatic relations with mainland China.

What: International relations

When: August 24, 1992

Where: Seoul, South Korea, and Beijing, China

Who:

ROH TAE WOO (1932-), president of South Korea from 1988 to 1993

DENG XIAOPING (1904-1997), senior leader of the People's Republic of China

LEE SONG OCK (1934-), South Korea's foreign minister

CHIANG KAI-SHEK (1887-1975), political leader of Taiwan

RICHARD M. NIXON (1913-1994), president of the United States from 1969 to 1974

Establishing Diplomatic Relations

South Korea's president Roh Tae Woo, using Foreign Minister Lee Song Ock as an intermediary, established diplomatic relations with the People's Republic of China (PROC), announcing the opening of relations on August 24, 1992. Until that time, the South Korean government, centered in Seoul, had maintained diplomatic relations only with the island government established by Chiang Kai-shek in 1949 on Taiwan.

Like China, Korea is a divided country. Each pair of countries has two separate governments that vie for diplomatic recognition by the world's major countries. PROC maps before 1992, however, showed a single Korea and a single China, with P'yongyang and Beijing as their capitals.

The establishment of diplomatic relations with the PROC government meant that South Korea's diplomatic relations with Taiwan were breached. Beijing would not permit South Korea to recognize the dissident island government as legitimate, nor would Taiwan countenance South Korea's maintenance of diplomatic relations with both entities.

News of Seoul's intended recognition of the PROC was first leaked to the press in mid-August of 1992 by the foreign minister of Taiwan. By August 22, most major newspapers in the United States carried stories of the anticipated accord between the two countries, an event that occurred two days later.

As part of the agreement, the PROC continued to recognize North Korea as a legitimate political entity. The Chinese convinced the South Koreans to permit such recognition by arguing that this was a step toward unification and toward helping control the production of nuclear weaponry in North Korea. The possibility of North Korea producing nuclear weapons concerned the entire world.

South Korea had much to gain economically from the accord with the PROC. In 1991, even without the accord, trade between the two countries had reached $5.8 billion, much more than China's trade with North Korea and almost double South Korea's trade with Taiwan, which reached about $3 billion in the same year.

When the accord was announced, Taiwan responded by closing its embassy in Seoul, leaving it with embassies in and diplomatic recognition by only twenty-eight countries. It discontinued direct air service between Taipei and Seoul. Nevertheless, Taiwan could not afford to sever completely its strong economic ties with South Korea, the economy of which was rebounding strongly.

2457

AP/Wide World Photos

Roh Tae Woo.

Two Koreas, Two Chinas

As communism spread through Asia during the late 1940's, the PROC, which espoused a pure and unwavering communism, drove the rightist government of Chiang Kai-shek from the mainland to the island province of Formosa, renamed Taiwan, south of mainland China. With some United States support, Taiwan flourished economically, thriving under a free enterprise system that mainland China could not adopt if it was to remain consistent with communism.

In the same way that the PROC and Taiwan were split by the struggle between communism and free enterprise, Korea, which shared a substantial border with mainland China, faced a political upheaval in the early 1950's that led to the Korean War. The United States sided with South Korea, recognizing its government centered in Seoul. The PROC sided with North Ko-

rea, officially recognizing its government in P'yongyang.

As the conflict grew, the United States sent military advisers to South Korea, provided the country with arms and economic support, and eventually involved its own military forces in the conflict. The PROC provided personnel and arms to North Korea, which espoused the brand of communism the PROC was then promoting.

A cooling of relations between Taiwan and the United States occurred in the early 1970's as President Richard M. Nixon met in 1972 with PROC heads of state and brought about détente between mainland China and the United States. Nixon's visit to China created a breach between Taiwan and the United States.

As China softened many of its policies and gradually moved from communism toward a free enterprise system, South Korea became an important trading partner, thereby paving the way for the establishment of diplomatic relations between the two countries. The PROC could not afford to establish such relations, however, at the cost of discontinuing its diplomatic relations with North Korea, one of its nearest neighbors.

During the early 1990's, the United States was concerned about the possible production of nuclear arms in North Korea, a suspicion that much of Asia shared. Using this fear as a wedge, the PROC was able to manipulate the situation to its advantage. It convinced the Roh Tae Woo government that South Korea had everything to gain by permitting China to continue its diplomatic ties with North Korea. China could then pressure the P'yongyang government to permit inspection of its military and industrial facilities. This was perceived as a step toward controlling the spread of nuclear weapons.

Consequences

Five weeks after South Korea and the PROC reached an accord, South Korea's president Roh Tae Woo visited Beijing to urge PROC leaders to press North Korea to resume arms negotiations. China's president Yang Shangkun, however, urged restraint, arguing that to pressure North Korea too strongly might discourage reunification of the two countries. Nevertheless, China supported South Korea's policy of urging North Korea to forgo the production of nuclear weapons as a means of normalizing relations with the West.

Taiwan, meanwhile, experienced an immediate decrease in its tourist business, as South Koreans selected other travel destinations. New trade relations with mainland China stimulated the South Korean economy. Although relations between North Korea and the PROC cooled immediately after China's recognition of South Korea, in September, 1994, the PROC showed its support of North Korea by withdrawing from the Armistice Commission that tried to pressure the North Korean government to discontinue producing nuclear products.

R. Baird Shuman

Canadians Discuss Constitutional Reform

> *In a meeting in Charlottetown, Prince Edward Island, the Canadian prime minister and the ten provincial premiers formulated a new basis for Canadian national unity, but the accord later failed with Canadian voters.*

What: National politics; Political reform
When: August 27, 1992
Where: Charlottetown, Prince Edward Island, Canada
Who:

BRIAN MULRONEY (1939-   ), prime minister of Canada from 1984 to 1993

ROBERT BOURASSA (1933-   ), premier of Quebec from 1973 to 1976 and 1985 to 1994

JACQUES PARIZEAU (1930-   ), leader of the Quebec separatist movement

Once More, the Constitution

On August 27, 1992, Canadian prime minister Brian Mulroney announced that he and the ten provincial premiers had reached a settlement on constitutional reform following a summit in Charlottetown, Prince Edward Island. This was the third time in ten years that Canadians had tried to resolve issues dividing them and redefine themselves as a nation.

The Charlottetown accord promised major revisions in the Canadian constitution. The Canadian Senate previously had been an appointed chamber more resembling the British House of Lords than its elected American counterpart. The Charlottetown accord made it an elected chamber and gave it a considerable amount of power that promised to change the operation of Canadian politics in a fundamental manner. The Senate was given the right to amend the legislation of the lower, but traditionally more powerful, House of Commons. Such amendments could be overwritten only by a 60 percent margin in the lower house, or 70 percent in the case of any bill dealing with natural resources. If between 50 percent and 60 percent of the lower house voted to override a Senate-amended bill, the bill automatically would go to a conference committee comprising members of both houses. The purpose of these provisions was to make sure that the provinces, with equal representation in the Senate regardless of population, had more of a voice in the federal government, especially as far as natural resources were concerned.

Whereas most Canadian constitutional debates had dealt with relations between English and French Canadians, the Charlottetown accord formally recognized those who had been in what is now Canada before either the English or French arrived. Native Canadians had shown increasing political awareness and skill, such as Elijah Harper, who had been crucial in the 1990 failure of the Meech Lake accord, the last attempt before Charlottetown to come to a constitutional settlement. In this accord, native Canadians were recognized as a distinct people with considerable tribal rights, especially over natural resources, as well as being given substantial veto power over federal and provincial actions affecting their interests.

The major reason for the Charlottetown conference was continued discontent in the French-speaking province of Quebec. The Charlottetown accord had at its heart the recognition of Quebec as a "distinct society" empowered to preserve its own particular culture and way of life. Although Jacques Parizeau, the leader of the Quebec separatist movement, was not satisfied by the clause in the accord related to the distinct society, Robert Bourassa, the more moderate premier of the province, accepted it. Other provinces frequently complained that Quebec received special and disproportionate treatment. For this reason, the "distinct society" clause was embedded

2460

in a set of resolutions that gave more power to the provinces and less to the federal government. The recognition of Quebec's cultural and linguistic autonomy thus was associated with a reinvention of Canadian federalism that had long appealed to elements in the western provinces, which were usually the most fiercely opposed to Quebec's nationalism.

The Charlottetown accord did not fully satisfy any interest group, but it gave some sort of reward to each of them. Even though it was a compromise adopted under pressured conditions, it promised to appeal to a Canadian populace anxious to have these matters settled.

One Country, Several Nations

For more than two centuries, English and French Canada have coexisted uneasily. Despite massive English predominance in economic and political power, French language and culture managed to hold their own in the country. French influence was concentrated in the province of Quebec.

In the beginning and middle of the twentieth century, it looked as if Quebec would become wholly assimilated within a united and bilingual Canada. The growing urbanization and modernization of Quebec promised to end Quebec's sense of separateness and particularity. Exactly the opposite happened. Quebec grew more buoyant and self-confident, and it produced a new political elite capable of articulating the necessity of Quebec's independence. Although a referendum held in May, 1980, failed to produce an endorsement of independence by Quebecers, it was clear to national Canadian leaders that the problem of Quebec's coexistence with the rest of Canada remained to be settled.

Canadian prime minister Brian Mulroney placed his power behind the Meech Lake accord of 1987. This agreement had as its principal goal amendment of the Canadian Constitution Act of 1982 in order to provide for the recognition of Quebec as a "distinct soci-

ety." After a long process of negotiation with the ten provincial premiers, the Meech Lake accord ultimately failed to be adopted because of objections by native Canadian leaders as well as by the premier of Newfoundland, Clyde Wells.

The tension between the two founding cultures of Canada, the English and the French, was complicated by the growing power of native Canadians as well as continued immigration to Canada, especially from Asia and the Caribbean. Canada, which for so long had been two nations, now contained several within its borders. Furthermore, Canadian regionalism had become heightened. Political opinion in the western provinces was opposed to the eastern provinces, which it saw as seats of entrenched power. Mulroney

Robert Bourassa.

AP/Wide World Photos

and the provincial premiers sought to address these potentially divisive ideas in the Charlottetown accord.

Consequences

The Charlottetown accord was rejected overwhelmingly by Canadian voters in a referendum on October 27, 1992. A casual observer might have reacted to this by assuming that Canada was at an end and that the several interest groups that had tried to work out an accord at Charlottetown were doomed to divide the country. Most Canadians, though, treated the outcome of the referendum as a reflection on the personal unpopularity of Mulroney and his handling of the troubled Canadian economy. Although the unresolved problems of Canada's several nations within one country remained urgent, Canada looked to survive as a single entity for the near future. A referendum on independence for Quebec, held on October 30, 1995, failed by a vote of 49.4 percent to 50.6 percent.

Nicholas Birns

Iraq Is Warned to Avoid "No Fly" Zone

In order to stem Saddam Hussein's military attacks against rebellious Shiite Muslims in southern Iraq, on August 27, 1992, the United States, Great Britain, and France established a "no fly" zone for Iraqi aircraft below the 32d parallel.

What: International relations; Military conflict
When: August 27, 1992
Where: Iraq
Who:
GEORGE HERBERT WALKER BUSH (1924-), president of the United States from 1989 to 1993
SADDAM HUSSEIN (1937-), president of Iraq from 1979
JOHN MAJOR (1943-), prime minister of Great Britain from 1990 to 1997
FRANÇOIS MITTERRAND (1916-1996), president of France from 1981 to 1995

The "No Fly" Zone

On August 27, 1992, the United States, Great Britain, and France made a joint decision to establish a "no fly" zone in southern Iraq. The decision not to allow Iraqi aircraft south of the 32d parallel was controversial. There were those who doubted the humanitarian motives of the allies. The Iraqis saw the change as a plot designed to break up their country. Iranian leaders wondered about U.S. president George Bush's motives—did he support the "no fly" zone simply to give himself a boost in public opinion polls and improve his chances for reelection? Such doubts were echoed in newspaper editorials within the United States. Egypt, Syria, and several Persian Gulf states worried that Iraq would eventually splinter and leave Iran in control of the area. The Arabs wanted to keep the eastern flank of their world intact, but not necessarily under the control of Iran's fundamentalist Shiite government.

The allied leaders dismissed such concerns about the "no fly" zone. Their stated motives were simple—they were merely responding to the de-fiance of Saddam Hussein, the political leader of Iraq. Hussein, the Allies argued, had repeatedly defied requests of the international community and violated the cease-fire agreements still in effect from the Persian Gulf War of 1991. Hussein, for example, deployed antiaircraft missiles in northern Iraq, refused to accept his postwar border with Kuwait, blocked humanitarian aid convoys throughout the countryside, and interfered with United Nations (U.N.) weapons inspectors trying to investigate suspected nuclear, biological, and chemical facilities.

As a result of these violations and many others, the United States, Great Britain, and France decided to issue an ultimatum and then strike if Hussein did not comply. The allies' military options were limited by the fact that U.N. representatives remained in Iraq. If his three foes attacked him, Hussein could take hostages from among more than one thousand U.N. representatives. The "no fly" zone policy seemed to be a better option. It would help the Shiites in southern Iraq and possibly inspire the overthrow of Hussein from within the country.

U.N. Resolution 688, which passed in April, 1991, required Iraq to ensure the human and political rights of all of its citizens. Without any legal blessing from the United Nations, the United States and its two allies, Great Britain and France, responded to Hussein's aggression and adopted a "688 strategy" to defend the rights of Iraqi citizens. The strategy went into effect on August 27, 1992, and established an Iraqi "no fly" zone south of the 32d parallel. The zone covered 47,500 square miles, or about one-fourth of Iraq's land area. The goals of the strategy were simple: to protect the Shiites, although not from Iraqi ground forces already in the area; to stir up a mutiny in the south; and possibly to inspire a palace coup in Baghdad against Hussein. The strategy was not immediately successful in achieving its three goals, although allied fighter planes continued to keep the skies clear of Iraqi aircraft below the 32d parallel.

The Politics of the Decision

After the Persian Gulf War, American, French, and British aircraft continued to control the skies over the northernmost portion of Iraq. Their goal was to protect the local Kurdish population from the wrath of Hussein. The Kurds were an ethnic minority that rebelled against the Iraqi dictator after the Gulf War. They tried to establish their own state, but the so-called butcher of Baghdad had other ideas. He wanted to reassert his control over northern Iraq and force the Kurds into submission.

To prevent this from happening and also to prevent the creation of a Kurdish state, the United States, Great Britain, and France established a "no fly" zone north of the 36th parallel. The Allies vowed to attack any Iraqi aircraft or ground troops that entered the zone and tried to attack the Kurds. The policy appeared to work. The Kurds remained protected, and the Iraqi regime did not bully its way back into northern Iraq.

In the case of southern Iraq, however, things were different. The Allies did not establish a "no fly" zone there after the Gulf War, even though local Shiite Muslims also rebelled against Baghdad. The uprising failed, but rebels continued to operate in the vast marshes where the Tigris and

Euphrates rivers meet. Hussein wanted to crush these rebels once and for all and turn the area into a "dead zone." He deployed 100,000 troops and began a strafing and bombing campaign against the estimated 50,000 Iraqi Shi'ites living in the marshes.

The Iraqi campaign did not go unnoticed. Max van der Stoel, a U.N. human rights inspector, reported on August 11, 1992, that Hussein was killing his own people. This was a violation of U.N. Resolution 688. The United States and its two allies, Great Britain and France, responded to Hussein's aggression with their "688 strategy."

Consequences

The "no fly" zone policy failed to accomplish its most important goals. Because it did not require Iraqi ground forces to retreat north of the 32d parallel, troops stayed in southern Iraq. They drained wells and defoliated marshlands, and they then used tanks and artillery to wipe out completely the Shiite Muslim rebels. The United States, Great Britain, and France may have controlled the skies of southern Iraq, but there was no real resistance to Hussein on the ground. Further, there have been few mutinies and coup attempts against his government. By protecting his supporters from the effects of the U.N. economic blockade, Hussein bought their loyalty, at least for the near future, and remained firmly in power into the next century.

Peter R. Faber

Non-Aligned Movement Meets in Indonesia

The leaders of the Non-Aligned Movement, a loose and mainly Third World alliance, met to discuss its future and agreed that it should continue to represent their interests despite the end of the Cold War.

What: International relations
When: September 1, 1992
Where: Jakarta, Indonesia
Who:
FIDEL CASTRO (1926 or 1927-   ), prime
    minister of Cuba from 1959 to 1976;
    president of Cuba from 1976
HOSNI MUBARAK (1928-   ), vice
    president of Egypt from 1975 to 1981;
    president of Egypt beginning in 1981
SUHARTO (1921-   ), president of
    Indonesia from 1967 to 1998

Summit of the Non-Aligned Movement

On September 1, 1992, the Non-Aligned Movement (NAM) began its first summit following the collapse of the Soviet Union. The NAM's leaders met in Jakarta, Indonesia, to discuss the continuing relevance of their alliance in the new, post-Cold War era.

To many outsiders, it appeared that the NAM no longer had any real purpose in the 1990's. It had been founded nearly four decades earlier, in 1955, as part of an effort by Third World countries to insulate themselves from the Cold War, which sometimes seemed to require every nation to choose either the Soviet or the Western side. Its other purpose was to raise issues of Third World economic development, decolonization, and social justice onto the agenda of international politics, which NAM members saw as disproportionately dominated by Cold War interests and issues.

With the end of the Cold War, there was no longer any great international divide among the superpowers and therefore no need to be "non-aligned." By 1992, the Soviet Union no longer represented an alternative to the West. In turn, this meant that Western countries began to perceive the Third World as having less strategic importance. As one result, there was increasing pressure on NAM members to adjust their outlook and policies toward Western approaches and concerns, in particular by adopting free market and democratic strategies of economic and social development.

Many Third World governments and elites had invested considerable time and effort in the NAM and were reluctant to see it fail. The majority of NAM member governments believed that it was even more important than previously to present the West with a united front of Third World concerns. Western economic aid and political attention were drawing away from the Third World and toward the emerging market economies and democracies in Eastern Europe and the former Soviet Union. Therefore, despite internal divisions and diverging paths to development and domestic government, the member states of the NAM agreed at their Jakarta meeting that the NAM should continue to represent their collective interests in the post-Cold War era.

The Issue of Alignment

The NAM began at a conference at Bandung, Indonesia, in 1955. It was founded in response to the perception that both sides in the Cold War were for the most part ignoring Third World concerns as well as seeking to force the generally poor and weak nations of Africa, Asia, and Latin America to choose sides. The three major figures in the NAM's early years were Gamel Nasser, dictator and prime minister (later president) of Egypt; Jawaharlal Nehru, democratic prime minister of India; and Tito, president and dictator of Yugoslavia. Throughout its first decades, the NAM's major themes and concerns were anticolonialism, opposition to racism (most notably, apartheid in South Africa), and formal neutrality toward the Cold War. In later years, it brought together Third World states that wanted to raise economic and social issues onto the agenda of international organizations, where most larger powers were otherwise preoccupied with Cold War security matters.

The NAM promoted the New International Economic Order in the 1970's. This was a campaign that included demands for guaranteed prices and markets for Third World commodities, heavy regulation of multinational corporations, free technology transfers to developing countries, and reform of bank rules regarding development loans. The program was resisted by many Western governments, and there followed a decade of rhetorical confrontation amid a crisis of deepening Third World debt.

During the Cold War, the NAM suffered from a chronic lack of unity born of the following factors. First, some of its members were not truly nonaligned, supporting either the United States or the Soviet Union. The Philippines, for example, backed the United States, and Cuba was solidly in the Soviet camp. These states sometimes acted as Trojan horses within the NAM for the varying agendas of the superpowers. Second, differences in levels of development, conflicting ideologies, and traditional regional animosities and territorial disputes among members were too large for the NAM to present a genuinely united negotiating front. Third, the rise of the OPEC (Organization of Petroleum Exporting Countries) oil cartel and the breakthrough of

several countries into the ranks of developed economies (for example, South Korea and Singapore) exacerbated differences among members and further divided the movement.

By the mid-1980's, the debt crisis led to a new pragmatism about mutual interests between the NAM and the West. The end of the Cold War advanced this new convergence of interests. Thus, by the 1992 summit the original purpose of the NAM was irrelevant and economic differences among member states were greater than ever.

Consequences

Perhaps out of sheer diplomatic inertia more than anything else, the NAM survived the end of the Cold War. It remained to be seen whether the Jakarta declaration of common intent and purpose would survive in the longer term, as member states scramble and compete for access to Western-dominated capital and export markets and contend in different ways with internal demands for political and social reform. The movement was still intact early in the twenty-first century; however, it still seemed possible that the NAM could yet break apart under the pressure of such competition and internal conflicts, but it may survive for some time as a forum wherein Third World countries discuss and jointly express collective concerns not otherwise prominent on the agenda of world politics.

Cathal J. Nolan

World Court Resolves Honduras-El Salvador Border Dispute

> *The World Court issued a decision, accepted by both Honduras and El Salvador, resolving a border conflict that had led to frequent violence in the past.*

What: International relations;
International law
When: September 11, 1992
Where: The Hague, The Netherlands
Who:
JOSE SETTE-CAMARA (1920-   ),
presiding judge of the five-judge
World Court panel
RAFAEL LEONARDO CALLEJAS (1943-   ),
president of Honduras from 1990 to
1994
ALFREDO CRISTIANI (1947-   ), president
of El Salvador from 1989 to 1994

The World Court's Decision

On September 11, 1992, the World Court (International Court of Justice) issued a decision ending a 130-year border dispute between Honduras and El Salvador. The court awarded almost two-thirds of the disputed 170-square-mile region to Honduras. The peaceful resolution of this dispute ended a disagreement that had led to violence in the past and helped spark a brief war in 1969.

The five-judge panel, headed by Jose Sette-Camara of Brazil, sat through more than fifty court sessions and examined more than twelve thousand documents dating back to the Spanish colonial era before reaching its verdict. Honduras gained control of a region along the Goascorán River and most of the territory along two other rivers. El Salvador received a smaller region of land along its border with Guatemala and control of two islands on the Pacific Coast. Judge Camara called the case the most complicated in World Court history.

A few hours after the decision, President Rafael Leonardo Callejas of Honduras met with El Salvador's president Alfredo Cristiani to celebrate the decision and praise the peaceful solution to the dispute. President Callejas explained that economic realities required both sides to accept the decision because close relations were necessary to promote Central American survival. El Salvador's first reaction had been critical of the World Court's view, arguing that the court had not paid enough attention to original Spanish colonial maps. President Cristiani vowed that his country would go along with the decision, however, because El Salvador had promised to accept the ruling and would not go back on its word. The few thousand people living in the area would receive dual citizenship so that they could move or stay, as they chose.

Origins of the Dispute

Disagreements over boundaries between Honduras and El Salvador date back to the 1850's. The Honduran-Salvadoran border remained undetermined in part because the region was very mountainous and had a small population. In 1861, a conflict over land resulted in several local villages taking up arms. A commission established by both governments failed to resolve the issue. After another short conflict in 1884, another commission was set up. It drew a map, but the Honduran legislature refused to accept the boundary. Conflict in the region broke out frequently in the twentieth century. The most noteworthy example was the 1969 Soccer War.

This conflict resulted from an increase in emigration from El Salvador into Honduras. Thousands of agricultural workers left El Salvador during the 1960's and entered Honduras seeking

2467

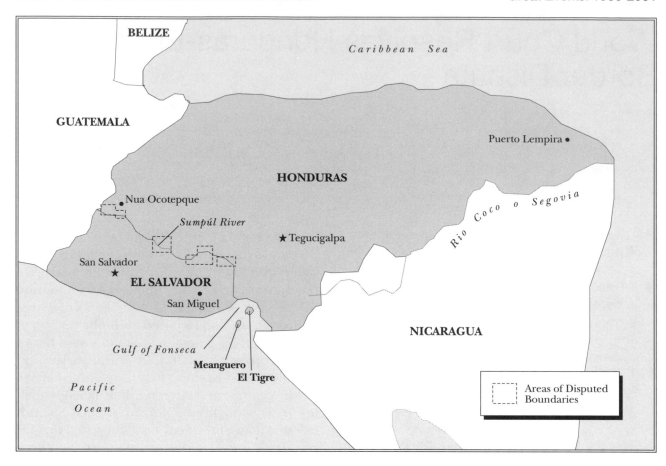

employment. El Salvador's problems resulted in part from a skewed distribution of land ownership. About 250 families owned 95 percent of the country's land, meaning that most Salvadorans could not support themselves by independent farming and were forced to work for wages. Honduran anger toward the growing number of immigrants led to the passage of laws limiting the number of foreign-born workers in any business to 10 percent of the payroll and outlawing ownership of land by anyone from El Salvador. Diplomatic relations between the two countries were broken in 1967 after an incident in which Hondurans took forty Salvadoran soldiers in the disputed area hostage. They remained in Honduran jails until American president Lyndon B. Johnson helped resolve the issue during a visit in June, 1968.

Nationalist tensions increased in 1969 when the two countries' soccer teams faced each other in the World Cup playoffs. Violence followed the first two games, played in each country's capital city. El Salvador won the third and decisive game,

played in Mexico City. On hearing the news, Honduran mobs in many cities savagely beat Salvadoran immigrants. In El Salvador, the smaller number of Hondurans were likewise beaten. Within hours, the governments broke diplomatic relations and called out their armies. El Salvador demanded reparations on behalf of its citizens in Honduras.

On July 14, the Salvadoran army invaded Honduras. It quickly won the war. Four days later, a cease-fire was arranged by the Organization of American States (OAS), though Salvadoran troops remained on Honduran soil until late in August. More than two thousand soldiers died in the conflict, more than 60 percent of them Honduran. The OAS sent a multilateral force into the region to patrol the border. It remained for six years.

Occasional fighting continued to afflict the area, and a treaty ending the war was not signed until October, 1980. This treaty failed to settle the boundary question, however, instead creating a binational border commission to look into the issue. If this committee could not reach a conclusion before 1985, both countries agreed to send the dispute to the World Court the following year and to abide by its decision.

The 1980 treaty reflected both nations' concern with the success of communist-led guerrillas, calling themselves the People's Revolutionary Army, in the disputed areas. Settlement of the border question was necessary to deprive the guerrilla units of the relatively safe zone from which they launched their attacks against both governments. The new agreement allowed both sides to send their armies into the region and to conduct joint operations against the rebels. Hundreds of people were killed in the area during these campaigns, and the rebels eventually were defeated.

Consequences

Both sides adhered to the World Court decision, demonstrating that international judicial decisions can have an impact on international conflicts if both sides are willing to accept the verdict. An extremely complicated border dispute was finally resolved peacefully. Many previous World Court decisions had been ignored because the court had no method of enforcing its findings. The El Salvador-Honduras settlement resulted from the willingness of both nations to abide by the verdict as a matter of honor. The outcome of this process may inspire other nations to make similar commitments in the future.

Leslie V. Tischauser

Peace Accord Ends Mozambique Civil War

President Joachim Chissano and resistance leader Afonso Dhlakama signed a peace treaty ending Mozambique's sixteen-year civil war.

What: Civil war
When: October 4, 1992
Where: Rome, Italy
Who:
JOACHIM CHISSANO (1939-   ), president of Mozambique from 1986
AFONSO DHLAKAMA (1952-   ), leader of Renamo (Mozambique National Resistance)

The End of Civil War

For sixteen years, government forces in Mozambique had battled Renamo (Mozambique National Resistance) troops, who claimed to be fighting for democracy and free enterprise. The long civil war ended in October, 1992, with the conclusion of the General Peace Accord, which called for a cease-fire, demobilization of soldiers, and multiparty elections.

The United Nations (U.N.) sent approximately eight thousand peacekeepers to oversee the agreement. The U.N. operation, called UNOMOZ, faced the difficult task of demobilizing soldiers on each side prior to elections. This was essential to avoid repetition of the situation in Angola, another former Portuguese colony, where failure to demobilize the warring factions led to renewed fighting after democratic elections failed in 1992.

Democratic elections in Mozambique were originally scheduled to take place within a year of the General Peace Accord. The prelude to Mozambique's first multiparty elections was long and complicated, and the election date had to be postponed. Troop demobilization was delayed as both sides tried to maintain a military advantage in case of a breakdown in the peace accord. Renamo stalled while it tried to transform itself from a guerrilla movement into a political party

and sought to gain assurances that it would play a role in governing the country, whatever the outcome of elections.

Elections finally were scheduled for October 27-28, 1994. One day before the elections began, Renamo's leader, Afonso Dhlakama, threatened to boycott them, but international pressure forced Renamo to reconsider. Western diplomats also encouraged Mozambique's president, Joachim Chissano, to form a government of national unity with Renamo in order to avoid the winner-take-all formula that had contributed to electoral failure in Angola. Chissano resisted this pressure.

The U.N.-monitored electoral process cost an estimated $1 billion. The elections were judged to be substantially free and fair, with more than 85 percent voter turnout. Renamo received a surprising 38 percent of the popular vote, giving it 112 seats in the 250-seat parliament. Frelimo (Front for the Liberation of Mozambique), the ruling party, got 48 percent of the popular vote and 129 seats in the parliament. The remaining 9 seats went to the Democratic Union party. Renamo won a majority of the popular vote in five of the eleven provinces in the country.

Renamo's leader, Dhlakama, was less successful in the presidential poll, receiving only 34 percent to President Chissano's 53 percent. On the strength of the Frelimo party's electoral victory, Chissano appointed a government made up entirely of his Frelimo supporters, ensuring continued political divisions.

Independence and Civil War

Mozambique gained its independence from Portugal in 1975, following a thirteen-year guerrilla struggle waged primarily by Frelimo. At independence, a Frelimo government came to power that espoused Marxist doctrine and implemented unpopular policies. The new government also represented a potential challenge to white minority regimes in Rhodesia and South Africa. By 1977, Frelimo was challenged by Renamo. The rebels of Renamo were organized by the Rhodesian security forces to destabilize the Frelimo regime and prevent the use of Mozambique as a staging area for attacks against Rhodesia's white-minority regime.

When Rhodesia became independent Zimbabwe in 1980, South Africa took over as Renamo's patron and provided assistance to the rebels in their increasingly brutal war with the Frelimo government. South Africa had a number of motives for supporting Renamo. Mozambique was committed to ending apartheid in South Africa and allowed the African National Congress to operate within its territory. South Africa wanted to prevent this and at the same time demonstrate its contention that black majority regimes were incapable of governing effectively. South Africa's leaders also were convinced that Frelimo's Marxist government represented a Communist threat to South Africa.

In 1984, Mozambique concluded the Nkomati Accord with South Africa. This accord called for the South Africans to stop their assistance to Renamo in return for the Frelimo government's agreement to prohibit antiapartheid rebels from the African National Congress from operating out of Mozambique. Despite the accord, South Africa continued to support Renamo, and the civil war continued. By 1992, there had been approximately 1 million casualties, some 1.7 million refugees had fled the country, and another 2 million to 3 million people were displaced by the war. The sixteen-year civil war left Mozambique one of the poorest countries in the world.

The end of the Cold War helped to set the stage for the end of Mozambique's civil war. The Frelimo government's Soviet backing ended, and South Africa withdrew its support from Renamo. It became apparent that a military solution to the conflict was unlikely. Momentum to-

Joachim Chissano.

AP/Wide World Photos

ward negotiations built, and peace talks began in Rome, Italy, in July, 1990.

Negotiations were protracted because each side raised difficult issues. Frelimo wanted a cease-fire, and Renamo wanted the withdrawal of Zimbabwean troops that had been protecting the rail corridor linking Zimbabwe and Mozambique's capital, Maputo. Renamo also demanded Kenyan mediation in the talks as a way to gain greater international recognition and counter charges of atrocities committed by its troops. Plans for the demobilization of troops also proved difficult to formulate. In the meantime, the Frelimo government implemented policy reforms that met many of Renamo's demands, including multiparty elections and renunciation of Marxism. In response, Renamo sought to maintain a military challenge in order to remain relevant.

2471

Consequences

Despite the success of Mozambique's first multiparty, democratic elections, the country continued to face formidable obstacles. The country remained one of the poorest in the world, and its economy was in shambles. The lack of economic opportunity, particularly for demobilized soldiers, and the presence of large numbers of weapons left over from the war contributed to rising crime rates. Ethnicity was a factor in the election campaign, raising the potential for future ethnic and regional tension. Political tensions between Frelimo and Renamo also continued, and other political parties formed. At the first session of the new parliament, Renamo members walked out in a dispute about the election of a parliamentary chair. A further legacy of the long civil war was the presence of an estimated two million land mines scattered throughout the country, threatening civilians and complicating vital agricultural production.

Robert J. Griffiths

British Queen Elizabeth Begins Paying Taxes After Castle Fire

A $120 million fire at Windsor Castle led Great Britain's Queen Elizabeth to reassess her financial relationships with the British government.

What: National politics
When: November 20, 1992
Where: London, England
Who:
ELIZABETH II (1926-   ), queen of Great Britain from 1952
PETER L. BROOKE (1934-   ), British minister responsible for the national heritage

The Windsor Castle Fire

On November 20, 1992, Windsor Castle, Queen Elizabeth's main residence, suffered a disastrous fire. The next day, Peter L. Brooke, the British minister responsible for the national heritage, announced that the government would pay the expected $120 million cost of restoring the castle. The public response to Brooke's announcement was unexpectedly negative.

Television and radio phone-in shows, newspaper editorials, and public opinion polls left no doubt that the general public wanted the queen to use her personal assets to restore the castle. The issue of restoring the burned-out castle seemed to prompt Britons to question the entire cost of maintaining the monarchy. One poll found that three-fourths of the British population believed that the cost of the monarchy should be reduced. For example, at that time the queen received support in an amount equivalent to $11.8 million per year. Her husband, Prince Philip, received an additional $540,000 annually, and the queen's mother received a stipend of $960,000. Several other members of the royal family received lesser amounts. Moreover, these amounts were tax-free, as was the investment income the queen received on her personal portfolio. In general, there was,

according to the newspapers, an impression of an out-of-touch government pandering to a wealthy and out-of-favor royal family.

Less than a week after the fire, Queen Elizabeth attempted to win back public favor by announcing that she would voluntarily begin paying income taxes, beginning in April, 1993, and would personally absorb the cost of supporting five family members. The latter activity would reduce government expenditures by $1.9 million annually. The financial impact of the decision to pay taxes was somewhat uncertain. Although many unofficial estimates placed the queen's personal wealth as high as $13 billion, Buckingham Palace officials suggested to reporters that her personal holdings amounted to about $100 million. They estimated that this latter amount would generate annual income of about $10 million. At the highest tax rate, 40 percent of income, the queen's expected tax bill would amount to approximately $4 million. The queen's actual tax returns are to remain well-guarded secrets; only a tiny group of tax officials will see the tax return forms.

Prince Charles supported his mother's decision and announced that he would also pay taxes at the 40 percent rate. Prince Charles previously had been paying a voluntary 25 percent tax. His additional taxes were expected to amount to $950,000 per year.

Monarchy Cost and Marital Problems

The prospect of paying $120 million to restore Windsor Castle was not the sole reason for the public outcry leading to the queen's decision to pay income taxes. The fire only brought to a head long-standing concerns about the cost of the monarchy and the lifestyle of the royal family. Recent marital troubles within the royal family

2473

A fire smolders in Windsor Castle.

had caused controversy among Britons, who had been brought up to believe that the monarchy could do no wrong. One newspaper called the queen the head of the world's most dysfunctional family. Princess Anne, the queen's daughter, recently had been divorced; the duchess of York, the queen's daughter-in-law, had been photographed cavorting in the nude with her Texas consort; and Prince Charles and Princess Diana were lurching toward a formal separation.

The tabloid newspapers emphasized the fallibility of the royal family. The British population was willing to support an infallible monarchy, but not fallible human beings. The queen acknowledged the problems at a luncheon given in her honor a few days after the Windsor Castle fire. She commented in her speech that 1992 had been an "annus horribilis," a horrible year. The speech, which was unusually personal, closed with a plea to critics to look at her family with "a touch of gentleness, good humor, and understanding."

Part of the controversy over the queen's tax-exempt status emanated from a book, published in early 1992, titled *Royal Fortune*. The author, Phillip Hall, estimated the value of the queen's stock portfolio at about $800 million. Hall also discussed the history of the monarch's tax-exempt status. In the early years of the British income tax, the monarchy paid its share. For example, Queen Victoria began paying in 1842. Victoria's grandson, King George V, negotiated a partial exemption in 1910. Queen Elizabeth's father, King George VI, was awarded a full exemption in 1937. Queen Elizabeth's decision to begin paying taxes merely restored what was a requirement prior to 1910.

Consequences

The queen's gesture of paying taxes may not have made much of a contribution to the nation's treasury, but the public relations value to the royal family was tremendous. One publication called the queen's decision the most dramatic gesture of her forty-year reign. Some called it a bold move toward a more accountable and more accessible monarchy. One government minister stated that the queen's decision would help bridge the gulf that was threatening to grow between her and her subjects.

Later events did more toward reducing the costs of supporting the monarchy. In August, 1993, the queen opened Buckingham Palace for public tours. The $16 admission charge was an effort to raise funds for the Windsor Castle restoration. The tours were expected to generate more than $6 million per year. Following the tour, visitors could purchase souvenirs. In late 1994, the queen allowed oil drilling to take place at Windsor Castle in another attempt to generate funds to offset the cost of supporting the royal family.

The burning of Windsor Castle was costly to the British government, but in many respects the disaster may have been a boon to the monarchy. The queen was forced to recognize her role as leader of the empire. The royal family had been hurt by the media's treatment of the monarchy, but the queen was able to use the fire as the springboard for a public relations coup.

Dale L. Flesher

United Nations Asks for End to Cuban Embargo

> *The United Nations General Assembly passed a nonbinding resolution calling for an end to the United States' embargo on Cuba.*

What: International relations
When: November 24, 1992
Where: New York City
Who:

GEORGE HERBERT WALKER BUSH (1924-   ), president of the United States from 1989 to 1993

FIDEL CASTRO (1926 or 1927-   ), prime minister of Cuba from 1959 to 1976; president from 1976

ALCIBIADES HIDALGO BASULTO, Cuban delegate to the United Nations

THOMAS L. RICHARDSON (1941-   ), British delegate to the United Nations

ALEXANDER F. WATSON (1939-   ), American delegate to the United Nations

The Resolution Passes

On November 24, 1992, the General Assembly of the United Nations (U.N.) overwhelmingly passed a nonbinding resolution calling for the United States to end its thirty-year embargo on Cuba. The vote was 59 in favor and 3 opposed, with 79 abstentions. Only the United States, Israel, and Romania voted against the resolution. Among the nations voting in favor were Canada, France, Mexico, Spain, Brazil, Venezuela, and New Zealand. Great Britain and the rest of the European Community abstained.

The United States' allies, which usually vote with that country, either voted for the resolution or abstained as a protest against a planned extension of the embargo. The Cuban Democracy Act of 1992 signed by President George Bush extended the embargo of Cuba to cover foreign subsidiaries of U.S. companies. Many nations saw the attempt to extend U.S. law beyond that country's borders as an infringement upon their sovereignty and upon their right to trade with any country they chose. The allies of the United States believed that they could vote against the United States because of the breakup of the Soviet Union, the end of the Cold War, and the fact that the resolution did not require the United States to take any action.

The Cuban delegate to the United Nations, Alcibiades Hidalgo Basulto, said that the embargo had cost Cuba $30 million and was "the most serious of the diverse forms of aggression that the United States has carried out against Cuba." Britain's U.N. representative, Thomas Richardson, criticized the United States' application of jurisdiction beyond its territory and characterized it as a "violation of a general principle of international law and sovereignty of independent states." U.S. representative Alexander Watson replied that Cuba was simply trying to use lofty sentiments to involve the international community in one aspect of its bilateral relations with the United States.

Cuba introduced the resolution because its economy was in shambles as a result of curtailed aid from the countries of the former Soviet bloc and their move to capitalism and market economies. The vote on the resolution clearly demonstrated the lack of support among the nations of the world, including the allies of the United States, for the thirty-year embargo on Cuba and the attempt to extend it.

The Embargo on Cuba

The embargo, imposed on trade with Cuba by the United States, began in 1960, one year after Fidel Castro came to power. It was the result of the opposing interests of the two nations. Castro

2476

had adopted policies to end political intervention and to break the economic domination of Cuba by the United States.

The United States invested in Cuba as early as the 1880's and increased its control after Cuba became independent in 1898. In addition to the sugar industry, Cuba's largest economic activity, the United States controlled insurance, banking, mining, and utilities.

Beginning in 1960, Castro started to nationalize industries and to confiscate other holdings of United States citizens. President Dwight D. Eisenhower responded by eliminating the Cuban sugar quota, which granted Cuba a percentage of U.S. sugar imports at prices higher than the international market price. The loss of the U.S. market was a serious economic blow to Cuba.

Castro retaliated by seizing all properties owned by U.S. citizens, accusing the United States of various crimes against Cuba and Cubans, and using vituperative language. President

Eisenhower on October 20, 1960, issued an embargo on all commodities except medicines, medical supplies, and a long list of foods. Foods subsidized by the U.S. government's Commodity Credit Corporation or from that corporation's stocks were prohibited. Corn, corn products, wheat, rice, and soybeans were included in the embargo. Technical information denied to communist countries was banned, and the sale, transfer, or charter of U.S.-owned ships to Cuba was prohibited unless approved by the Maritime Commission. The embargo became an important part of U.S. foreign policy toward Cuba.

The United States' attempt to persuade other nations to cooperate in the embargo met with limited success at first. Latin American nations and U.S. allies in Europe became increasingly unwilling to accept the policy.

Castro began to turn to the Soviet Union and the Soviet bloc for military and economic aid and for trade. As a result of Cuban dependence

Cubans protest the U.S. blockade of their nation.

on the Soviet Union, Cuba became a pawn for the Soviets in the Cold War between the United States and the Soviet Union. The Cuban Blockade and the Cuban Missile Crisis, along with Cuban military involvement in Africa, resulted from that dependence.

The Cuban Democracy Act of 1992, the occasion for the United Nations General Assembly resolution, was a continuation of U.S. economic policy toward Cuba. The act was passed following the breakup of the Soviet Union and, according to the United States Department of State, was intended to meet the new situation. The act was designed to hasten and facilitate the establishment of democracy in Cuba at a time when the future of Castro and Cuba seemed uncertain.

Consequences

Passage of the United Nations General Assembly resolution on the U.S. embargo of Cuba did not require the United States to lift the embargo. The embargo continued, but the attempt to extend it to subsidiaries of U.S. companies abroad antagonized the nations of the world, including the allies of the United States, and was not successful. Many people believed that the embargo harmed U.S. interests and drove Cuba into the Soviet orbit. The attempted extension of the embargo in 1992 did little to harm Cuba but damaged the international reputation of the United States, which continued to resist efforts to lift the embargo through the decade that followed.

Robert D. Talbott

Militant Hindus Destroy Mosque in India

> *In a defiant challenge to the secular principles enshrined in the constitution of independent India, militant Hindus stormed and destroyed the Babri Mosque, leaving behind a trail of Hindu-Muslim violence that engulfed India's major cities.*

What: Religion; Civil strife; Ethnic conflict
When: December 6, 1992
Where: Ayodhya, India
Who:
P. V. NARASIMHA RAO (1921-   ), prime minister of India from 1991 to 1996
LAL KISHAN ADVANI (1927-   ), leader of the opposition Bharatiya Janata Party (Indian People's Party)
VISHWANATH PRATAP SINGH (1931-   ), prime minister of India from 1989 to 1990

Destruction of the Mosque

On December 6, 1992, the Vishwa Hindu Parishad (VHP, or World Hindu Council) and the Rashtriya Swayamsevak Sangh (RSS, or National Volunteer Force), in league with the Bharatiya Janata Party (BJP), India's national opposition party, organized a rally to press for the demolition of a historic mosque in the northern Indian city of Ayodhya, in the state of Uttar Pradesh. The protest march escalated into violence as angry crowds made up of thousands of militant Hindus scaled barricades and succeeded in destroying the mosque with bare hands and sledgehammers.

The mosque, known as the Babri Masjid, was built in 1528 by the Emperor Babar, founder of the Mughal dynasty in India. The dispute over the mosque centered on the Hindu claim that the mosque was built by Babar after the demolition on that location of an ancient Hindu temple dedicated to Lord Ram. Ayodhya is venerated by Hindus as the birthplace of Lord Ram and is an important Hindu pilgrimage site.

In the wake of the destruction of the Babri Masjid, Hindu-Muslim riots broke out in major Indian cities such as Bombay, leaving more than twelve hundred people dead. In response to the events in Ayodhya, the BJP-led government of Uttar Pradesh resigned. The Indian national government under the control of the ruling Congress Party banned the VHP and the RSS for their complicity in the Ayodhya attack. Lal Kishan Advani, leader of the national opposition BJP, resigned and was taken into government custody. The BJP since 1986 had made common cause with the VHP and the RSS on a Hindu revivalist agenda calling for reassertion of Hindu primacy in India.

The destruction of the mosque epitomized the Hindu-Muslim tension that had plagued India since its independence from Great Britain in 1947. In that year, two states—India and Pakistan—were created from the British Indian Empire. Pakistan, a Muslim majority state, was founded on the principle of Muslim identity. India was founded on the principle of secularism, which has as its basis the separation of church and state. The growing strength of Hindu revivalism posed a threat to the secular foundations of India's democracy.

Muslim-Hindu Conflict

Hindu revivalists in India have claimed that Indian secularism in practice has often resulted in government policies that appease Muslims at the expense of the Hindu majority. This view has found a strong resonance among Hindus since the mid-1980's, for several reasons. The BJP platform became more popular in the light of the rapidly declining political fortunes of the secular-minded ruling Congress Party, whose leaders have been widely perceived as being corrupt, and the lack of credible secular alternatives. The growth of Punjabi and Muslim militant separatism in the states of Punjab and Kashmir has fueled Hindu insecurity.

2479

AP/Wide World Photos

Militant Hindus stand on a dome of the Babri Mosque in Ayodhya.

Finally, and more directly related to the Muslim question, many Hindus perceive that the ruling Congress Party has pandered to the Muslim minority. Such perceptions were fueled by isolated cases such as the Shah Bano episode of 1985, when a court order awarding a modest alimony to a Muslim woman deserted by her husband (in accordance with the country's secular law) was greeted with stern protests from Muslim clerics. In response, the Indian government took steps to permit Islamic law to prevail.

The mosque in Ayodhya came to symbolize the struggle of Hindu revivalists against secular-minded Hindus and Muslims. The controversy over the mosque predates Indian independence. Under the British, who duly noted competing Hindu claims to the site, Muslims were allowed to worship inside the mosque and Hindus were permitted to worship outside. After independence in 1947, the Indian government declared the

mosque closed to both communities. In 1949, riots erupted following a rumor that an image of Lord Ram miraculously had appeared inside the mosque. The image actually had been placed there by some Hindus.

In 1984, the VHP began agitating to have the site opened to Hindus. The movement was soon joined by the BJP, which saw in the issue a chance to improve its political fortunes nationally. A 1986 district court decision ordered the disputed site to be opened to the public in a move that allowed Hindus to gather and worship there. This decision sparked Hindu-Muslim violence all over India. In October, 1990, Advani, leader of the BJP, decided to march to Ayodhya and begin construction of a Hindu temple on the site of the mosque. He was arrested before he could enter the state of Uttar Pradesh, under the orders of Prime Minister Vishwanath Pratap Singh of the Janata Dal Party. Singh insisted that an interim court order securing the status quo at the site would be enforced. This action by the prime minister resulted in the resignation of his coalition government when the BJP withdrew its support. The events of December, 1990, resulted in yet another round of Hindu-Muslim violence that left more than three hundred people dead.

The publicity surrounding these events gave the BJP a boost that allowed it in 1991 to win control of the state government of Uttar Pradesh, where Ayodhya is located. It also catapulted the BJP into the role of the main opposition party at the national level.

Consequences

The BJP became the principal opposition party in India's national parliament as a result of its success in the 1991 elections. It won 119 of 545 seats, up from a mere 2 seats in 1984. The BJP's emphasis on Hindu revivalism raised important questions about the health of India's secular democracy. Some Indians argued that Indian democracy would rise to this challenge, but others feared that the emergence of Hindu nationalism would alter the secular foundations of the Indian republic.

Vidya Nadkarni

International Troops Occupy Somalia

In the wake of widespread anarchy and starvation, the United Nations voted for the first time to interfere in a country's internal affairs in order to protect the lives of its citizens.

What: International relations; Human rights

When: December 9, 1992

Where: Somalia

Who:

MOHAMMED FARAH AIDID (1934-   ), former general of United Somali Congress troops and head of the Somali National Alliance

ALI MAHDI MOHAMMED (1940-   ), president of the United Somali Congress and leader of the Somali Salvation Alliance

MUHAMMAD SIAD BARRE (1919 or 1921- 1995), socialist military dictator of Somalia from 1969 to 1991

GEORGE HERBERT WALKER BUSH (1924-   ), president of the United States from 1989 to 1993

BILL CLINTON (1946-   ), president of the United States from 1993 to 2001

The International Community Enforces Order

In the wake of massive starvation and confiscation of supplies of international relief agencies, in late November of 1992, U.S. president George Bush committed up to twenty-four thousand American troops to a Somalian operation. The United Nations Security Council unanimously endorsed the proposal on December 3, and six days later eighteen hundred United States Marines arrived in the Somali capital of Mogadishu at the head of the multinational United Task Force (UNITAF), which eventually numbered more than thirty-seven thousand troops. Total U.S. troop strength reached twenty-eight thousand. UNITAF's purpose was to safeguard humanitarian deliveries, initiate a low-key political process, and establish working relationships with all Somali factions and groups.

Early UNITAF operations were successful. Military protection for food deliveries and a good harvest in January reduced the threat of famine, and the warlords signed a cease-fire in Addis Ababa, Ethiopia, on January 15, 1993. The cease-fire stabilized the political situation in the countryside.

The second phase of active intervention began when formal command of Operation Restore Hope was transferred to the United Nations on May 4, 1993. The United Nations Operation in Somalia (UNOSOM) II was the first operation in which the United Nations asserted the right to intervene in a country's internal affairs in order to protect the lives of that country's citizens. Based upon Security Council resolution 814 of March 26, 1993, which authorized a sweeping nation-building program, UNOSOM II went far beyond the initial humanitarian aims of the UNITAF mission. Comprising more than thirty-three thousand troops and costing $1.5 billion annually, it was the most expensive peacekeeping operation in U.N. history.

Sporadic rioting continued in Mogadishu. On June 5, Somali rebels thought to be acting under the direction of Mohammed Farah Aidid attacked U.N. forces, killing twenty-four Pakistani troops. By mid-1993, UNOSOM had become increasingly concerned with capturing Mogadishu strongman Aidid. In an unsuccessful raid on his headquarters on October 3-4, 1993, two U.S. helicopters were shot down. Eighteen soldiers were killed, and seventy-eight were wounded. The outcry in the U.S. Congress led President Bill Clinton to promise to withdraw most American troops by March 31, 1994.

In November, 1993, the U.N. Security Council voted to end its manhunt for Aidid in order to encourage negotiations. Discussions between rival

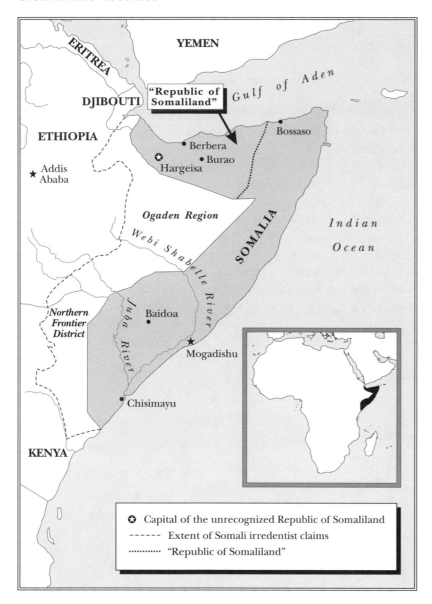

were frequently at war with their Christian Ethiopian neighbors throughout the Middle Ages. Regional relations were further fractured as the Somalis themselves were eventually divided among five distinct territories. France colonized the northern port of Djibouti in the 1860's, Britain occupied most of northern Somaliland in 1884, and Italy established control of southern Somaliland in 1889. The Somali Ogaden, west of British and Italian territories, was seized during the 1890's. Many Somalis also lived in northeastern Kenya, occupied by the British in 1895. During World War II, Italy was driven from Somaliland. It was granted a ten-year trusteeship in 1950 to prepare its former colony for independence.

In 1960, the former British and Italian regions combined to form the Somali Republic. After almost a decade of democratic rule in which Somalia received increasing amounts of economic aid from China and the Soviet Union, in 1969 the military-backed Supreme Revolutionary Council assumed power under Muhammad Siad Barre. The division of Somalis among Ethiopia, Kenya, and Djibouti had been a constant source of regional conflict. The conflict erupted into open warfare between Somalia and Ethiopia during 1977 and 1978.

Russia had been supplying both countries with arms and assistance since independence. All Russians were ordered out of Somalia late in 1977. After initial Somali successes, Ethiopian troops with the help of a Cuban army of ten thousand soldiers drove them from the Ogaden in March, 1978. In January, 1991, Siad Barre was driven from power by forces of the United Somali Congress (USC), destroying all effective central government. As the country disintegrated into dozens of armed factions ruled by local warlords, the for-

warlords yielded no settlement, though representatives for Aidid and Ali Mahdi Mohammed signed agreements on January 16 and March 24, 1994, renouncing violence and promising to work together to establish a new government. On the following day, the last major contingent of U.S. forces left Somalia, putting the peacekeeping mission in the hands of nineteen thousand UNOSOM troops. Fifty U.S. Marines remained in order to protect the forty U.S. diplomats in the country.

A History of Division

For hundreds of years, Somalia had been an area of international tension. Islamic Somalis mal economy collapsed. Much of the fighting was between USC factions led by President Ali Mahdi Mohammed and General Mohammed Farah Aidid. A U.N.-brokered cease-fire in February, 1992, had little effect. The five hundred U.N. peacekeepers who arrived in September to oversee distribution of humanitarian aid were too few in number to halt the massive looting of supplies. In the north, the independent Somaliland Republic, which had declared its independence in May, 1991, began to fracture along ethnic lines.

The drought of 1991 intensified problems. Starvation was rampant, and international relief efforts were often thwarted by the theft of food shipments by irregular troops of the various factions. By mid-1992, the International Red Cross estimated that 1.5 million of southern Somalia's 4.5 million people were in immediate danger of starvation. By the end of 1992, more than 300,000 people in northern and southern Somalia combined had died. UNOSOM I (April-December, 1992) failed in its diplomatic attempt to reconcile warring Somali factions.

Consequences

The prospects for political stability in Somalia appeared no better in late 1995 than they had been three years earlier, when the United States decided to intervene. Intervention by the United States and the United Nations saved hundreds of thousands of lives in the short term but produced no coherent policy for resolving factional disputes. The March 24, 1994, agreement had called for an April 15 meeting to lay the groundwork for a peace conference and creation of a national legislature, but that meeting was repeatedly postponed and eventually canceled. Under the protection of twenty-two hundred American and Italian marines, the last fifteen hundred UNOSOM II troops left Mogadishu on March 10, 1995.

John Powell

Israel Deports 415 Palestinians

On December 17, 1992, Israel deported an unprecedented number of Palestinians who were accused of membership in Hamas.

What: International relations
When: December 17, 1992
Where: Israel
Who:
YITZHAK RABIN (1922-1995), prime
    minister of Israel from 1974 to 1977
    and 1992 to 1995

Israel Deports Palestinians

On December 17, 1992, Yitzhak Rabin, prime minister of Israel, authorized deportation of 415 Palestinians from the West Bank and Gaza Strip despite an international outcry. The deportees were thought to be members of Hamas, an Islamic fundamentalist organization advocating active resistance against Israeli military occupation of the West Bank and Gaza Strip.

Although Israel had in the past deported Palestinians, the number involved in this incident was unprecedented. Palestinians feared that Israel was embarking on a campaign of population transfer. That fear was grounded in the rhetoric of Israeli extremist parties during elections.

The deportations came on the heels of a Hamas kidnapping and killing of an Israeli soldier. Subsequent events showed that the deportations had not been planned well. Some of the deportees had no connection to Hamas, and some were in need of medical attention.

The deportations were intended to calm Israeli tension. Some analysts believed that Rabin wanted to deal a blow to Hamas and thus enhance the status of the Palestine Liberation Organization (PLO) as a negotiating partner. Whatever strategy Rabin may have had, the deportations quickly backfired.

Jordan previously had declared that it would no longer accept deportees, leaving Lebanon as the destination of previous deportees. Lebanon declared that it would not accept these busloads of deportees. The Palestinians were sent off their buses in a desolate area of southern Lebanon and forced to march north to the incongruously named Marj al-Zuhur (Plain of Roses), a snow-covered area in the mountains. Relief organizations provided basic necessities such as tents, blankets, and food while a flurry of diplomatic activity commenced in an attempt to solve the crisis. Israel refused to let relief workers ship supplies through Israeli territory. The makeshift camp quickly became the site of demonstrations, press conferences, and interviews.

Israel appeared to be losing popular support in its peace negotiations. The PLO was embarrassed in the eyes of its people. Palestinian negotiators refused to resume the eighth round in the largely inconsequential peace talks until a satisfactory solution could be found. International condemnation of the deportation was not uncommon. The United Nations Security Council convened to discuss the matter and passed Resolution 799, which called on Israel, in part, to respect the applicable provisions of the Fourth Geneva Convention, to cease such measures of collective punishment, and to repatriate all the deportees. Israel, on December 28, allowed ten of the deportees to return upon discovering that they were not in fact members of Hamas.

The solution came at the prodding of the newly installed U.S. administration of President Bill Clinton. The United States and Israel agreed that exile would be for one year instead of two and that a phased repatriation would be put into effect until all the deportees had arrived home. Although this plan fell short of Palestinian demands and of the demands made in Resolution 799, the eighth round of peace talks resumed in Washington, D.C., in April, 1993.

Palestinian Opposition

Israel occupied the West Bank and the Gaza Strip following the June, 1967, Six-Day War. The PLO resisted the occupation and demanded Palestinian statehood. It gradually gained sympathy and support among Palestinians, who organized themselves on secular nationalist grounds. Islam, although it was the religion of the majority, was never regarded as an ideological basis for the resistance during the first decade of the occupation. Official Islamic leadership was tied to Jordan and was financed by the conservative regimes of Saudi Arabia and Kuwait.

Islamic activism increased during the late 1970's and early 1980's, spurred by the lack of success of the nationalists, by the Islamic revolution in Iran, and by the emergence of militant activism in Egypt. This activism was focused on "re-Islamization" of society, targeting such issues as alcohol consumption, veiling of women, and public morality. Newly founded Islamic groups, along with the Muslim Brotherhood, constituted an open challenge to the PLO, Palestinian Communists, and other leftists. They continued to receive financial backing from Saudi Arabia and Kuwait. Because at this stage they rarely took anti-Israeli stands, they operated under the auspices of the Israeli military, which hoped to split the Palestinian national movement. When the Islamic Jihad became more involved in the resistance, it clashed with the Israeli military, which largely suppressed it by the mid-1980's.

Palestinians began the *intifada*, an uprising against the Israelis, in early December, 1987. Elements loyal to the PLO took the initiative and formed the Unified National Leadership of the Uprising (UNLU), which organized the uprising by issuing clandestine leaflets that contained demands, directives, and instructions on how to conduct demonstrations, general strike days, and other forms of civil disobedience. By late 1988, the Islamists had regrouped and formed the Movement of Islamic Resistance, better known as Hamas, which began to issue its own leaflets, sometimes with different general strike days and locations. In the aftermath of the Persian Gulf War (1991), the Madrid Conference, and the activation of the peace process between Israel and the PLO, Hamas stood out as an active force among Palestinians who either criticized or opposed the talks.

Consequences

Despite opposition from some Palestinian groups, the peace talks between Israel and the Palestinians continued. Several rounds were held, but they produced little. This may have allowed Hamas to recruit more sympathizers. A secret channel of talks between Israel and the PLO led to the Oslo Agreement and the signing of the Declaration of Principles on the White House lawn on September 13, 1993. Later agreements moved toward establishment of a Palestinian national authority over Gaza and Jericho, then later other areas in the West Bank. These talks did not prevent atrocities from being committed by extremists on both sides. Despite the severity of those acts, and also because of them, peace talks continued. Rabin was shot to death on November 4, 1995, by Jewish law student Yigal Amir.

Mahmood Ibrahim

Kim Young Sam Is Elected South Korean President

In a three-way race, dissident Kim Young Sam won the 1992 presidential election with a 42 percent plurality.

What: National politics
When: December 18, 1992
Where: South Korea
Who:

KIM YOUNG SAM (1927-), president of South Korea from 1993 to 1998

KIM DAE JUNG (1924-), major contender in South Korea's 1992 presidential race

CHUNG JU YUNG (1914-2001), important contender in South Korea's 1992 presidential race

ROH TAE WOO (1932-), president of South Korea from 1988 to 1993

SYNGMAN RHEE (1875-1965), president of South Korea from 1948 to 1961

PARK CHUNG HEE (1917-1979), president of South Korea from 1963 to 1979

Kim Young Sam Wins

On December 18, 1992, Kim Young Sam was elected president of South Korea by a 42 percent plurality in a three-way race. Kim was the candidate of the Democratic Liberal Party (DLP). A political dissident, he had been imprisoned in 1961 for opposing the military coup that controlled South Korea for two decades, had been expelled from Parliament in 1979, and had endured house arrest in 1980 and 1982. He was banned from politics in 1982.

The chief contenders in the 1992 presidential election were Kim Young Sam, Kim Dae Jung, and Chung Ju Yung. The founder and owner of Korea's largest and most influential conglomerate, the Hyundai Group, Kim Young Sam formed a third party, the United People's Party, whose standard-bearer he became at the age of seventy-seven.

Comparisons have been made between Chung and Ross Perot. Although Chung captured a mere 16 percent of the overall vote—another 8 percent was divided among several minor candidates—this 16 percent was crucial in the outcome of the election. Kim Dae Jung, running on the Democratic Party ticket, received 34 percent of the vote, giving Kim Young Sam a clear victory, by a margin of 8 percent.

In 1987, as the presidential candidate of the Reunification Democratic Party, Kim Young Sam lost to Roh Tae Woo, candidate of the Democratic Justice Party. Roh Tae Woo served the single five-year term permitted under South Korean law. Kim enhanced his chances of winning the 1992 election by merging his party with Roh Tae Woo's ruling party, renamed the Democratic Liberal Party.

This combination of a political merger and of the candidacy of a significant third-party opponent ensured victory for Kim Young Sam. Kim was a popular symbol of the struggle against the authoritarian South Korean governments of former presidents Syngman Rhee, who served from 1948 until 1961, military strongman General Park Chung Hee, who served from 1963 until his assassination in 1979, and Lieutenant General Chun Doo Hwan, who seized power in 1979 and was reelected president under the new constitution of 1981, serving until 1988.

Partition, Authoritarianism, and Rebellion

Annexed by Japan in 1910, Korea existed as a Japanese colony until 1945. As World War II neared its end, the allied governments assured Korea of independence from Japan after the war. This promise resulted in partitioning of the country, which occupies a peninsula six hundred miles long between the Sea of Japan and the Yellow Sea.

The Soviet Union occupied Korea north of the 38th parallel, and the United States occupied the part of Korea south of that line. The two portions were named the Republic of Korea (South Korea) and the Democratic People's Republic (North Korea).

Rich in such minerals as gold, coal, iron ore, tungsten, and graphite, Korea occupies a crucial position in Asia's economy. The richest mineral deposits are in North Korea. South Korea has abundant supplies only of tungsten and graphite. Because 80 percent of the land is not suited to agriculture, Korea is dependent on the sea for much of its protein. The land that is suited to agriculture is carefully cultivated and produces substantial crops of rice, soy beans, and grains. The partitioning of Korea resulted in two quite unequal entities.

South Korea's first president, Syngman Rhee, who served for thirteen years, was authoritarian. When General Park Chung Hee grasped the reins of government, he installed a military regime that became increasingly dictatorial and repressive. It ended abruptly in 1979 with Park's assassi-

Kim Young Sam.

nation. The coup that brought about Park's assassination was led by Lieutenant General Chun Doo Hwan, who seized the presidency on Park's death and was subsequently reelected under the new constitution in 1981. He served until 1988.

During these tumultuous years, Kim Young Sam emerged as a dissident leader. His opposition to the oppressive governments in power from 1961 until 1988 led to three arrests and to his being banned from politics in 1982. This ban, enforced until 1985, worked to Kim's political advantage. He had established a reputation as an advocate of the people, having begun his political career in 1954 as the youngest member of Korea's parliament.

Kim's twenty-three-day hunger strike in 1983 during his house arrest attracted international attention. The resulting publicity made him a strong contender in his later bids for the presidency. His participation in the 1988 election kept Kim Dae Jung from winning, with 40 percent of the vote going to Roh Tae Woo. By bringing about a consolidation between his own party and the ruling party, Kim ensured his victory in 1992. His margin of victory was enhanced by the presence of a significant third-party candidate on the ballot.

Consequences

On October 9, 1992, Park Tae Joon, who chaired South Korea's Democratic Liberal Party, resigned from that post because of ideological conflicts with Kim Young Sam. Eleven other key conservatives left the party at this time to form the New Korea Party, and others were on the brink of resigning.

Upon assuming the presidency, Kim appointed a cabinet of twenty-six members, most of them drawn from the academic community. He then began a program of changing his country's politics and economy drastically, beginning by firing politicians known for their corruption. Under Kim's leadership, interest rates were reduced and $2 billion was funneled into subsidized loans for small businesses. These two acts stimulated an economy already on the rebound because of increased trade with mainland China.

R. Baird Shuman

Wiles Offers Proof of Fermat's Last Theorem

Andrew Wiles presented his proof of the "Last Theorem" of Pierre de Fermat, which had defied solution by mathematicians for more than three and a half centuries.

What: Mathematics

When: 1993

Where: Cambridge, England, and Princeton, New Jersey

Who:

ANDREW J. WILES (1953-   ), an English mathematician working at Princeton University

RICHARD L. TAYLOR (1962-   ), a mathematician at Cambridge University and a former student of Wiles

FRED DIAMOND (1964-   ), a Cambridge mathematician, also a Wiles student

An Old Mystery

Pierre de Fermat (1601-1665) was a French jurist whose genius at mathematics led to his occasional nickname, "the father of number theory"—number theory being the study of the relations of whole numbers. In 1637, while rereading the *Arithmetica* of the third-century Alexandrian mathematician Diophantos, Fermat was struck by a discussion of the Pythagorean theorem, which states that, for a right triangle, the square of the hypotenuse is equal to the sum of the squares of the other two sides—or, in mathematical notation, $x^2 + y^2 = z^2$, where z is the hypotenuse and x and y are the other sides; x, y, and z are all integers. The most familiar whole-number example of the Pythagorean relationship is probably that in which $x = 3$, $y = 4$, and $z = 5$.

Generalizing the Pythagorean equation to $x^n + y^n = z^n$, Fermat noted in the book's margin that this equation had no whole-number solutions for any n larger than 2, and that he had a "truly marvelous" proof of this assertion—which was, unfortunately, too long to write in the margin. This deceptively simple equation, and the implied proof of Fermat's statement, came to be called "Fermat's Last Theorem" (FLT), because it was the last of a number of his mathematical assertions left unproven at his death, and the only one to resist solution by later mathematicians.

Fermat himself provided a proof of FLT for $n = 4$, and the Swiss mathematician Leonhard Euler did so for $n = 3$. By mid-nineteenth century, FLT was proved for all cases up to $n =$ one hundred, and in the 1980's computer calculations extended this to four million. Yet demonstrations of specific cases, however extensive, are not the mathematician's definition of "proof." Proof must be established for the absolutely general case, from which an infinite number of specific cases can be deduced.

The Solution

This was where the matter stood when Andrew Wiles began work on FLT in 1986. His approach was to establish x, y, and z as points on "elliptic curves" (curves with the general equation $y^2 = x^3 + ax^2 + bx + c$), then to make use of the "Taniyama-Shimura conjecture," which maps these points onto a non-Euclidian surface called the "hyperbolic plane." (An alternative explanation relates the elliptic curves to a group of "modular curves" that are related to complex number planes. Both explanations are oversimplifications of very difficult mathematical ideas.) If elliptic curve solutions existed that violated FLT, they could not be mapped onto the hyperbolic plane, and the Taniyama-Shimura conjecture would be violated. Thus, solutions that violate FLT are impossible, and FLT is proved by this contradiction.

The difficulty Wiles faced was that no one had proved the Taniyama-Shimura conjecture, even for the limited set of cases he needed. This was the problem that occupied him for seven years. He solved it by devising mathematical counting systems for a group of elliptic curves called

2489

"semistable" and their modular counterparts. He could then show a one-to-one correspondence between the two groups that proved the Taniyama-Shimura conjecture in the limited case that was sufficient for his FLT argument. This was the greater part of the material of the three lectures at Cambridge University from June 21 to June 23, 1993, in which Wiles presented his findings. In the third of these lectures, he announced his proof of FLT almost as an afterthought, as a corollary of the work discussed in the first two lectures. The audience of normally reserved mathematicians burst into spontaneous applause.

The lecture presentations were gathered and expanded into a two-hundred-page paper submitted to the journal *Inventiones Mathematicae,* and flaws were found by the six referees to whom the manuscript was sent. Most of the flaws were quickly repaired, but one serious flaw involving the unproven upper limit of a mathematical construct, the Selmer group, took nearly two years to straighten out. Wiles finally appealed to one of his former students at Cambridge, Richard L. Taylor, and together they found a way around the missing proof. Wiles's overall proof of FLT was published in revised form in the *Annals of Mathematics* in 1995; in the same issue, Wiles and Taylor published the Selmer group work as a separate article.

Consequences

Proof of FLT was the headline story of Wiles's achievement, because of the longtime intractability of the problem and the near-hero status accorded its conqueror. To the world of mathematicians, however, the real story was the limited proof of the Taniyama-Shimura conjecture, which is already on the way to a more general proof in the hands of another Wiles student, Fred Diamond of Cambridge University. The Taniyama-Shimura conjecture provides a bridge between two major areas of mathematics, algebra and analysis, and is expected to further the "Langlands program," a hoped-for grand unification of all mathematics. In the meantime, Wiles's work has drawn on so many areas of mathematics that it has opened the way to a host of other advances in the field, and research papers building on his ideas have begun to appear in many journals.

Robert M. Hawthorne, Jr.

"Velvet Divorce" Splits Czechoslovakia

In what came to be known as the "velvet divorce," Czechs and Slovaks peacefully negotiated an end to their seventy-four-year political union.

What: International relations; Political independence
When: January 1, 1993
Where: Czech Republic and Slovakia
Who:
VÁCLAV HAVEL (1936-   ), president of Czechoslovakia from 1989 to 1992; president of the Czech Republic from 1993
VÁCLAV KLAUS (1941-   ), prime minister of the Czech Republic from 1993
VLADIMIR MECIAR (1942-   ), prime minister of Slovakia, intermittently from 1990

Slovaks Gain Independence

On January 1, 1993, Slovaks celebrated the birth of their new republic in the streets of the city of Bratislava. The realization of independence for Slovakia brought to an end the awkward political relationship that had been forged between Czechs and Slovaks during World War I. As 1993 began, the governments of both Slovakia and the new Czech Republic pledged themselves to the principles of democracy and human rights as expressed in the Helsinki Accords of 1975 and monitored by the Conference on Security and Cooperation in Europe.

On January 26, 1993, Czechs elected Václav Havel as president of their new nation. Havel, a former playwright, had led prodemocracy forces in the struggle against the communist regime during the 1980's. Although Havel represented the national ideals of the Czech people, the new government was dominated by Václav Klaus, the leader of the majority party in the Czech parliament. In July, Klaus was named prime minister. He continued to advocate both the Czech Republic's formation of economic and military associations with the nations of Western Europe and integration into existing associations.

Slovaks elected economist Michal Kovac as president in February. Prime Minister Vladimir Meciar continued to dominate the political life of the new state. Both a former communist and a strident nationalist, Meciar attempted to slow the process of privatization in Slovakia. His fiery rhetoric and erratic behavior sparked fears of persecution among minority groups, especially ethnic Hungarians.

The Decision to Separate

The collapse of Czechoslovakia's communist government in December, 1989, paved the way for the reemergence of democracy. Dissident leader Havel attempted to establish a government dedicated to preserving civil rights and to developing a capitalist economic system.

The reintroduction of democracy led to a proliferation of political parties. Some embraced the reforms of the new government, but others called for a return to the state socialism of the past. Economic hardship in the eastern region of the republic sparked a wave of Slovak nationalism. The results of a national election held on June 5-6, 1992, revealed a growing political split between the Slovaks and the Czechs of Bohemia and Moravia. The Civic Democratic Party, led by Klaus, won a third of the votes cast for the federal parliament and nearly 30 percent of the votes in the Czech assembly. Klaus, a former finance minister, was the architect of Czechoslovakia's transformation from a socialist to a free market economy.

In Slovakia, the Movement for a Democratic Slovakia, led by former boxer Meciar, captured 37 percent of the vote in the Slovak parliament. Meciar pledged to slow the pace of economic reform and to give Slovaks greater control over their internal affairs. Although Meciar vowed to create a sovereign state in Slovakia, polls taken at

2491

three votes more than the three-fifths majority required by law. Two earlier votes on the measure had been defeated by a coalition of communists, Social Democrats, and conservatives.

On January 1, 1993, two new independent states legally came into existence. Although many Czechs regretted the end of the federation, separation was accomplished with a minimum of distress and without violence.

Consequences

The Czech Republic prospered after the division of the federation. Foreign capitalists invested heavily in Czech industry, and the unemployment rate remained at about 4 percent during 1993.

Slovakia was plagued by both political and economic instability. The Meciar government increased tensions between Slovaks and ethnic minorities by attempting to regulate the use of minority languages. The Hungarian language was eliminated from road signs, and a bureaucracy was created to screen names given to newborns. Attempts were also made to control the press. On March 11, 1994, Meciar and his cabinet were ousted from power amid charges that the prime minister was using his position to sell state-owned companies to his friends. Meciar returned to power in December, 1994, after his party captured more than a third of the vote in national elections.

On the international scene, the new Slovak state became involved in a protracted dispute with Hungary over the construction of the Gabcikova dam on the Danube River. The project was designed to provide Slovakia with an adequate source of electrical power and to control flooding. Although Hungarians agreed to the construction of the dam in 1977, the government in Budapest later sought to block the project, citing environmental concerns. The issue went before the International Court of Justice.

Thomas D. Matijasic

the time suggested that only about 20 percent of Slovaks desired independence.

When negotiations between the two groups stalled, Klaus proposed that the federation split unless Slovaks agreed to maintain a strong central government. Meciar quickly seized the offer. On June 20, Czech and Slovak officials signed an agreement to prepare to end the federation. President Havel advocated a popular referendum on the issue but pledged to work for a painless separation if that was the consensus of the electorate.

On July 3, Slovak delegates in the Federal Assembly blocked the reelection of Havel as president of the republic. Hundreds of Czechs gathered outside the capitol and jeered Slovak representatives as they left the building. Although his term did not expire until October, Havel chose to resign as president after the Slovak parliament issued a declaration of sovereignty on July 17.

The final bill of separation passed the Federal Assembly on November 25, 1992, after receiving

2492

Russia and United States Sign START II

START II continued the process of dismantling nuclear weapons systems in the only remaining superpower countries.

What: International relations; Military
When: January 3, 1993
Where: Moscow, Russia
Who:
GEORGE HERBERT WALKER BUSH
(1924-   ), president of the United
States from 1989 to 1993
BORIS YELTSIN (1931-   ), president of
the Russian Federation from 1991 to
1999
RONALD REAGAN (1911-   ), president
of the United States from 1981 to
1989
MIKHAIL GORBACHEV (1931-   ), general
secretary of the Communist Party of
the Soviet Union from 1985 to 1991

Continued Reduction of Nuclear Arsenals

On January 3, 1993, George Bush and Boris Yeltsin, at a U.S.-Russian summit in Moscow, signed the second Strategic Arms Reduction Treaty, known as START II. The treaty was designed to leave each side with a maximum of thirty-five hundred strategic warheads. The agreement called for the elimination of all heavy, land-based, multiple-warhead nuclear missiles to eliminate the possibility of a first strike. Each side would be left with a maximum of five hundred single-warhead, land-based, strategic missiles. Russia thus pledged to give up its heavy multiple-warhead SS-18 missiles, its most dangerous first-strike weapon. The United States, in turn, agreed to eliminate half of its first-strike weapons, its submarine-launched missiles. In all, the two sides agreed to reduce their strategic arsenals by 73 percent from 1991 levels. In 1991, the United States possessed 12,646 strategic nuclear warheads and the Soviet Union 11,012. Each had the firepower needed to devastate the other many times over.

The countries agreed to a schedule of arms reduction. By 2003, the United States would have a maximum of 500 land-based, 1,728 submarine-based, and 1,272 air-launched missiles. Russia would have no more than 500 land-based, 1,744 submarine-based, and 752 air-launched missiles.

In February, 1992, at the first U.S.-Russian summit following dissolution of the Soviet Union, Bush and Yeltsin had agreed to a round of nuclear disarmament. They produced a framework agreement that called for far deeper cuts in strategic nuclear weapons than had START I, signed in 1991.

At the very end of 1992, on the eve of the signing of START II, several sticking points remained. The United States sought the destruction of not only Russia's heavy SS-18 and SS-19 missiles but also their silos. For economic reasons, the Russians wanted to retain the silos as well as the ten-warhead SS-19 missiles, which they agreed to refit with single warheads. The elimination of the heavy multiple-warhead SS-18's remained the basis of the conservatives' complaint in Moscow that the United States had gotten the better part of the agreement. The United States, anticipating the accusations, granted concessions to Yeltsin so that he could defend the treaty. It agreed to permit Russia to convert 90 (of 154) SS-18 silos for use by single-warhead SS-25's and to convert 105 (of 170) multiple-warhead SS-19's to single-warhead missiles. The United States would abolish or modify its own land-based, multiple-warhead Minuteman III and MX missiles.

From SALT to START

By the late 1970's, the Soviet Union was well on its way to drawing even with the United States in nuclear arms. Ronald Reagan, the Republican Party's presidential candidate in 1980, stressed the need for a reduction of the number of strategic warheads. In 1982, as U.S. president, Reagan first proposed to General Secretary Leonid Brezhnev a START treaty to supersede the Strategic Arms Limitation Talks (SALT) treaties of 1972 and 1979. The SALT treaties had limited the deployment of certain strategic weapons but at the same time had made it possible to continue to stockpile strategic weapons launchers not covered by the treaties. Thus the strategic nuclear arms race, although now bound by certain constraints, continued.

Despite twelve rounds of formal negotiations, thirteen meetings of foreign ministers, and six summit conferences, the superpowers failed in their complicated attempts to work out a balanced reduction of U.S. and Soviet strategic weapons. There was too much mistrust to find an agreement. By the summer of 1991, however, relations between the Soviet Union and the United States had undergone remarkable changes. The Berlin Wall had fallen, the Cold War had come to an end, and General Secretary Mikhail Gorbachev had announced a phased withdrawal of all Soviet troops from Eastern Europe by 1995. The time had come to reverse the U.S.-Soviet strategic nuclear arms race.

On July 31, 1991, at a summit meeting in Moscow, Bush and Gorbachev signed the first START treaty. It called for a reduction by 1998 of the number of strategic missiles to sixteen hundred and that of warheads to six thousand. At the end of 1991, the Soviet Union had ceased to exist and its successor state, Russia, inherited the bulk of the Soviet Union's nuclear weapons as well as its treaty commitments. On October 1, 1992, the U.S. Senate ratified START I. The Russian parliament followed suit on November 4, 1992. This set the stage for further commitments to arms reduction.

Consequences

With START I and START II, Russia completed a train of events that Gorbachev's "new thinking" had set into motion as early as 1987. It moved from the primitive concept of "the more the better" to the principle of "reasonable sufficiency." In 1993, Yeltsin echoed Gorbachev's position of a "minimum-security level" of nuclear warheads and spoke of a commitment to address the needs of Russia's citizens, half of whom, he admitted, lived below the poverty line. The U.S. government abandoned the quest to maintain nuclear superiority and was thus able to continue to cut back its military spending in an attempt to come to grips with its budget deficit.

START II committed the signatories to return to the number of strategic warheads they had possessed in the late 1960's and early 1970's, before they began to refit their missiles with multiple warheads. The treaty was a tacit recognition that the stockpiling of strategic weapons and escalating nuclear arms race had undercut, rather than enhanced, the national security of the superpowers. Neither side, however, was able to reap an immediate peace dividend because the dismantling of the missiles and the discarding of the nuclear matter proved to be nearly as expensive as the initial buildup. To ease Russia's burden of the dismantling process, the U.S. Congress allotted $800 million in assistance.

Harry Piotrowski

Clinton Becomes U.S. President

Bill Clinton's inauguration as president of the United States ended twelve years of Republican control of the executive branch.

What: National politics
When: January 20, 1993
Where: Washington, D.C.
Who:
BILL CLINTON (1946-   ), president of
   the United States from 1993 to 2001
GEORGE HERBERT WALKER BUSH
   (1924-   ), president of the United
   States from 1989 to 1993

Clinton Takes Office

On January 20, 1993, Bill Clinton was sworn in as the forty-second president of the United States. In a three-way race against incumbent Republican George Bush and independent Ross Perot, he had won a substantial electoral victory. Clinton seemed to be well positioned to produce new programs and initiatives during the first days of his administration. Not only had the election ended twelve years of Republican control of the executive branch, but the Democrats also were in control of both houses of Congress. Many observers believed that the coalition assembled by President Ronald Reagan—right-wing Republicans together with Republican and Democratic centrists—had finally become unraveled as a result of the weakened American economy.

Clinton's campaign promises included restoration of economic health, deficit reduction, health care reform, campaign spending reform, and support for the North American Free Trade Agreement (NAFTA). Taken as a whole, Clinton's program seemed to be well calculated to expand or restore the power of the Democratic Party in national politics. To his dismay, the new president found within a few months of his inauguration that there were severe limits on what he could accomplish. Solid Republican opposition

to his proposals, coupled with the defection of many Democrats whose interests were harmed by Clinton's program, prevented him from achieving much in the early days of his administration. His efforts to find compromises acceptable to Congress often made him appear to vacillate, further weakening his political influence.

By the end of the first two years of Clinton's term, he had achieved only two major legislative victories—passage of his budget and approval of NAFTA. In the latter instance, he found it necessary to appeal to Republicans for legislative support. His health care reform program became mired in controversy over costs and who would pay them.

Both before and after the election, Clinton's greatest political challenge came from opponents trying to dig up scandals from his past. A variety of charges were made, many without evidentiary foundation. President and Mrs. Clinton were charged with conflicts of interest and financial improprieties in connection with the Whitewater real estate development in Arkansas. In 1993, Clinton also was charged with propositioning and sexually harassing a former Arkansas state employee, Paula Jones. Jones, with the financial support of conservative groups opposed to the president, filed suit against him. Although Clinton was able to delay the hearing until after his term as president had ended, his political reputation was harmed again.

Clinton also faced strong factional opposition among Democrats. He had campaigned for office as a "New Democrat," a centrist. In order to prevail with Congress and the bureaucracy, however, he found it necessary to adopt a much more liberal stance as president than he had as a candidate. His new positions on gay rights, abortion, and minority issues were not generally popular with the public. The new positions also made the

Library of Congress

President Bill Clinton takes the oath of office.

president appear to be waffling on a variety of campaign promises. The administration's abandonment of the promised middle-class tax cut and welfare reform was particularly damaging.

The 1992 Campaign

Clinton's victory in 1992 can be traced to events in the late 1960's that reshaped both major American political parties. Since the Great Depression, the Democratic Party had been the country's dominant political force. A coalition of Southern white politicians and Northern urban and labor leaders was formed. The Democrats could rely on votes from large cities, members of minority groups, unionized workers, Southerners, Catholics, and Jews. The Republicans were led by Northeastern and Midwestern Protestants. They could usually count on support from the business community and from middle- and upper-class suburban voters.

Two powerful political struggles of the 1960's brought about reconstruction of the Democratic coalition. Opposition to the war in Vietnam induced many liberal Democrats to attack and ultimately weaken the influence of big-city Democratic machines and labor leaders. Liberal activists for a variety of other causes began to support special interest groups rather than the party itself. The Civil Rights movement weakened the Democratic Party in the South. Although liberals could control many congressional districts, the party's strength in presidential campaigns was diminished. A pattern of Republican presidents and Democratic Congresses was established. Before Clinton, only Jimmy Carter had been able to overcome the trend.

Consequences

President Clinton's apparent inability to prevail in the legislative process or to become popu-

2496

lar enough to silence critics of his personal behavior made him appear weak and vacillating. In politics, appearance is sometimes as important as reality, and the Democrats were faced with a leader who appeared weak. At the midterm elections of 1994, the Democrats suffered a stunning reversal, losing control of both houses of Congress for the first time in more than forty years.

Clinton's move toward the political center following the 1994 elections did little to reestablish his reputation for principled and steadfast politi-

cal leadership. Many of the crucial decisions of the second half of Clinton's term were legislative rather than administrative. Because congressional rather than executive action was required, the presidency appeared to be far less relevant to the political process than before. This political situation had been practically unknown in the United States since before the Great Depression. It established the 1996 elections as critical to the future of American politics.

Robert Jacobs

Terrorists Bomb World Trade Center

In one of the most destructive acts of terrorism in U.S. history, the World Trade Center in New York City was shaken by a car bomb that killed six people and injured more than one thousand.

What: International relations; Terrorism
When: February 26, 1993
Where: New York City
Who:
SHEIKH OMAR ABDEL-RAHMAN (1938-   ), fundamentalist Muslim cleric
EMAD SALEM (1950-   ), bodyguard for Abdel-Rahman and FBI undercover agent
MAHMUD ABOUHALIMA (1959-   ), suspected mastermind behind the bombing
MOHAMMAD A. SALAMEH (1968-   ), Islamic fundamentalist arrested for the bombing

Terrorism in the United States

On February 26, 1993, a massive car bomb exploded in the hundred-story World Trade Center in New York City. Six days later, Mohammad A. Salameh, a twenty-five-year-old man described as an Islamic fundamentalist, was arrested.

During the investigation of the blast site, officials found part of a van that they believed contained the bomb. Identification of this piece of steel showed that it was part of the chassis of a van that had been rented in New Jersey a few days before the bombing. On March 4, when Salameh went to the rental agency seeking a refund of the money he had paid to rent the van, he was arrested by Federal Bureau of Investigation (FBI) agents.

Investigators soon learned that Salameh was a close follower of Sheikh Omar Abdel-Rahman, a Muslim cleric who held Islamic services in a storefront mosque in Jersey City, New Jersey. The sheikh, blind and stricken with diabetes, had preached that violent action is an acceptable way to overthrow governments in countries that have large Islamic populations but are not ruled by strict Islamic law. His goal was to overthrow secular regimes in Islamic countries, and he had professed a desire to bring harm to Western nations that support moderate governments.

Abdel-Rahman had been under suspicion for terrorist activities for several years. He slipped into the United States in 1990 after successfully receiving a tourist visa at the United States embassy in Khartoum, Sudan. Abdel-Rahman's name was on a "watch list," meaning that he was not to be granted admission into the United States. Officials in the embassy in the Sudan failed to check the list and thus granted Abdel-Rahman a visa to enter the country.

Upon his arrival, Abdel-Rahman applied for political asylum. This meant that he could not be forced to leave the United States until his request for asylum had been reviewed, a procedure that takes several months.

Abdel-Rahman, who was born in Egypt, was a fierce opponent of Egyptian president Hosni Mubarak. At the time of the bombing in New York, Egypt was trying Abdel-Rahman *in absentia* on charges of inciting illegal demonstrations against the government.

Further investigation by American officials into the sheikh's activities found evidence of at least four additional terrorist plans. These called for placing homemade bombs in at least four other New York City locations and targeting for death United States senator Alfonse D'Amato. Federal agents believed that Mahmud Abouhalima, an Egyptian-born cabdriver and follower of Abdel-Rahman, was the mastermind behind the World Trade Center explosion.

Muslims Want Control

Followers of the religion of Islam are called Muslims. Islam was founded by the prophet Muhammad and is one of the most widespread reli-

2498

gions in the world. Messages that Muhammad said came directly from God are written in the Koran, the Muslim holy book.

For many years, fundamentalist Muslims have been active in their desire to make Islamic countries live under the strictest Islamic laws. These ancient laws do not conform to modern society. They demand severe penalties such as cutting off the hand of a thief. Islamic law makes women subordinate to men, forbids them to appear in public without veils over their faces, and prevents them from exercising many personal rights granted to men. Islamic law also does not permit consumption of alcohol. Fundamentalist Muslims believe that those who die in a jihad, or holy war, are martyrs and ensured a place in paradise.

Fundamentalist Muslim aggression and desire for control came into public notice in 1979 when another Muslim cleric, Ayatollah Ruhollah Khomeini, led a revolution to overthrow the government of Iran. After overthrowing the progressive government of the shah of Iran, Muslim radicals stormed the American embassy in Tehran on November 4, 1979, and took fifty-two American citizens into custody. After 444 days of negotiations and a failed American military action to rescue the captives, the Americans were released, immediately after U.S. president Ronald Reagan was sworn into office.

The toppling of Mohammad Reza Shah Pahlavi gave fundamentalists firm control of Iran and a power base from which to direct other fundamentalist activities. On October 6, 1981, Egyptian president Anwar el-Sadat was assassinated by a Muslim fundamentalist. Sadat was a moderate who had opened negotiations with the Jewish state of Israel and was on friendly terms with many Western governments.

Investigations into the murder of Sadat turned up evidence that Abdel-Rahman may have been involved in, or at least encouraged, the assassination. That, however, was never proved.

Much of the evidence against Abdel-Rahman in the World Trade Center bombing came from Emad Salem, a man who served as a bodyguard to Abdel-Rahman and was an informant for the FBI. On July 8, 1993, the sheikh was taken into custody by agents of the Immigration and Naturalization Service, to be held pending resolution of his appeal of an order for deportation.

Emergency vehicles fill the street near the twin towers of the World Trade Center in New York after an explosion in the underground parking garage.

2499

In March of 1994, Mahmud Abouhalima, Mohammad Salameh, and two others were found guilty on thirty-eight charges stemming from the World Trade Center bombing.

Consequences

The bombing of the World Trade Center was an awakening for the United States. Never before had terrorists made such a bold strike within that country. Terrorist strikes had become commonplace in the Middle East and even in Europe. The blast in New York was a warning that America was no longer immune to terrorism.

As a result of the bombing, security measures were increased in airports and in all places considered to be potential terrorist targets. Security was tightened in public buildings located in major cities, and intelligence operations were focused more on terrorism, with special emphasis on Muslim fundamentalist activities.

On October 1, 1995, Abdel-Rahman and nine others were convicted on forty-eight counts of conspiracy in connection with plans to bomb the United Nations headquarters and other buildings and to assassinate various political leaders. Some of those charges were connected with planning, though not actually performing, the bombing of the World Trade Center. Abdel-Rahman received a sentence of life in prison.

Eight and a half years after the bombing, the World Trade Center was the target of a vastly greater terrorist attack. On September 11, 2001, terrorists hijacked four jetliners and converted two of them into suicide bombs that they used to level both towers.

Kay Hively

World Meteorological Organization Reports Ozone Loss

In the most dramatic report of ozone depletion since the phenomenon was first reported, the World Meteorological Organization announced a rapid decline in ozone levels in the Northern Hemisphere.

What: Environment
When: March 5, 1993
Where: Geneva, Switzerland
Who:
Zou Jingmeng (1929-   ), president of the World Meteorological Organization
Godwin Olu Patrick Obasi (1933-   ), secretary-general of the World Meteorological Organization
Daniel L. Albritton (1936-   ), cochairman of the Global Ozone Research and Monitoring Project
Robert Tanner Watson (1922-   ), cochairman of the Global Ozone Research and Monitoring Project

The Danger Moves North

On March 5, 1993, the World Meteorological Organization, an agency of the United Nations based in Geneva, Switzerland, announced that levels of ozone in the atmosphere above Northern Europe and Canada had fallen 20 percent below normal during the winter of 1992-1993. The report followed a similar announcement made on November 13, 1992, that ozone levels in the same areas were 12 to 20 percent below normal during the winter and spring of 1991-1992.

The finding was confirmed by Canadian scientists, who announced on March 9, 1993, that ozone levels over the Canadian cities of Toronto and Edmonton were the lowest in 30 years and that ozone levels over Canada had dropped 25 percent since 1980. American scientists with the National Aeronautics and Space Administration also reported similar findings, recorded by the Total Ozone Mapping Spectrometer aboard the satellite Nimbus-7, which had been measuring global ozone levels since 1978. The satellite records showed that after several years of slow decline, ozone levels over the Northern Hemisphere had dropped suddenly to record lows.

Most scientists blamed the sudden loss of ozone on the massive eruption of Mount Pinatubo on the Philippine island of Luzon from June 14 to June 16, 1991. The volcano released enormous amounts of debris into the atmosphere, accelerating the destruction of ozone by chlorine from chlorofluorocarbons.

This ozone loss was recorded in the atmosphere above heavily populated regions in Canada, the United States, Europe, Russia, and China. Scientists estimated that every 1 percent decline in ozone levels would increase the amount of ultraviolet light reaching the Earth's surface by 1.3 percent. Although they advised against undue alarm, they suggested that light-skinned persons take reasonable precautions against excessive exposure to sunlight to lessen the chance of developing skin cancer.

Chlorofluorocarbons and Ozone

Chlorofluorocarbons were invented in 1928 by industrial engineer Thomas Midgley, Jr., for use in refrigeration. By 1932 they were used in air conditioning and by 1941 in aerosol sprays. These chemicals were nonflammable, noncorrosive, and nontoxic. Their use soon became widespread.

Chlorofluorocarbons were considered to be harmless until 1973, when chemists F. Sherwood Rowland and Mario Molina theorized that they slowly migrate to the stratosphere, where they are broken down by ultraviolet light to release chlorine, which destroys ozone. Ozone, a molecule containing three oxygen atoms, is created by the action of ultraviolet light on ordinary oxy-

2501

gen, which contains two atoms. Ozone is unstable and is constantly formed and destroyed in the stratosphere. It absorbs much of the sun's ultraviolet light, preventing it from reaching Earth's surface, where it would be damaging to living organisms.

The potential danger of chlorofluorocarbons remained highly controversial throughout the 1970's and the early 1980's. Some environmentalists demanded a ban on chlorofluorocarbons. Many industrialists insisted that they were safe and that their loss would cause economic hardship.

In March, 1977, the United Nations Environment Program held the first international conference on chlorofluorocarbons in Washington, D.C. It was followed by a second conference in Munich, Germany, in December, 1978. Delegates could agree only that more study was needed. In April, 1983, the governments of Norway, Sweden, and Finland drafted a plan to ban all chlorofluorocarbons in aerosols and to restrict their use in other products. This proposal was known as the Nordic Annex. The United States and Canada supported the Nordic Annex, but most European countries opposed it and instead suggested a limit on chlorofluorocarbon production. Delegates met again in Vienna, Austria, on March 22, 1985, but negotiators failed to resolve the conflict. The Vienna Convention for the Protection of the Ozone Layer only required participating nations to share information and to develop specific restrictions at a later date.

The debate took a sudden turn on May 16, 1985, when Joseph Farman, head of the Geophysical Unit of the British Antarctic Survey, announced that ozone levels above Antarctica had dropped by as much as 40 percent below normal at the beginning of each spring since 1982. On September 16, 1987, delegates met in Montreal, Canada, to sign the Montreal Protocol on Substances That Deplete the Ozone Layer. The Montreal Protocol froze chlorofluorocarbon production at 1986 levels starting in 1989, followed by a 20 percent reduction by 1994 and a 30 percent reduction by 1999. Meanwhile, American expeditions to the Antarctic in 1986 and 1987 established that chlorine from chlorofluorocarbons was responsible for the ozone loss.

On March 15, 1988, the Ozone Trends Panel, a large international research project started in 1986, announced that ozone levels had dropped about 2 or 3 percent in the Northern Hemisphere. Meeting in London, England, in June, 1990, delegates agreed to eliminate all use of chlorofluorocarbons by 2000. At a meeting held in Copenhagen, Denmark, in November, 1992, the deadline was moved to 1996.

Consequences

The announcement by the World Meteorological Organization of serious ozone depletion in the Northern Hemisphere made it even more imperative to stop chlorofluorocarbons from entering the atmosphere. E. I. du Pont de Nemours and Company, the world's largest producer of chlorofluorocarbons, had responded to the Ozone Trends Panel by announcing on March 24, 1988, that it would phase out production. On March 9, 1993, four days after the World Meteorological Organization report, it announced that the chemicals would no longer be made by the end of 1994. Scientists predicted that ozone would reach its lowest levels in the early twenty-first century, then slowly return to normal over the next several decades.

Rose Secrest

Afghan Peace Pact Fails to End Bloodshed

A peace plan to end Afghanistan's long and bloody civil war was signed in Pakistan by eight rival military factions, but differences among the groups were unresolved and violence resumed.

What: Civil war
When: March 7, 1993
Where: Islamabad, Pakistan
Who:
GULBUDDIN HEKMATYAR, a Pashtun tribal leader
BURHANUDDIN RABBANI (1940-   ), president of Afghanistan from 1992 to 1996
ABDUL RASHID DOSTAM, Uzbek chief in control of the most disciplined militia in northern Afghanistan
AHMED SHAH MASOUD (1954?-   ), minister of defense and a sworn enemy of Hekmatyar
NAWAZ M. SHARIF (1949-   ), prime minister of Pakistan from 1990 to 1993; broker of the peace plan

A Shaky Peace Treaty

On March 7, 1993, the Afghan government and leaders of eight rebel factions met in Islamabad, Pakistan, to sign a peace pact aimed at ending the fighting that had killed an estimated five thousand people and wounded thousands more. The conflict had forced an estimated 750,000 people to flee Kabul, the capital of Afghanistan.

In March, 1993, Prime Minister Nawaz Sharif of Pakistan invited the chiefs of Afghanistan's leading factions to Islamabad to negotiate an end to their bitter fighting. Representatives of eight of the ten major groups participating in the conflict attended. After six days of bitter wrangling, they agreed to a cease-fire in which all heavy weapons were to be moved out of range of Kabul and placed under a central military command.

The representatives also agreed that President Burhanuddin Rabbani would share power with Gulbuddin Hekmatyar, an extremist Pashtun leader who would take the office of prime minister. Hekmatyar was a bitter opponent of the Rabbani government and had played a key role in the destruction of Kabul. Observers found it difficult to imagine that Rabbani and Hekmatyar could jointly select government ministers on whom both could agree, but they promised to do so.

The peace conference also resulted in agreement that an assembly would be elected within eight months to draw up a constitution and that a parliamentary election would be held before June 28, 1994, the end of Rabbani's term of office. The agreement gave the Afghan leaders collective responsibility for establishing a national army and conducting elections.

At the invitation of King Fahad of Saudi Arabia, the Afghan leaders left for Mecca on March 9 to pray for peace in Afghanistan and to bless the Islamabad agreement. The prayers and international pressure failed to bring two of the most important actors to the agreement. The first was the Uzbek faction under the leadership of General Abdul Rashid Dostam, who formerly had been the main prop of the Soviet-backed regime. Dostam, arguably the most powerful man in northern Afghanistan, was recognized by many of the factions as the closest person to being neutral of those involved in the intertribal feuding. Some, however, could not forgive him for his role as a Soviet protégé. Also missing from the peace negotiations was Defense Minister Ahmed Shah Masoud. Rabbani wanted to keep Masoud in that position, but Masoud was bitterly opposed by Hekmatyar. Hekmatyar argued that no single faction should control both the presidency and the ministry of defense, and other tribal leaders agreed.

No one at the negotiations represented the many ordinary Afghans who wanted neither an

Islamic government nor a Marxist government. The people of Kabul who had not already fled the country—most of them professionals—lived amid constant rocket shelling, with shortages of food, medicine, and hospital facilities. They had witnessed the withdrawal of all foreign aid and abandonment of foreign embassies as diplomats found it impossible to function within the country.

The peace pact failed to end the bloodshed in a civil war that had lasted fourteen years. At least ten groups, with differences based on tribal ethnicity and variations in Islamic ideology, began new violent struggles for power. Although the peace pact was supported by the governments of Pakistan, Iran, and Saudi Arabia, observers were not hopeful that hostilities would cease.

Cold War Roots

In 1979, the Soviet Union responded to a request for help in suppressing an internal revolt in Afghanistan. The request came from Prime Minister Babrak Karmal, who had been installed by the Soviets as leader of the new Marxist government that replaced the country's monarchy. Because the Soviets considered Afghanistan to be within their sphere of influence, they sent in troops under the guise of assistance.

Burhanuddin Rabbani.

The United States perceived the troop movements as unwarranted aggression. Afghanistan became a battlefield in the Cold War, with the Soviets waging warfare against the revolutionaries and the Americans clandestinely arming and training these "freedom fighters." These guerrillas primarily were tribal Muslim fundamentalists and, as such, received financial backing from the Saudi government. Pakistan, which had recently transformed a secular government to Muslim fundamentalism, also offered support. Pakistan became the primary conduit for Western arms being shipped to Afghanistan. After the fall of the Soviet Union, the communist regime of Afghanistan fell, to be replaced by an Islamic government on April 28, 1992. Rival factions began fighting for power.

Consequences

A twenty-two-member cabinet was named on May 20, 1993, with cabinet positions divided among the ten major factions. The May cease-fire agreement called for the ministry of defense to be run by a commission, under Rabbani's direction, rather than by Masoud. An interim constitution was approved on September 27.

The apparent peace did not last. On January 1, 1994, forces controlled by Hekmatyar and Dostam coordinated a rocket and artillery assault on Kabul. At the end of the year, Kabul remained divided into sectors controlled by various factions. The central government retained almost no authority. Rabbani refused to surrender the presidency as scheduled, and the Supreme Court extended his term for six months, into December of 1994.

Dostam's previous support of the Soviet government led to close relations between him and the Uzbekistan government. Saudi Arabia supported factions it saw as counterweights to Iranian influence. Pakistan gave refuge to 1.5 million Afghan refugees and offered a permanent home to members of the Pashtun ethnic group, traditional leaders in Afghan politics. The rest of the world watched and waited, hoping that rival factions would reach some internal agreement. Instead, a movement known as the Taliban took control of the country in 1996 and began a five-year reign of extreme Islamic fundamentalism.

Laina Farhat

AP/Wide World Photos

Egyptian Government Troops Attack Islamists

> *Government forces killed twenty-six militants in Cairo and Aswan as part of a massive crackdown against Islamists advocating overthrow of the secular government of President Hosni Mubarak.*

What: Civil strife; Religion
When: March 10, 1993
Where: Egypt
Who:
HOSNI MUBARAK (1928-   ), president of Egypt from 1981
HASSAN AL-ALFI, minister of the interior
NAGUIB MAHFOUZ (1911-   ), winner of the 1988 Nobel Prize in Literature

Crackdown on Islamists

The Egyptian government announced on March 10, 1993, that its troops had attacked and killed twenty-six Islamists in shoot-outs in Cairo and Aswan. These incidents were the latest in a series of confrontations between the government and Islamist groups advocating the overthrow of the secular government of President Hosni Mubarak and the establishment of an Islamic form of government. The most prominent of these groups was al-Ikhwan al-Muslimun, or the Muslim Brotherhood.

A day before the attack, the government put more than forty militants on trial. The militants, however, were undaunted and continued their attacks against government officials and civilians. On June 9, a bomb exploded on the Giza road, killing one Egyptian and wounding fourteen other people, five of whom were British tourists. On June 20, another bomb exploded on a busy Cairo street, killing seven people and wounding fifteen others.

The war between the government and the Islamists continued. The government hanged seven Islamists on July 9 and five more on July 18. Fifty-three militants were put on trial on August 8, and the militants responded with an ambush of a senior police official in Qena, in upper Egypt, the next day. On August 19, they ambushed Interior Minister Hassan al-Alfi in the heart of Cairo. Although al-Alfi escaped, four people died in the attack, and fifteen others were wounded.

President Mubarak declared on October 12, 1993, that the Islamist forces were being subdued. By then, the battle between the two sides had claimed the lives of at least two hundred people. Islamists continued to target government officials, civilians, and tourists and tourist installations. The attacks on tourists apparently were attempts to hurt the Egyptian economy.

The battle continued in 1994. On February 1, for example, government forces killed seven suspected militants. In retaliation, terrorists exploded a bomb in a passenger train, injuring eleven people, six of whom were foreign tourists. March 2 witnessed a new government offensive, with the detention of approximately one thousand suspected militants. In response, the Islamists ambushed and killed a leading government undercover operative.

On October 14, members of the Muslim Brotherhood stabbed Naguib Mahfouz, winner of the 1988 Nobel Prize in Literature and critic of the extremists. Violence continued as the attackers were hunted down and killed. By November, 1994, more than four hundred people had died in the conflict.

Roots of Activism

The roots of Islamic activism in Egypt go back to the country's struggle against the British. After several failed attempts by secular nationalists to gain independence for Egypt, a group advocating the return to "pure Islam" was founded in 1928 by Hasan al-Banna. The group, al-Ikhwan al-Muslimun (the Muslim Brotherhood), gained wide acceptance during the 1930's and 1940's. It played a significant role in the 1952 Egyptian revolution that toppled the pro-British monarchy of King Farouk and brought Colonel Gamal Abdel Nasser to power. Soon afterward, however, relations between the Muslim Brotherhood and Nasser soured as the government adopted secular nationalist policies and sought to subordinate religion to its interests. After members of the Muslim Brotherhood were implicated in an assassination plot against Nasser in 1954, several leading members were either executed or jailed. The organization was declared illegal, and its large membership went either underground or into exile.

The most vehement attacks against the Muslim Brotherhood came in the mid-1960's, especially after the group's leader was accused of sedition, arrested, and later executed. By then, the Muslim Brotherhood had established numerous chapters throughout the Arab world.

As long as Nasser dominated Egyptian politics, the Muslim Brotherhood remained subdued. When Anwar el-Sadat came to power in 1971, he was much less popular than Nasser. In search of political allies and to create a counterforce to the Nasserites, Sadat legalized the Muslim Brotherhood and allowed it to publish newspapers and magazines. This strategy appeared to be successful. Islamist groups, especially among university students, challenged Nasserites and leftists, and Sadat acquired a firmer grip on power.

Eventually his policies led to a clash with his allies in the Islamic camp, leading to the emergence of a more activist, militant wing. Three broad policies advocated by Sadat caused the break. The first was his "Open Door" economic policy, which benefited a limited number of Egyptians and led to a wider gap between the rich and the masses of the poor, and eventually to the eruption of massive bread riots in Cairo in early January, 1977. The second was Sadat's unwillingness to institute Islamic law. Finally, Islamists objected to Sadat's unprecedented visit to Jerusalem to meet with Israeli leaders in November, 1977. This visit shocked many Islamist groups as a betrayal of Islam and the Palestinian cause.

The peace initiative notwithstanding, discontent with Sadat's domestic and international policies mounted among secular and Islamist groups. The latter found a model in the success of the Islamic revolution in Iran. More groups were founded, such as the Islamic Jihad. In October, 1981, members of the Islamic Jihad assassinated President Sadat while he was viewing a military parade held in celebration of the October War.

Despite the subsequent mass arrests, trials, and execution of some of their members in the early 1980's, Islamist groups resumed their activities. In 1990, they assassinated Rifaat al-Mahgoub, the speaker of the assembly. They also continued to target foreign tourists in an effort to hurt the government's international standing as well as to cripple the tourist industry, a major source of hard currency. This wave of violence seemed to culminate in the attempted assassination of President Mubarak on June 26, 1995, while he visited Ethiopia.

Consequences

Despite the 1995 assassination attempt against Mubarak, it seemed that the wave of insurgency, which by the summer of 1995 had claimed more than 750 lives, was beginning to taper off. For Egypt to hold off an Islamist takeover, the government would need to provide solutions to the problems of unemployment, lack of housing, and inadequate education and health care. The government also needed to democratize politics and deal with bureaucratic inefficiency and corruption. Those problems provided fertile ground for the recruitment of the discontented, who directed their frustration against the government and sought to replace it with another that promised them a better future.

Mubarak's competent management of Egypt's foreign policy continued to win support from the United States and other western nations into the late 1990's. This close friendship with the United States did not, however, lead Egypt to support all U.S. proposals. For example, in November, 1997, despite strong U.S. pressure, Egypt boycotted a U.S.-backed Middle East economic conference with Israeli representatives because of the lack of progress in the Israeli-Palestinian peace process.

Mahmood Ibrahim

North Korea Withdraws from Nuclear Non-Proliferation Treaty

> *In one of the most troubling events of the nuclear age, North Korea announced its withdrawal from the Nuclear Non-Proliferation Treaty, apparently intending to become a new nuclear power.*

What: Military; International relations
When: March 12, 1993
Where: P'yongyang and Yongbyon, Democratic People's Republic of Korea (North Korea)
Who:
KIM IL SUNG (1912-1994), president of North Korea from 1972 to 1994
KIM JONG IL (1942-   ), Kim Il Sung's son and successor

The NPT Withdrawal Announcement

On March 12, 1993, the Democratic People's Republic of Korea (North Korea) became the first country to try to withdraw from the Nuclear Non-Proliferation Treaty (NPT), an international agreement that limits the spread of nuclear technology. The treaty allows countries to acquire nuclear technology for peaceful purposes such as generating electric power but not for weapons. In return, the countries that already possess nuclear weapons promise to share knowledge of nonweapons nuclear technology. Because the treaty requires ninety days' advance notice for a member to drop out, the other members had three months to persuade the North Koreans to change their minds.

The North Koreans' most important reason for pulling out of the NPT was that it required them to permit inspectors from the International Atomic Energy Agency (IAEA) to look at two sites. The IAEA suspected that these sites contained evidence that North Korea secretly had developed an atomic bomb, in violation of the NPT. When North Korea gave the IAEA an inventory of its nuclear facilities in 1992, it neglected to list the two sites. When the IAEA dis-

covered them through satellite reconnaissance, it demanded "special inspections" to assess them, inspections that the North Koreans refused to allow. When the IAEA threatened to go to the United Nations Security Council to seek sanctions against North Korea for refusing to cooperate, the government of President Kim Il Sung announced that it was pulling out of the NPT and the IAEA. If North Korea had actually done this, it would have been able to pursue a nuclear program beyond any international controls.

The Kim Il Sung government defended its decision by accusing the IAEA of acting as an arm of the U.S. Central Intelligence Agency (CIA), the source of the satellite information. The North Koreans demanded that the IAEA be impartial and believe their assurances that they were not secretly building nuclear weapons. They pointed out that the IAEA had never demanded special inspections of Israel, South Africa, Pakistan, or Iraq, all of which actually had secret weapons programs. They said this proved that the IAEA was treating North Korea unfairly.

North Korea's Nuclear Program

North Korea became a communist country in the late 1940's as a result of Korea's liberation from Japanese colonial rule and division into U.S.- and Soviet-occupied territories. In 1950, North Korea invaded South Korea, whose government had been sponsored by the United States and the United Nations, but it failed to unify the country under communist rule. The Korean War ended in July, 1953, after a United Nations army led by the United States intervened to defend the south.

China, the Soviet Union, and other socialist countries helped North Korea recover and create a strong economy. The United States and the West helped South Korea rebuild and eventually outpace North Korea in productivity and living standards. Both sides have maintained huge armed forces to deter attack, and each has had a defense treaty providing protection, including a "nuclear umbrella." The Soviet Union's nuclear forces that could defend North Korea were in Siberia. The United States defended South Korea with nuclear forces on Navy ships and on bases in South Korea and on Okinawa. When the Cold War ended in the early 1990's, North Korea lost most of its foreign allies and its defense treaty with the Soviet Union, making it the only nation in the region without a nuclear umbrella to protect it.

One of President Kim Il Sung's most important ideas was his concept of *juch'e*, or self-reliance. In nuclear matters, this meant using North Korea's own uranium to develop a self-sufficient nuclear power industry and, apparently, a nuclear weapons capability. North Korea's nuclear program began with a small research reactor in 1965 and was supplemented in 1986 by a bigger reactor designed to produce electricity in the town of Yongbyon.

In 1989 the Yongbyon reactor was shut down for maintenance, and some of its partially spent uranium fuel was removed. This could have been made into plutonium, which is used in atomic bombs. To make plutonium, the North Koreans would have needed special reprocessing facilities that the world community thought it lacked. In 1989, however, North Korea was found to be building a radiochemical laboratory in Yongbyon that would have provided the capability. Construction stopped when North Korea ratified the NPT and accepted IAEA inspections in 1992. The discovery of the two "hot spots" later that year, however, suggested that North Korea already had created some plutonium and had concealed it from the IAEA, perhaps in an undiscovered laboratory. This evidence of bad faith was worrisome to the international community as a whole, particularly North Korea's neighbors, South Korea and Japan.

Consequences

During the ninety-day waiting period after North Korea announced its withdrawal from the NPT, the other NPT members frantically sought a way to persuade North Korea to continue abiding by the treaty. The United States, which out of loyalty to South Korea had never had high-level diplomatic contacts with North Korea, began talking with top North Korean officials to persuade them to reconsider. With the eighty-year-old President Kim Il Sung in virtual retirement, this diplomatic victory was seen in North Korea as a triumph for Kim Jong Il, the president's son and successor, as he tried to build up his reputation. On the last day of the waiting period, North Korea agreed to stay in the NPT and continue discussions. In July, 1993, the two sides began working on a plan to replace the North Korean reactors with safer models less capable of producing weapons-grade plutonium.

The United States wanted North Korea to agree not to produce nuclear weapons. It saw the prospect of a North Korea with nuclear weapons as a threat, not only to the security of Northeast Asia, to its global efforts to limit the spread of nuclear weapons. U.S.-North Korean negotiations included American commitments to replace North Korea's old and dangerous nuclear reactors, to supply North Korea with fuel oil until new reactors were completed, and to set up liaison offices in both nation's capitals as a preparatory step to full diplomatic relations.

While many observers thought that North Korea got the best of the United States, the North Koreans took the negotiations seriously and continued negotiating while Kim Il Sung was dying in the summer of 1994. That fall in Geneva, the agreement was signed, averting the potential for a renewed hostilities in Korea, this time possibly involving nuclear weapons.

Donald N. Clark

Elections Increase Speculation About China's Future

The approaching death of China's aged leader, Deng Xiaoping, and the elections of Jiang Zemin as president and Li Peng as premier stimulated worldwide concern about China's future.

What: National politics
When: March 31, 1993
Where: Beijing, China
Who:
DENG XIAOPING (1904-1997), China's most powerful leader
JIANG ZEMIN (1926-   ), former Communist Party general secretary who was elected president from 1993
LI PENG (1928-   ), conservative former minister who was premier from 1987 to 1998

Deng's Last Selections?

After his political reinstatement in 1977, Deng Xiaoping became China's supreme leader among the country's elite collective leadership. Despite holding no official positions after the 1970's, Deng ruled by virtue of personal prestige and powerful connections. These allowed him to choose, or to sanction the choice of, presidents, premiers, and ministers, all of whom nominally were elected by the nearly three thousand delegates of the Communist Party's National People's Congress (NPC).

Within the realm of Chinese government, where personal connections and alliances can matter more than official positions, Deng's choices for top governmental offices provided sinologists and journalists with important indications of stresses and divisions within the nation's only political party, the Chinese Communist Party (CCP). They also stirred speculations about China's future leadership. Deng was eighty-nine years old and in ill health in March, 1993. His selections of Jiang Zemin as the next president and Li Peng as the next premier were expected to be his final ones. They also were expected to clarify the question of who might succeed him. Whoever did so would determine the degree of political stability that prevails among China's 1.2 billion people, the fate of the CCP, and the direction taken by China's remarkably reformed and expanding economy.

Reflecting Deng's wishes, the NPC on March 31, 1993, with customary unanimity elected Jiang Zemin as president. In addition, it named Li Peng as premier. More than 250 NPC delegates opposed Li. Relative to the rest of China's aged leadership, Jiang and Li, both in their sixties, were young. Both were entitled to hold office for five years, although if they fell out of favor with the supreme leader, they were subject to instant dismissal.

The Tides of Succession

Ensuring peaceful successions to authority has always been a major problem for authoritarian regimes. At the time of his death in 1976, Mao Zedong, who had led the Chinese Communist Revolution of 1949, was revered almost as a god. His leadership role was difficult to fill. China's difficulties were complicated further by the death, the same year, of Zhou Enlai, China's remarkable foreign minister and, like Mao, one of the founders of the CCP. Zhou was a figure who might otherwise have won the political support required to lead China. Instead, Mao's wife and the so-called Gang of Four assumed control. The Gang of Four's four-year rule was ended by a series of coups. Subsequently, in turn, Hua Guofeng, Hu Yaobang, and then Zhao Ziyang and Hu Qili rose to and fell from powerful positions.

Behind the scenes throughout these episodes, Deng Xiaoping as deputy premier and, more important, as CCP secretary attempted to create via-

2509

AP/Wide World Photos

Chinese president Jiang Zemin (right) and U.S. president Bill Clinton, in Seattle in November, 1993.

ble new leaders and an enduring political succession during the 1980's. China's leadership nevertheless continued to reflect deep instabilities within the CCP. During the first forty-four years of communist rule, from 1949 to 1993, about half of the members of the CCP's Politburo Standing Committee, which has been synonymous with China's post-1976 "collective leadership," were ousted for their mistakes. Such were the troublesome preludes to the 1993 elections of Jiang and Li.

Upon his 1993 election, President Jiang Zemin was considered by Western observers to be in a weak political position. Superficially, that was not immediately apparent. Jiang, who earned a high reputation as mayor of Shanghai, previously had

been the only official since the 1970's to occupy China's top state, party, and military positions simultaneously. Western observers and sinologists believed that he had made too many enemies and mustered only modest political support.

Li's position as premier appeared even weaker than Jiang's as president. Li, who like Deng had been a protégé of Zhou Enlai, was the son of a revolutionary martyr who was killed by Nationalists. Subsequently, he was trained in Moscow as an engineer and rose to authority as a bureaucrat in the ministry of power, then in the education ministry, eventually taking his place as a member of the CCP's Central Committee. Li was considered a hard-line communist who opposed the ongoing success of China's economic reforms, which had thrown open the country to international investments, instituted capitalist practices, and decentralized state planning and enterprises.

Li furthermore was closely identified with the infamous Tiananmen Square massacre of June 3, 1989. After his denunciations—as a former and much-disliked state education minister—of students and other prodemocracy dissidents, thousands were shot to death by the army, ostensibly on his orders. Li accordingly was often characterized as arrogant or incompetent, and as the most unpopular official in China. Despite even Deng's criticisms of him, however, Li still managed to be elected premier for the first time in 1988.

Consequences

By mid-summer of 1995, Deng Xiaoping's life expectancy had been reduced to weeks. Deng's daughter was about to publish a book dedicated to, but not uncritical of, her father. She estimated his remaining life in days, leading some observers to ask if she and the remaining Deng clan were positioning themselves to seize power. The Chinese foreign ministry, at the same time, reported Deng to be in "excellent" health. In the meantime, Jiang Zemin appeared to have consolidated his presidential power by moving old cronies and allies from his Shanghai political base—figures such as Wu Bangguo and Jiang Chunyun—into Central Committee and vice ministerial posts.

Although Premier Li was obliged to make public statements favoring economic reform, and although he tried to distance himself from the Tiananmen massacre, he remained immensely unpopular. In 1995, he was reported to be in ill health. He recovered fully but was replaced in office by Vice Premier Zhu Rongji in 1998.

Clifton K. Yearley

Branch Davidian Cult Members Die in Waco Fire

> *More than eighty Branch Davidian cult members and their leader, David Koresh, died in a fire in Waco, Texas, on April 19, 1993, after a lengthy standoff with government officials.*

What: Civil rights and liberties; Religion
When: April 19, 1993
Where: Waco, Texas
Who:

DAVID KORESH (VERNON WAYNE HOWELL, 1959-1993), leader of the Branch Davidian cult

JANET RENO (1938-   ), U.S. attorney general in 1993

WILLIAM SESSIONS (1930-   ), director of the FBI in 1993

The Raid in Waco

On April 19, 1993, the Federal Bureau of Investigation (FBI) stormed the Branch Davidian compound in Waco, Texas, after a fifty-one-day standoff. The result was a fire that killed the people the agency had hoped to save.

The FBI had been waiting for the residents of the compound to surrender after an initial exchange of gunfire on February 28. Progress seemed to be made when more than thirty adults and children left the stronghold during the following four weeks. Unfortunately, many of the members remained within the walls when the FBI struck, despite repeated attempts for a peaceful resolution.

The FBI employed floodlights and constant loud music in an attempt to disrupt the compound, along with cutting off electricity. These and other tactics failed to force the Davidians to turn themselves over to authorities. On the day of the final attack, FBI agents used tanks fitted with battering rams to break holes in the walls of buildings, then filled the buildings with a type of tear gas. Agents watched for signs of surrender, then repeated the gassings. By noon, the build-

ings were ablaze. More than eighty people died. Nine escaped the blaze and were captured; they were later sentenced to forty-year prison terms.

Branch Davidian leader David Koresh and his followers held an apocalyptic vision, a religious belief that they would die in a violent clash with evil. When the FBI increased the force used against them, the Davidians' confidence grew. The cult members were certain that this was the predicted conflict and that their deaths would fulfill the will of God.

Although Koresh's followers were given opportunities to cooperate, they often made promises and then failed to meet deadlines. On April 14, Koresh claimed he needed only two more weeks to complete important work. He was supposedly decoding a religious prophecy called the "Seven Seals."

The FBI was frustrated by the delays and presented an attack plan to Attorney General Janet Reno. She asked many questions, expressing concern for the safety of children in the compound. After she discussed her decision with President Bill Clinton, Reno gave her permission for the attack.

Did They Deserve to Die?

The Branch Davidian cult had its beginnings within the Seventh-day Adventist church. It had made its home in Waco since 1935 without any incident. Major changes became apparent in May of 1992. A United Parcel Service employee making a delivery to the Davidian gun business saw fifty hand grenades fall out of a package. He notified the authorities, and the Bureau of Alcohol, Tobacco, and Firearms (ATF) ultimately became involved.

The ATF began investigating the group's purchases of guns and explosives. By the end of the year, the ATF estimated that the cult had built up an arsenal of approximately three hundred weapons. These weapons were supposedly purchased for resale through the group's business, The Mag Bag. The cult had several sources of income, including a yard service, an automotive store, and The Mag Bag. The gun business came under suspicion because of unauthorized conversion of semiautomatic weapons into machine guns. The group also had failed to pay taxes and register some of its weapons.

The ATF also became interested in investigating drug charges and allegations of sexual abuse involving the group. Only after the raid was it discovered that the drug charges had been against former leader George Roden, not against Koresh. In fact, there had been no evidence of drug involvement since Koresh's takeover. Accusations of sexual abuse against children were never proved, despite several visits by public officials.

Based on its suspicions, on February 28, 1993, the ATF decided to take action. It attempted to storm the compound even though information on the raid had been leaked to the press and someone had gotten the information to Koresh. The raid was neither a surprise nor a success. Four agents and six Davidians were killed, and the compound remained in the hands of the cult members.

Some sources say the violent attack was unnecessary. They claim that Koresh was often seen in the community and arrest warrants easily could have been served against him. Rumors circulated that the ATF conducted the raid and tipped off the media in order to showcase its Special Response Team and prevent budget cuts. Others say the agency was within its rights because of the number of weapons held in the compound.

The FBI, under the leadership of William Sessions, became involved after the February attack. Various methods were attempted to achieve a surrender, but nothing worked. Koresh made demands and the FBI complied, yet he never emerged. The FBI's final plan, designed to pre-

Fire consumes the Branch Davidian compound in Waco, Texas.

AP/Wide World Photos

2513

vent a threatened mass suicide, had the opposite effect. Agents were to gas the building, negotiate, gas the building again, and negotiate further until the compound was empty. They began their movements with a telephone call and loudspeaker announcements describing the plan. Fires soon broke out and were fanned by strong winds. The blaze left the agents powerless.

The cause of the fire remained unknown. Tanks used by the FBI may have ignited flammable materials, or cult members may have set the fires as a means of mass suicide.

Survivors of the fire and relatives of sect members who died in the fire filed a $695 million wrongful-death suit against the federal government. In July, 2000, a five-member federal advisory jury found that the government had not been responsible for the deaths at Waco. However, the jury's decision was only advisory, and U.S. district judge Walter Smith was left to make the final decision. In September he confirmed the jury's verdict, thereby exonerating the government of wrongdoing.

Consequences

The Waco confrontation led to questions concerning the methods used by government agencies. Mismanagement and lack of communication were listed as factors contributing to the loss of life. Traditional law-enforcement techniques, including a "show of force," were closely scrutinized. It was expected that future conflicts would be handled quite differently.

Another consequence involves the role of the media. Media involvement and leaks to the press may have affected the February 28 raid. This raised the issue of balancing the public's right to know against the government's need for secrecy to carry out its plans and protect the lives of its agents.

Finally, cults and other nontraditional groups came under closer scrutiny. Each militia, religious sect, or organization would thereafter be viewed with suspicion. Their freedom thus might come into conflict with the goal of public safety, as government officials could choose to limit individual freedoms to protect the larger society.

Suzanne Riffle Boyce

IRA Bomb Explodes in London

The Irish Republican Army detonated a bomb in the heart of London's financial district, killing one person, injuring several dozen people, and causing an estimated billion dollars in damage.

What: Civil strife; Terrorism; International relations
When: April 24, 1993
Where: London, England
Who:

JOHN MAJOR (1943-   ), prime minister of Great Britain from 1990 to 1997

GERRY ADAMS (1948-   ), president of Sinn Féin beginning in 1983

FRANCIS MCWILLIAMS (1926-   ), lord mayor of the city of London from 1992 to 1993

A Spectacular Explosion

On April 24, 1993, the largest bomb ever to explode on Great Britain's mainland was set off by the Irish Republican Army (IRA) in the financial district of London, known as "the City." The bomb contained more than two tons of ammonium nitrate fertilizer and was hidden in a dump truck parked on the street. The explosion left a crater fifteen feet deep, shattered windows for several blocks, and caused widespread structural damage to several buildings. Particularly hard hit was the National Westminster Tower, which at fifty-two stories was the second tallest skyscraper in Britain.

The human toll of one person dead and thirty-six injured could have been much worse. The IRA struck on a Saturday, when the City is largely deserted. Furthermore, the IRA followed its usual pattern of giving several coded telephone warnings to the police shortly before the blast so people could be evacuated from the danger area.

Prime Minister John Major's immediate response was that Londoners would not give in to terrorism and that London's City would be back to work as usual on the following Monday. Sir Francis McWilliams, the lord mayor of London,

made similar statements. Behind this brave rhetoric was the fear, openly voiced in the press, that the explosion jeopardized London's status as one of the world's leading financial centers. Only a year previously, the IRA had planted another bomb in the City, resulting in loss of life and substantial damage to the Baltic Exchange building. It was feared that further terrorist attacks would drive up property insurance rates, prod police into conducting time-consuming security checks of cars and people coming into the area, and drive away foreign firms afraid of becoming innocent victims in any further attacks.

A Divided Ireland

This bombing was only one more event in the troubled history of Anglo-Irish relations. For centuries, the Irish had endured British domination, but by the early twentieth century, Great Britain had reconciled itself to granting home rule to Ireland. Irish Protestants, living mainly in the northeastern corner of Ireland, an area popularly known as Ulster, refused to be forcibly incorporated into a united Catholic Ireland, where they would be a minority. As Protestants and "Loyalists," they wished to remain part of the United Kingdom. Consequently, the Government of Ireland Act of 1920 provided for the principle of partition. In 1921, the six counties of the north were given a home-rule parliament in Belfast, and the union with Great Britain was maintained. The following year, the twenty-six counties of the south were given a parliament based in Dublin and formed the Irish Free State.

The problem centered on Northern Ireland. Approximately two-thirds of the population was Protestant and Loyalist, but Roman Catholics formed a large minority. They were the victims of blatant and organized discrimination. By the late 1960's, an effective Catholic civil rights movement emerged, and in 1969 terrible street clashes

2515

British prime minister John Major.

broke out between Catholics and Protestants. This forced Great Britain to send British troops in to restore order. Eventually they were perceived as an occupying army by the Catholic community.

This provided an opportunity for the IRA. Founded in 1919 to fight the British, the IRA had difficulty establishing a presence in Northern Ireland. In 1969-1970, the IRA split into two wings, the "Officials," who tended to be Marxist and more theoretical, and the "Provos," who were associated with violence and terror. The Provos waged a ruthless campaign in Northern Ireland against British authority and soon began operations on Great Britain's mainland. By the mid-1970's, they had carried out a number of attacks there, with the hidden bomb being a favorite weapon. By its very nature, the bomb was an indiscriminate device, and Britons were outraged every time children, the elderly, or innocent bystanders were among the victims.

To the British, members of the IRA were terrorists and cold-blooded killers, while the IRA saw itself as "freedom fighters" seeking to oust an occupying army from Irish soil and achieve a united Ireland. Further angering the British were the clever propaganda tactics of Gerry Adams, the president of Sinn Féin, the political arm of the IRA. Adams effectively presented the Irish cause not only in Great Britain but abroad as well, especially in the United States.

During this bloody campaign, the IRA was seen as largely successful. For example, in the ten years previous to the City bombing, thirty-five people in England were killed, more than five hundred were injured, and billions of dollars were lost in property damage. In addition, there were two daring but unsuccessful attempts on the lives of British prime ministers. Most disheartening from the British perspective was the low rate of convictions for these offenses. To the general public, it appeared that Great Britain's security forces were losing the battle with the IRA.

Consequences

In September, 1994, the IRA declared a unilateral cease-fire, and the following month Protestant paramilitary units in Northern Ireland responded similarly. The British army in turn modestly reduced the level of its armed forces there. For the first time since 1969, the violence had stopped. By May, 1995, talks between representatives of the British government and Sinn Féin had begun. Later that month, Prince Charles, heir to the British throne, made a historic visit to Dublin. All of this gave grounds for cautious optimism.

It was unclear what role the bombing campaign played. Some IRA sympathizers argued that the British government was driven to the bargaining table by the organization's successful tactics. Other observers believed that the IRA finally realized that its bombing campaign was counterproductive and unlikely to sap the morale of the British nation. Whatever the reason for the cessation, people were relieved that perhaps the nightmare of sectarian strife and indiscriminate bombings might finally be coming to an end. The IRA, however, claimed responsibility for a bombing in February, 1996, breaking the cease-fire.

David C. Lukowitz

Tamil Rebels Assassinate Sri Lanka's President

A suicide bomber killed Sri Lankan president Ranasinghe Premadasa during a May Day celebration.

What: Assassination; National politics
When: May 1, 1993
Where: Colombo, Sri Lanka
Who:
RANASINGHE PREMADASA (1924-1993), president of Sri Lanka from 1988 to 1993
DINGIRI BANDA WIJETUNGA (1922-   ), prime minister of Sri Lanka from 1989 to 1993 and successor to Premadasa as president
KUMARAKULASINGHAM WEERAKUMAR (1970-1993), suicide killer of Premadasa

The May Day Bombing

During the annual May Day rally, President Ranasinghe Premadasa died when a suicide bomber triggered an explosion. The blast killed Premadasa, the bomber, and twenty-two others who were watching or participating in the parade. The Liberation Tigers of Tamil Eelam (LTTE), a rebel guerrilla group, were believed to be responsible for the killing, although they denied responsibility.

The bomber was Kumarakulasingham Weerakumar, a twenty-three-year-old member of the Tamil ethnic minority and a resident of Gurunagar, a small village in the LTTE-controlled Jaffna peninsula on the northern part of the island. He strapped explosives around his waist and rode a bicycle up to the vehicle carrying Premadasa. The bomb was triggered by a radio transmitter held in his hand as he pedaled the bicycle. Premadasa had stepped down from the vehicle shortly before Weerakumar approached him. The bomb killed both instantly.

Although the LTTE denied responsibility for the bombing, police believe that the group was behind it. The LTTE carried out a similar bombing in India in 1991, when a female suicide bomber presented former Indian prime minister Rajiv Gandhi with a bouquet of flowers. A remote control device activated a bomb in the bouquet, killing Gandhi and those around him. In addition, fragments of a cyanide capsule were found in Weerakumar's neck. Members of the LTTE are known for their vow to take cyanide rather than surrender. All LTTE fighters carry a cyanide capsule around their necks. Although members of the United National Party (UNP), President Premadasa's political party, were angry with him and might have killed him, the police insisted that the LTTE committed the killing despite any hard evidence.

Weerakumar had moved to Colombo a year earlier and had befriended Premadasa's personal valet. His friendship with a member of Premadasa's staff gave him access to the president's house and apparently allowed him to get past security personnel on the day of the killing.

Leaders of the UNP met on the day of the killing to select a new leader, whose name was submitted to and approved by the parliament on May 4. Prime Minister Dingiri Banda Wijetunga was selected to lead the country. Wijetunga was a senior statesman of his party. His reputation for hard work overshadowed his lack of charisma and leadership abilities. The obvious choices to replace Premadasa, the younger Ranil Wickremasinghe and Gamini Dissanayake, refrained from contesting the office in the interest of party unity. Both placed the interests of political stability ahead of their own political ambitions and avoided a possibly divisive battle for the presidency. They were both young and knew that elections would be held within two years. Four days

2517

after the assassination, the UNP-dominated parliament overwhelmingly elected Wijetunga to serve the rest of Premadasa's term of office.

A History of Conflict

The LTTE has been fighting the Sri Lankan government since 1983 to create an independent state for the Tamil-speaking ethnic minority in Sri Lanka. The Tamils, who are divided into two ethnic groups, the Sri Lankan and the Indian Tamils, compose 18 percent of the country's population. The Sinhalese, on the other hand, compose 74 percent and thus dominate the political system.

The Tamils claim that the government, dominated by the Sinhalese ethnic group, has discriminated against them. The civil war has been confined largely to the north and east of the island, where the Sri Lanka Tamils are the majority of the population. On occasion the war has spilled into the capital city of Colombo, as it did on May Day. On several occasions the LTTE has assassinated government leaders in Colombo.

The Tamils are separated from the Sinhalese by language and religion. The Tamils speak a Dravidian (South Indian) language and are mostly Hindus. The Sinhalese are mostly Buddhist and speak an Indo-Aryan language unrelated to Tamil. The approximately 13 million Sinhalese of Sri Lanka are the only speakers of Sinhala in the world and are fearful that the 3.2 million Tamils on the island will ally themselves with the 50 million Tamils in India and overrun the culture of the Sinhalese. The Sinhalese have controlled the Sri Lankan government to protect and promote their religion and culture, often at the expense of the minority groups on the island.

The assassination provided the first test of the 1978 constitution's provisions for replacing a dead leader. The power to replace the president lies with Parliament. The election of the new president went smoothly, with little opposition, as all political factions among the Sinhalese united to maintain political stability and to ensure a smooth transition.

Consequences

An important element of all democracies is their ability to withstand the trauma of an unexpected death of their head of government. The killing of Premadasa provided a test of Sri Lanka's commitment to democratic values. The political transition to a new leader was smooth, and the political system survived the test. The assassination also sent a shock wave through the government by illustrating how successful the Tamil guerrillas could be in targeting and killing important political leaders. The guerrillas made it clear that they had the ability to carry out any action they desired and that a military settlement of the civil war would be difficult.

Robert C. Oberst

Paraguay Holds Democratic Presidential Election

Voters in Paraguay were able to choose a president freely in elections monitored by international observers, although there were charges of fraud.

What: Civil rights and liberties; National politics
When: May 9, 1993
Where: Asunción, Paraguay
Who:
JUAN CARLOS WASMOSY (1938-), candidate of the ruling Colorado Party
DOMINGO LAÍNO, (1935-), candidate of the opposition Liberal Party
GUILLERMO CABALLERO VARGAS (1940-), candidate of the National Encounter Party
JIMMY CARTER (1924-), president of the United States from 1977 to 1981 and leader of a team of international observers of Paraguay's election

First Fair Elections

On May 9, 1993, the people of Paraguay participated in the first open and fair presidential elections in the country's history. To ensure that the Colorado Party, which controlled the government and the electoral machinery, would allow fair elections, the three leading candidates requested election monitors from the Democratic Institute for International Affairs. The institute's delegation was led by former U.S. president Jimmy Carter. In addition, SAKA, a consortium of private groups that was to provide an independent vote count, requested observers from the Latin American Studies Association. In all, there were several hundred election observers, including a team from the Organization of American States.

On election day, more than one million Paraguayans quietly went to the polls. Several incidents marred the election process. Shortly after the polls opened, rifle fire damaged the transmitter of the only opposition television station. In the afternoon, as the government announced the victory of its candidate, the state-owned telephone company cut the wires to the headquarters of SAKA as it was about to make an independent tally of the votes Also on election day, the government ordered international borders closed to Paraguayans returning from neighboring countries, denying them the opportunity to vote.

Carter raised strong objections to these incidents. After several hours of negotiations, the telephone lines were restored to SAKA.

Election results were later certified independently. The government candidate, Juan Carlos Wasmosy, won with 38.9 percent of the vote. The leading opposition candidate, Domingo Laíno of the Liberal Party, received 31.5 percent, and Guillermo Caballero Vargas came in third with 27.1 percent. Carter and other observers agreed that incidents of fraud and intimidation had not altered the final outcome and declared the elections to be fair. Even though Wasmosy won with a plurality and a 7 percent lead over his closest rival, the two main opposition parties received nearly 60 percent of the vote, making the newly elected president the first in recent history without an automatic majority in the country's legislature.

After the Coup

General Alfredo Stroessner had ruled Paraguay from 1954 until he was ousted in 1989. His power rested on a coalition of the armed forces, government bureaucracy, and the Colorado Party. Although it provided stability to a country that had seen two civil wars in the twentieth century and twelve presidents between 1936 and 1954, Stroessner's regime was both repressive and corrupt.

In 1989, a combination of pressures for change among the Colorado Party's leadership and from a society stifled by corruption led to a military coup that overthrew Stroessner. General Andres Rodríguez, who led the coup, scheduled new elections three months after Stroessner's departure. Opposition parties complained that they did not have enough time to mount effective campaigns, but elections were held as scheduled on May 1, 1989. Rodríguez was elected to a four-year term as president of Paraguay. On May 26, 1991, the electoral process was expanded as voters went to the polls to choose mayors and city councils for the first time in Paraguay's history. A constituent assembly was elected in December, 1991, to write a new constitution for Paraguay.

In each of these elections, Paraguayans experienced opportunities to vote freely, but not always fairly. There was still a close relationship between the Colorado Party and the government bureaucracy that it controlled. Opposition politicians complained that government workers were campaigning for the Colorado Party and that the military recruited voters and transported voters to the polls. Nevertheless, increasing numbers of Paraguayans exercised their right to vote, with the turnout rising from 53 percent of the electorate in 1989 to more than 71 percent in 1991.

As General Rodríguez's term came to an end in 1993, intense struggles occurred within the ruling Colorado Party. Some within the party opposed the candidacy of Wasmosy because he was a businessman without political experience. No clear choice emerged until a December, 1992, primary. Wasmosy lost by a narrow margin, but the outcome was overturned by a party panel in March, only two months before the elections.

Several opposition parties sensed the opportunity for victory over a badly divided Colorado Party. The Authentic Radical Liberal Party (PLRA) nominated Domingo Laíno, an economist who had served in the constituent assembly that drafted the new constitution. The National Encounter Party (EN) was an alliance of reformers from within the Catholic Church, a trade union federation, and some Christian Democrats. They named Vargas, a lawyer and businessman, as their presidential candidate. Eleven other candidates ran for president.

The elections of May 9, 1993, saw an increase in political campaigning and extensive use of radio and television. Observers for the Latin American Studies Association concluded that total campaign spending by the three major contenders reached $27 million. The voter turnout, although down slightly from that in 1991, was nearly 70 percent of the electorate and demonstrated the continued support of Paraguayans for their open political system.

Consequences

The election of Wasmosy represents a break in Paraguay's history of fraudulent elections that automatically returned the ruling party to power. The people of Paraguay had the opportunity to choose their president freely in May, 1993. The high rate of voter participation demonstrated enthusiasm for the democratic process. The strength of the new opposition parties indicated that the people of Paraguay had a choice at the polls. The violations of electoral laws and the intimidation of voters were serious offenses but did not alter significantly the results of the elections, according to observers.

James A. Baer

Demonstrations Publicize China's Occupation of Tibet

When demonstrators in Lhasa protested against inflated prices charged by Chinese shop owners, military authorities responded by using tear gas and plastic bullets.

What: Human rights; Civil rights and liberties
When: May 23, 1993
Where: Lhasa, Tibet
Who:
LI PENG (1928-   ), prime minister of China from 1987 to 1998
TENZIN GYATSO (1935-   ), fourteenth Dalai Lama

Protests Against the Chinese

By noon on May 23, 1993, several hundred demonstrators occupied the Barkhor, which circles the seventh century Jokhang temple in the center of Lhasa, Tibet's capital. Some chanted slogans related to rent increases and rising prices; others shouted "Chinese out of Tibet." Many carried pictures of the Dalai Lama, Tibet's leader in exile Within two hours, the crowd of as many as four thousand people became disruptive. Chinese soldiers fired tear gas for two hours until stability was restored.

Protests continued sporadically for the next five days. On June 1, several Tibetans were arrested for raising their national flag. The Dalai Lama publicly asked Prime Minister Li Peng of China to exercise restraint. The official response offered by Tibet's governor, Gyalcon Norbu, was that dialogue with the Dalai Lama was conditional upon his acceptance that Tibet is an integral part of China, an idea that many Tibetans including the Dalai Lama reject. The turmoil of May 23-28 suggests that economic growth cannot make Tibetans forget the grievances they have endured since the 1950 invasion by China.

In the late 1980's the Chinese government placed fewer restrictions on new settlements and entrepreneurship throughout the People's Republic of China, which since 1951 has included Tibet, known as Xizang. As a result, thousands of Han Chinese fortune seekers from central regions have entered Lhasa annually and set up business. Approximately fifty thousand of Lhasa's sixty thousand registered private businesspeople are from central China. Although immigrants are expected to apply for residence permits and business licenses, Chinese authorities in Lhasa bend the rules in order to promote economic development. The Chinese government invests heavily in Tibet, compared with other provinces, following the conventional wisdom that relative prosperity reduces political opposition.

By all accounts, this colonization process has been a sad reminder of how little occupied countries gain from takeover by a superpower. Because of low literacy levels and poor technical training, Tibetans cannot compete with the Han Chinese; thus, they are actually victimized by a booming economy. The fourteenth Dalai Lama, living in exile in northern India, denounced this social domination as a form of cultural genocide.

Chinese Rule

Since the Chinese invasion of Tibet, an estimated one million Tibetans have died of unnatural causes. In the early 1980's, the Beijing government ostensibly placated Tibetans after twenty years of harsh policies by lifting a ban on public worship and permitting tourism. Systematic abuses continued. Through the 1980's and into the 1990's, human rights violations and appalling acts of torture and degradation were reported, especially after widespread rioting in 1987 and again in 1989, when martial law was imposed for thirteen months.

National Archives

The Dalai Lama (left) greets Indians and the press after fleeing Tibet.

Paramedic teams have tried to sterilize populations of entire villages. Pregnancies were forcibly aborted, and newborn babies were given lethal injections. Buddhist nuns, who became the champions of the freedom movement, were often detained in prisons without charges against them. They were sentenced without trial to years in prison, and many were badly beaten and violated with electric batons. In 1994, Amnesty International reported that 628 prisoners were held in Tibetan jails because of political beliefs. Among them were 182 women and 45 people under the age of eighteen.

China's claim to Tibet goes far back in history. As an isolated nation, Tibet relied on Chinese support against Mongol and Ottoman influence from the thirteenth to the eighteenth century, when the Manchu Dynasty subsumed Tibet as a protectorate. In the 1940's, Chiang Kai-shek, leader of the Nationalist Chinese, condemned Tibet as a "heretic kingdom." The communists came to power in 1949 under Mao Zedong, who promised to "consolidate the motherland." Tibetans were supposedly liberated by Chinese troops who would protect them from the forces of imperialism.

The Dalai Lama did not fall for Mao's propaganda. He fled Tibet after the Chinese invasion of 1950 but returned in 1954 for discussions with Mao in Beijing. Perhaps no two ideologies were less compatible. The Dalai Lama talked about the possibilities for a synthesis of Buddhism and socialism while Mao reiterated that religion is poison to the masses. The Dalai Lama wisely concluded that Mao and the Chinese represented the enemy.

Soon afterward, the Chinese army collectivized Tibet. Prominent families were forced to confess to crimes against the people. Their property and possessions were confiscated, and some were executed on the spot. The warlike Khampa tribe resisted the Chinese intervention. In 1959 the Dalai Lama, with his family and one thousand retainers, fled south to India, barely making it across the border. At least eighty thousand of those who remained behind were slaughtered.

This was only the beginning of the annihilation of the religious culture of Tibet. The atrocities reached a peak in 1966 and 1967, during the fanatical Chinese Cultural Revolution. Tibetan villages were burned, all of Tibet's fortifications were destroyed, and six thousand monasteries and temples were gutted. Artistic treasures worth approximately $80 billion were transported to Beijing. They made their way through Hong Kong to European auction houses and private collections. Thousands of woodblock sculptures representing centuries of compiled data on Buddhist spirituality were burned.

Huge flocks of tame animals including ducks, geese, cranes, and songbirds were killed by Chinese people armed with machine guns, along with yaks, antelopes, wild donkeys, and blue sheep. The Chinese used these animals as food; Tibetans are primarily vegetarians. Trans-Himalayan trade by nomadic caravans was virtually eliminated. The forests of Eastern Tibet were destroyed.

With the death of Mao's follower Zhou Enlai in 1976, China began to reassess its Tibetan policy. Overt destruction of Tibetan culture and resources was publicly criticized, but the influx of Han Chinese was encouraged, especially through intermarriage with Tibetan women.

Consequences

Few reliable statistics document the aftermath of the 1993 insurrection. During the outbreak, the movement of foreigners, especially journalists, was severely restricted. Several demonstrators certainly were killed and many more arrested, but conclusive evidence is hard to find. Nuns continued to die in prison, and the Barkhor was carefully watched by secret police.

A planned hydroelectric project promised electricity and irrigation in the future, but as with many other Chinese developments, the immediate beneficiaries will be the Han Chinese. In Lhasa, bulldozers, picks, and shovels razed the artifacts of thirteen hundred years of Tibetan history and architecture. Two-thirds of the old city was leveled. According to Chinese plans, only the immense Potala, or Winter Palace, the Jokhang temple, and a handful of old homes will be retained.

The demonstrations of 1993 may not have been entirely fruitless, as they turned international attention toward Tibet. China's most favored nation status with the United States may fall into jeopardy as more people become aware of China's modernization of Tibet.

Robert J. Frail

Armenia and Azerbaijan Sign Cease-Fire Agreement

> *After five years of war, the governments of Armenia and Azerbaijan agreed to a cease-fire, but the peace process quickly broke down and fighting resumed without a long-term solution in sight.*

What: Military conflict
When: May 26, 1993
Where: Moscow, Russia
Who:
ABULFAZ ELCHIBEY (1938-   ), president of Azerbaijan from 1992 to 1993
SURAT HUSEYNOV (1959?-   ), rebel leader who seized power in Azerbaijan in June, 1993
LEVON TER-PETROSYAN (1945-   ), president of Armenia from 1990 to 1998
AYAZ MUTALIBOV (1938-   ), president of Azerbaijan from 1991 to 1992

The Cease-Fire Agreement

A lengthy and destructive war between Armenia and Azerbaijan began in 1988. Deep antagonisms between these two neighbors, especially over control of the Nagorno-Karabakh region, contributed to their mutual suspicion, confrontation, and violence. Several nations and international organizations attempted to bring about a cease-fire between the combatants and begin meaningful peace negotiations to resolve the dispute. Russia, Iran, the United States, Turkey, the United Nations, and the CSCE (Conference on Security and Cooperation in Europe) made these periodic efforts. Before 1993, however, all attempts to mediate the crisis and end the war failed.

The international community in the spring of 1993 again encouraged negotiations between the warring parties. This encouragement included United Nations Resolution 822, adopted in April, 1993, recommending a cease-fire. Finally, on May 26, Armenian and Azeri negotia-

tors agreed to stop the fighting and begin negotiations for a peace treaty. The agreement included a sixty-day cease-fire and Armenia's military withdrawal from the Azeri town and district of Kelbajar, an area captured the previous month in a major Armenian offensive. The Azeris agreed to end their successful trade and transportation embargo of Armenia, a policy that significantly blocked access to essential supplies, including food and fuel, for the Armenian population and economy. The Nagorno-Karabakh government reluctantly accepted the agreement on June 15, fearing a settlement opposed to its interests.

The May 26 cease-fire agreement was to be followed by formal peace negotiations in Geneva, Switzerland, in June under CSCE auspices. To assist in the process, a CSCE delegation visited the area to implement the agreement through the deployment of peacekeeping monitors and supervision of the withdrawal of military forces. Continued mutual suspicion, complicated by the outbreak of civil war within Azerbaijan and the overthrow of President Abulfaz Elchibey in early June, prevented the treaty's implementation. United Nations Resolution 853, adopted on July 29, expressed grave concern about the apparent collapse of the peace process.

Uneasy Neighbors

Armenia and Azerbaijan, located in the Caucasus Mountain region of the former Soviet Union, have long cultural and political traditions. These factors help explain the underlying misunderstandings and occasional conflicts between the two societies. Armenia's Christianity dates from the fourth century C.E., during Armenia's existence as a separate kingdom. Following several centuries under Ottoman and Turkish rule, a portion of Armenia came under Russian control

Disagreements about the status of Nagorno-Karabakh, along with accounts of mistreatment of its citizens by Azeris, led to the outbreak of hostilities between Armenia and Azerbaijan in early 1988.

The disintegration of the Soviet Union in 1991 compounded the difficulty of reaching a negotiated settlement. Neither side won enough decisive victories to end the conflict, although Armenia generally achieved greater military success. Armenian forces succeeded in May, 1992, in capturing Azeri territory providing direct access to Nagorno-Karabakh, and two corridors allowed transportation of needed supplies and reinforcements to the enclave. Military and civilian deaths between 1988 and late 1993 were estimated at fifteen thousand to eighteen thousand. Estimates of civilian refugees displaced by the fighting range as high as one million.

Internal conditions within Armenia and Azerbaijan deteriorated. Economies of both nations were adversely affected, and political tensions led to the overthrow of two Azeri presidents: Ayaz Mutalibov in March, 1992, and Abulfaz Elchibey in June, 1993. In both instances the Azeri presidents fell as the result of public hostility. Citizens believed that the war had not been waged energetically and that negotiating with Armenia was a sellout. In Armenia, opponents of President Levon Ter-Petrosyan resisted any negotiated or compromise peace. Ter-Petrosyan survived this opposition in 1993, largely because of the successful Armenian operations against Azeri territory earlier in the year.

and was declared a republic of the Soviet Union in 1920. Azerbaijan is a Muslim society, with historical and religious ties to the Islamic faith in the Middle East and Iran. Russia absorbed parts of this region by the early twentieth century, establishing a Soviet republic in 1920.

A more immediate problem affecting their relationship came from the creation of the "autonomous Soviet republic" of Nagorno-Karabakh in 1923. This area, approximately the size of Delaware, is an enclave located completely within Azerbaijan's borders. Its population of 200,000 is composed primarily of Christian Armenians or those identifying with Armenia. Local government officials in Nagorno-Karabakh, supported by public opinion in the region, sought to have the area legally transferred to Armenian control during the 1980's. The Armenian government in Yerevan also claimed the enclave as Armenian territory. Citizens of Azerbaijan, called Azeris, saw these efforts as hostile interference in its internal affairs and an attempt to gain both power and territory for its neighbor to the west. The Azeri government in Baku therefore opposed the Nagorno-Karabakh separatist movement.

Consequences

Suspicion between the two states and civil unrest within Azerbaijan in mid-1993 prevented implementation of the May 26 agreement. Rebel leader Surat Huseynov took power in Baku in June, overthrowing President Elchibey, and became prime minister of Azerbaijan. The Azeri-

Armenian conflict continued throughout 1993 and into 1994. Although both nations later signed additional cease-fire agreements, no formal negotiations took place to settle the Nagorno-Karabakh issue and permanently end the lengthy dispute. In December, 1996, the Armenians rejected a settlement based on self-determination and security for Nagorno-Karabakh within Azerbaijan.

In February, 1998, Ter-Petrosian was forced from office by the Armenian military and replaced by the Karabakhian prime minister, Robert Kocharian. Ter-Petrosian had expressed willingness to recognize formal Azerbaijani sovereignty over Nagorno-Karabakh if the region were accorded autonomous self-rule. His position was similar to one advanced in October, 1997, by Arkady Gukasian, who succeeded Kocharian as president of Nagorno-Karabakh.

In the late 1990's, the question of Nagorno-Karabakh continued to prevent the establishment of normal relations between Armenia and Azerbaijan. Moreover, it continued to bedevil the lives of thousands of Azerbaijani who were driven from their homes inside and outside the enclave and fostered political instability in Armenia and Azerbaijan.

Taylor Stults

Bomb Damages Florence's Uffizi Gallery

The Uffizi Gallery blast was one of several suspected Mafia bombings that occurred in response to the Italian government's crackdown on organized crime.

What: Law; Crime
When: May 27, 1993
Where: Florence, Italy
Who:

JOHN PAUL II (KAROL WOJTYŁA, 1920-   ), bishop of Rome and leader of the Roman Catholic Church from 1978

GIOVANNI FALCONE (1939-1992), chief prosecutor of Mafia figures

SALVATORE RIINI (1931-   ), purported boss of all bosses of the Sicilian Mafia

The Uffizi Gallery Is Bombed

Just after 1 A.M. on May 27, 1993, a two-hundred-pound bomb concealed in a stolen Fiat exploded, gravely damaging the west wing of the Uffizi Gallery in Florence, Italy. The car bomb killed five people, destroyed a part of the five-hundred-year-old structure, damaged the Corridoio Vasariano, and weakened the Buontalenti staircase, the gallery's old exit. The fireball destroyed the newly completed catalog of the Uffizi collection as well as the world's oldest assembly of agricultural research documents, housed in the Academia dei Georgiofilli. Four of the dead were from a single family, the caretakers and inhabitants of the Torre Della Pulci, a medieval tower behind the Uffizi Gallery.

The Uffizi is Florence's principal museum. Aside from the loss of life, perhaps the greatest tragedy was the permanent loss of three paintings, two by Bartolomeo Manfredi and one by Gerrit van Honthorst, seventeenth century followers of Caravaggio. Damage to artwork was extensive and varied. Sculpture was broken. Some thirty paintings had pigment blasted from their surfaces, others were slightly damaged, and still others were shredded by flying glass. In nearby rooms, bulletproof glass that had been installed to protect against vandalism protected the paintings of Titian, Raphael, Veronese, Caravaggio, and Michelangelo from flying debris. More than two hundred works had to be removed from their galleries because the blast had blown out skylights, exposing the works to the weather.

The day after the explosion, more than twenty thousand people marched through the city in mourning and anger, and as a demonstration against armed intimidation. No party was immediately prosecuted for the blast, but authorities strongly suspected Mafia involvement. The bombings were viewed as a backlash to a Milan police investigation that had resulted in the roundup of hundreds of alleged Mafia gangsters and bosses. The terrorist act was a blow against government attempts to diminish the power and control of the Mafia, to draw attention away from political corruption trials, and to break the growing popular will for reform. The Mafia targeted a world-famous museum to underscore the vulnerability of Italy's thirty-five hundred museums and hundreds of thousands of public monuments and, by extension, the foreign tourists visiting the treasures of Italian history.

The Uffizi bombing was one of five car-bomb attacks between May and August of 1993 in Rome, Florence, and Milan. The attacks left ten people dead and dozens wounded. Cultural sites, journalists who opposed the Mafia's efforts to destabilize the government, church figures, and properties all were among the targets. Two venerable Roman churches were bombed. The first was San Giovanni in Laterano, a seventeenth century baroque masterpiece by Francesco Borromini and the church where the pope serves as bishop. The second was San Giorgio in Velabro, one of Rome's oldest churches and the site where Romulus and Remus are said to have founded Rome. It was surmised that the churches were targeted because of the pope's call for resistance to organized crime. In the minds of Mafia members, that statement abrogated the church's unwritten hands-off policy toward their activities.

Nearly a year after the Uffizi bombing, Italian investigators confirmed original suspicions when they officially declared that the Mafia was behind the 1993 bomb attack. At that time, four organized crime suspects were named, but they remained at large.

The Mafia Retaliates

The sixteenth century Galleria degli Uffizi (the Gallery of Offices) in Florence, Italy, is one of the world's greatest repositories of Italian art. The Uffizi was designed by Giorgio Vasari to house the government offices of the Tuscan state. Vasari also built the so-called Corridoio Vasariano that stretches from the Uffizi across the Arno River, via the Ponte Vecchio, to the Pitti Palace. The corridor is nearly half a mile long and contains the world's largest collection of self-portraits. After Vasari's death, Bernardo Buontalenti carried on the work. The resultant structure consists of an elongated, U-shaped, three-story building that borders the Arno River on one side and is within sight of the Piazza della Signoria on the other. The Uffizi became a public museum in 1859 after Tuscany joined the Italian state.

In 1991, Milanese authorities had uncovered a vast network of corruption involving business executives and politicians who exchanged huge bribes for public works contracts. Nearly twenty-five hundred politicians and corporate leaders were implicated. Later, two leading businessmen associated with the scandal committed suicide. The Mafia retaliated by bombing cultural sites and assassinating government and church officials. In May, 1992, Italy's chief prosecutor of Mafia figures, Giovanni Falcone, was murdered. Two months later, another judge was killed. Salvatore Riini, the boss of all Mafia bosses, was believed to be involved and was arrested. It was also in May that Pope John Paul II, during a visit to Sicily, made the church a Mafia target when he urged people to resist the crime lords. In September, Giuseppe Puglisi was murdered. He was a priest in Palermo, Sicily, and an outspoken critic of the Mafia.

Consequences

The Uffizi bombing and others like it were designed to destroy irreplaceable items of cultural heritage. The attacks had little effect on tourism and served to galvanize Florentine resistance. The bombings led to the resignations

AP/Wide World Photos

Firefighters clear debris after a car bomb damaged the Uffizi Gallery in Florence.

of some intelligence officials and a shake-up of the national police organization. A voters' referendum called for a change in the way the country was run.

Thanks to the efforts of 150 restorers, custodians, administrators, and volunteers, the Uffizi Gallery reopened only twenty-four days after the bombing. Many of Florence's store owners contributed 2 percent of their net receipts to help defray the cleanup cost. The Italian citizens' positive response convinced the gallery's administration to complete thirty new display rooms. Completion had been stalled since 1990 because of budgetary and bureaucratic problems. The startlingly rapid turnabout so impressed the United Nations Educational, Scientific, and Cultural Organization that it was suggested that the Uffizi experience be used as a model for arts administrations worldwide.

William B. Folkestad

Chinese Immigrants Are Detained in New York

Following the grounding of the Golden Venture off Queens, New York, 276 illegal émigrés from mainland China were taken into custody.

What: International relations
When: June 6, 1993
Where: Rockaway Peninsula, Queens, New York
Who:
GUO LIANG CHI (1968-   ), leader of New York City's Fuk Ching gang

The Golden Venture Runs Aground

In the early morning hours of June 6, 1993, the *Golden Venture*, a 150-foot freighter, ran aground in the Atlantic Ocean a quarter of a mile off the coast of Queens, New York. On board were 285 Chinese people, most of them from the southern coastal province of Fujian, who were attempting to enter the United States without proper authorization. The ship had left Bangkok, Thailand, in February, crossing both the Indian and Atlantic oceans before reaching the United States.

In the course of the arduous trip, at least five of the émigrés were thought to have contracted tuberculosis. Living conditions on board were squalid. Each passenger was given only one bottle of water a week and one meal a day, consisting primarily of rice.

The *Golden Venture* arrived in U.S. waters in May but, after two attempts, was unable to connect with the smaller ships that were to smuggle the Chinese passengers ashore. On May 17, an agent of New York's Fuk Ching gang, which ran the smuggling operation, wrested control of the freighter from its crew of seven Burmese and six Indonesians. In an act of desperation, the *Golden Venture* was purposely permitted to run aground on June 6.

As the ship foundered fifteen hundred feet off the New York coast, some of its human cargo dived into the cold water, attempting to swim through six-foot waves to the shore. A rescue effort involving hundreds of police officers, firefighters, and members of the Coast Guard was launched immediately. The rescuers were shocked by the appalling sanitary conditions aboard the grounded ship.

By the time the rescue was completed, six of the Chinese émigrés had drowned. Of those remaining, 276 were remanded to the custody of the Immigration and Naturalization Service (INS) and dispatched to detention centers to await hearings on their requests for asylum. The INS announced its opposition to granting these requests.

Illegal Chinese Immigration

Illegal immigration from China to the United States has occurred for more than a century, particularly from the southern province of Fujian, home of the Chinese aliens aboard the *Golden Venture*. As early as the 1849 Gold Rush in California, crime syndicates in China lured people to the United States with promises of wealth and virtually limitless possibilities.

The system used in those early days differed little from the system employed by the current syndicates. They offer transportation to the United States, housing and jobs for people after they arrive, forged documents, and virtual immunity from deportation. In return, those who leave China pay a substantial down payment of the syndicate's fees and sign agreements to pay the remainder from the money they expect to earn upon arriving in New York City, San Francisco, or one of the other popular destinations for illegal aliens.

Once they reach their destinations, the Chinese usually realize that they have been deceived. They frequently work long hours in sweatshops, often for as little as a dollar an hour. Their income is barely enough to cover their living expenses, let alone repay the syndicate quickly and save enough money to bring other family members to the United States. They live in a condition of indentured servitude as virtual slaves, but they can do little to help themselves because of their tenuous legal status in the country.

An estimated fifty thousand to eighty thousand illegal Chinese aliens were smuggled into the United States each year in the late 1980's and early 1990's, paying fees of between $20,000 and $35,000 each. Only about 10 percent of these immigrants are apprehended by the INS and returned to China. In essence, then, the syndicates honor their claims of succeeding in helping people to escape from China. The rate of defection

from China accelerated sharply in the years following the Tiananmen Square massacre in 1989.

Illegal immigration to the United States from China far exceeds such immigration from most other countries. Chinese immigrants enter the country by land, sea, and air. Some are smuggled into Mexico and make their way over the U.S.-Mexico border.

On July 6, 1993, the United States Coast Guard stopped three ships smuggling a total of 658 Chinese nationals, mostly young males, in international waters off Mexico's northern coast. Mexican immigration officials finally permitted these three dilapidated ships to dock in the Pacific port city of Ensenada, where the immigrants were held in custody. Eventually they were returned to China in a chartered Mexican jet aircraft. Crew members of these three ships were arrested and charged with violations of Mexican immigration laws. Most of the crew members of

U.S. Coast Guard personnel intercept a ship carrying Chinese nationals off the coast of San Diego, California, in July 1993.

2531

the *Golden Venture* were charged with violating American immigration laws.

Consequences

An immediate consequence of the grounding of the *Golden Venture* was that nine crew members of the vessel were tried and convicted of smuggling. On December 3, 1993, they were each sentenced by the U.S. District Court in New York City to six months of imprisonment. These arrests and convictions did not, however, strike at the heart of the problem, the crime syndicates behind the smuggling.

On August 27, 1993, Guo Liang Chi, leader of New York's Fuk Ching gang, was arrested in Hong Kong. That gang was thought to have been involved in the *Golden Venture* debacle. Charged in the January deaths of two of his New York accomplices, Guo Liang Chi faced extradition to the United States. The following day, federal investigators arrested fourteen other Fuk Ching gang members in New York City on charges of conspiring to smuggle immigrants, kidnapping, and extortion.

R. Baird Shuman

United States Orders Trade Sanctions Against China

In an effort to stop the spread of nuclear weapons and technology, the United States imposed trade sanctions against China for two years for violating the Missile Technology Control Regime agreement.

What: International relations; Military
When: August 25, 1993
Where: Washington, D.C.
Who:

QIAN QICHEN (1928-), foreign minister for the People's Republic of China beginning in 1988

JAMES A. BAKER III (1930-), U.S. secretary of state from 1989 to 1992

GEORGE HERBERT WALKER BUSH (1924-), president of the United States from 1989 to 1993

BILL CLINTON (1946-), president of the United States from 1993 to 2001

WARREN CHRISTOPHER (1925-), U.S. secretary of state from 1993 to 2001

Trade Sanctions Against China

On August 25, 1993, the U.S. government announced that China and Pakistan had violated the Missile Technology Control Regime (MTCR) agreement. The problem occurred over the transfer from China to Pakistan of the technology to produce the M-11 missile. Qian Qichen, China's foreign minister, insisted that China had maintained compliance with the MTCR in its dealings with Pakistan. Although China admitted to selling missile technology to Pakistan, it maintained that the M-11 missiles were not involved.

According to U.S. laws, the Bill Clinton administration had no choice but to impose economic sanctions against China. Although the United States imposed measures against China, they were the weakest allowed by law. The State Department enacted a package of trade sanctions aimed primarily at the transfer of electronics with military uses. Covered under the two-year period of the sanctions were satellite equipment, rocket systems, computer systems related to warheads, and other software and computer equipment used in unmanned flight vehicles. The trade sanctions were projected to cost the United States more than $1 billion in trade with China.

These sanctions worried U.S. businesses involved in high-tech manufacturing. These corporations were concerned that other countries would take advantage of the void created in the Chinese market by the trade sanctions. Because the move was not triggered directly by an international agreement, other nations were not obligated to impose similar measures against China.

Even with pressure from corporate America not to enforce the sanctions, the United States was not willing to let China sell technology that could shift the balance of nuclear power between Pakistan and India. The MTCR agreement was meant to slow or stop the proliferation of nuclear weapons. The United States believed that the sanctions it legislated were an effective way of enforcing the agreement and that the agreement needed to be enforced.

Proliferation of Nuclear Weapons

The decision to impose trade sanctions against China stemmed from the MTCR, an agreement reached in 1987 after four years of negotiations. The countries involved in the initial signing of this policy were Canada, France, Germany (then the Federal Republic of Germany), Italy, Japan, the United Kingdom, and the United States. These seven nations wanted and encouraged all countries to join the security initiative. The People's Republic of China and what was then the Soviet Union were singled out as two possibilities for early inclusion in the MTCR.

The pact established a formal foreign policy and framework for restricting the exportation of missile systems and unmanned vehicle technology to Third World countries. The nations that were close to using this type of technology were Argentina, Brazil, India, and Pakistan. The MTCR divided equipment targeted for export control to Third World nations into two groups. Category 1 systems included ballistic missiles, space launch vehicles, sounding rockets, and other subsystems. Category 2 covered components that could be used to build missile and unmanned vehicle systems. The export of short-range missile systems that could be adjusted for longer ranges also was covered in the MTCR, but in vague terms. The items listed in Category 2 were allowed to be exported on a case-by-case basis, as long as the items could not be used in nuclear weapons.

In 1991, China agreed to abide by guidelines and parameters of the MTCR, even though it did not sign the initiative. Secretary of State James Baker III understood this as meaning that the M-9 and M-11 missiles and their technologies would not be exported to any Third World nation. It was this second missile type, the M-11, that eventually led to the trade sanctions of 1993. Furthermore, China's agreement to follow the MTCR had led President George Bush to grant China conditional most favored nation status in 1992. That status improved China's trading relationship with the United States, relieving a variety of trade restrictions between the two countries.

By March 25, 1993, twenty-three countries had signed the MTCR. These countries were Australia, Austria, Belgium, Canada, Denmark, Finland, France, Germany, Greece, Iceland, Ireland, Italy, Japan, Luxembourg, the Netherlands, New Zealand, Norway, Portugal, Spain, Sweden, Switzerland, the United Kingdom, and the United States. Additionally, Argentina and Hungary were invited to become partners in the MTCR. It was important that even though China was one of the first nations invited to join this agreement, it still had not officially signed the MTCR.

The Missile Technology Control Regime Plenary Session was held in Canberra, Australia, in March of 1993. Shortly after this meeting, the United States accused China of violating the guidelines set forth by the MTCR. A United States investigation found that China had sent M-11 missile technology to Pakistan. This technology was covered by the MTCR because it could be modified for use in long-range nuclear missiles. At this point, the U.S. government took action and placed economic trade sanctions on high-tech exports to China. The United States also placed less stringent sanctions on Pakistan for importing the missile technology.

Consequences

In October, 1994, the United States and China held talks to resolve the trade sanctions that were imposed on August 25, 1993. On October 4, the United States and China came to an agreement on how to control the proliferation of nuclear weapons. The first step was that the United States lifted all sanctions that had been placed on China in August of 1993. After these sanctions were removed, China agreed to abide by the MTCR and not export any missiles covered by that agreement. Moreover, China agreed that any missile technology that could be adapted to fall into one of the MTCR categories also would be banned from exportation.

David R. Buck

Nigerian President Babangida Resigns

> *After voiding the nation's first democratic elections in ten years, Nigerian president Ibrahim Babangida resigned in the face of a nationwide strike.*

What: National politics
When: August 26, 1993
Where: Abuja, Nigeria
Who:

IBRAHIM BABANGIDA (1941-   ), military leader of Nigeria from 1985 to 1993

ERNEST SHONEKAN (1936-   ), technocrat appointed as interim leader

MOSHOOD ABIOLA (1938-1998), Yoruba chief judged by many to have been legitimately elected president of Nigeria in June, 1993

SANI ABACHA, (c. 1944-1998), Babangida's succesor as military leader of Nigeria

Military Government Under Siege

General Ibrahim Babangida, Nigeria's military ruler since 1985, promised his people democratic elections many times, only to postpone them. He finally permitted elections to occur in June, 1993, under very restricted conditions. He then annulled the elections when it appeared that a candidate not of his approval, Yoruba chief Moshood Abiola, was the likely winner. This action unleashed a wave of protest, including strikes and demonstrations, that made Babangida's resignation inevitable.

Babangida, though a tyrant, was a skillful political leader. He knew his time as leader of the Nigerian people had ended. Pressure from abroad was also a prominent contributing element in Babangida's decision to resign. Although Nigeria, as the largest nation in West Africa, is a center of diplomatic and political influence in Africa itself, its relationship to the West is primarily an economic one. Nigeria is known in the West as a supplier of natural resources, particularly oil. Oil workers were among the groups that went on strike to protest the election's annulment, and the Nigerian petroleum industry had been brought to a near halt. Nigeria's trade with the outside world was drying to a trickle.

Babangida left office on August 26, 1993, without resolving the fundamental problem facing his country. He appointed only an interim leader, a passive technocrat and businessman named Ernest Shonekan who promised little to alleviate the country's problems. Many observers concluded that ethnic prejudice, not mere political ambition and hunger for power, lay behind Babangida's actions. Babangida, like most leaders throughout Nigeria's post-independence history, was a member of the Hausa tribe from the north, a people particularly dominant within the military. Babangida had managed the electoral process from beginning to end. The fact that Abiola had nevertheless won signaled that the Yoruba were threatening to become more politically conscious and to assume the mantle of leadership on their own behalf.

There were also religious aspects to this division. The Hausa are mainly Muslim, whereas the Yoruba, who live in the southwestern portion of Nigeria and are the most populous ethnic element within the country, are substantially Christian. Despite the fact that Abiola himself, though a Yoruba, was also a Muslim, it was thought that Babangida's objections to Abiola were based on religious and ethnic factors.

Few observers took Shonekan seriously as a leader. Although he was a successful businessman, his political experience was limited and his power base was insubstantial. Nobody thought that Babangida intended Shonekan as any sort of permanent successor. He was only a stopgap, a device for preventing Babangida's legitimate successor, Abiola, from assuming the power that the people had given him through democratic elections. For the most part, neither the world community nor any major group or institution within Nigeria acted as if Shonekan would be a durable presence on the Nigerian political scene. Most of the Nigerian public had grown jaded and cynical about the political process. Nigerians no longer expected Babangida or anyone associated with the military leadership to produce anything good for them. In recognition of success at achieving Babangida's ouster, however, the vast majority of strikes and protests were called off, and Nigerian public life returned to normal, albeit momentarily.

A Nation's Wasted Potential

By the early 1990's, Nigeria, long thought to hold the potential to be the dominant country in Africa and perhaps a major world power of the future, was in terrible shape. Economic depression and corruption plagued the nation, along with Babangida's extreme reluctance to relinquish his own grip on power despite promises of democracy. Babangida had been part of a military coup that had overthrown Nigeria's last elected government on New Year's Eve, 1983. In September of 1985, he had staged another coup against two of his fellow military leaders and assumed full power for himself.

Babangida promised a timetable for a gradual return for democracy, but this kept being delayed. Finally, he permitted elections in 1993, but they were sharply controlled by his government. Anyone deemed to have participated in corruption under the previous democratic regime was prevented from running. Furthermore, Babangida permitted only two political parties, invented by himself and designed to be slightly right of center and slightly left of center respectively. By designing only two political parties based on rather slim ideological divisions, Babangida claimed that he was trying to find a way around corruption and ethnic advocacy. Many suspected that he was trying to hold on to power as long as possible, as well as hoping for a continued supremacy of the Hausa over the Yoruba and the Ibo from the southeastern part of Nigeria. The Ibo had launched a ferocious war for independence in the late 1960's, declaring a separate Republic of Biafra. The revolt, and its eventual failure, was at the root of one of the many strands of tension within Nigeria's various communities. When the left-of-center candidate, Abiola, won against the right-of-center candidate, Bashir Tofa, Babangida decided that even this harnessed version of electoral politics was not for him. He annulled the vote, precipitating the crisis that led to his resignation of power.

Consequences

As expected, Shonekan's government did not last the year. In October, power was assumed by another general, Sani Abacha, a former colleague of Babangida who showed no signs of yielding power to democratic rule. When Abiola protested, he was imprisoned and, according to some accounts, subjected to torture and starvation under Abacha's harsh regime. Ironically, Abiola died in prison in July, 1998—only one month after Abacha's own unexpected death, apparently from a heart attack. An autopsy performed by an international team of medical experts concluded that Abiola's death was also due to natural heart problems, not foul play.

Nicholas Birns

Brazilian Police Massacre Rio Slum Dwellers

The slaughter of twenty-one inhabitants of a Rio de Janeiro slum followed shortly after the murder of eight street children and focused world attention on the brutality of Brazil's military police.

What: Civil rights and liberties
When: August 30, 1993
Where: Rio de Janeiro, Brazil
Who:

ITAMAR FRANCO (1931-   ), president of Brazil from 1993 to 1995

EMIR LARANGEIRA, state legislator and commander of the Ninth Battalion of military police

GILBERTO DIMENSTEIN, journalist who had denounced violence against children

Death Squads Attack a Shanty Town

On the night of August 29-30, 1993, a squad of hooded, armed men appeared in the Vigário Geral slum. They broke into smaller squads and spread out across the slum. In Corsican Plaza stood carts that some residents used to sell juice and other soft drinks in the city. The attackers machine-gunned the carts, killing a seventeen-year-old boy and an unemployed metalworker. Other assassins walked into the Caroço Bar, terrifying those inside. When some of the customers tried to leave, the murderers set off a bomb and then began shooting at the survivors. Seven people died.

Meanwhile, another squad of killers cut the telephone wires leading into Vigário Geral so that no one could summon help. The hooded gunmen then shot people they met in the street. When one woman's husband, Edmilson Costa, tried to rescue his family, the gunmen agreed to let them go if he took their place. The killers ordered the woman to run away without looking back. She heard her husband pleading with the men not to kill him. Shots rang out, and Costa was dead.

Hooded killers broke into the house of Gilberto Cardoso dos Santos. Some of the gunmen wanted to kill four small children found in the house. One killer took pity on the children and told them to run. The oldest, a girl of nine, led the escape, carrying with her a baby born two weeks earlier. Nine other family members were killed.

Before the assassins faded into the night, they had murdered twenty-one innocent people. Many victims, however, probably sensed that trouble was coming. The day before, a Sunday, several military policemen had come into Vigário Geral and tried to extort money from drug kingpin Flávio Pires da Silva. Four policemen died in a shootout. The massacre was the military police's revenge on Vigário Geral. By the following Friday, investigators had arrested five suspects, all members of the Race Horses, a death squad with ties to the Ninth Battalion of military police. Hoods and weapons were found in the suspects' homes.

Outlaw Policemen

The Vigário Geral killings stunned the world, in particular because they followed shortly after the Candelária massacre of July 22-23. On that night, about forty street children were sleeping in Pius X Square near Our Lady of Candelária church. Four hooded policemen appeared and asked for Marco Antonio da Silva, the children's leader. After identifying him, a policeman shot the boy in the head. Chaos erupted. When children screamed and tried to flee, the assassins' pistols sprayed bullets. Eight of the children died, and several suffered serious wounds. An inquiry showed that the murderers had links to the Ninth Battalion of military police. Authorities arrested three officers, including a lieutenant. The murderers apparently were retaliating against the boys for throwing rocks at a police car the day before.

Despite international outrage over the two massacres, Brazilians were shocked only by the number of victims. President Itamar Franco promised that the perpetrators would be punished, but he also publicly criticized Brazilian ambassadors in Europe and North America for not clearly defending Brazil against foreign criticism.

Brazilians understood that police committed murders, extortions, and other crimes not only in Rio de Janeiro but elsewhere in the country as well. Most victims were slum dwellers and street children. Not all Brazilians opposed such killings. A significant minority believed that the police were simply ridding the country of criminals. Many of the children became addicted to drugs or sniffing glue and supported themselves by selling drugs and robbing tourists. Some businesses paid off-duty policemen to kill troublesome street children who bothered customers. Police also extorted drugs and money from the boys.

The low wages paid to military police officers made corruption enticing. They shook down drug dealers and thieves in the slums for a share of the criminals' profits. In fact, police complicity allowed organized crime to survive. A famous case occurred in 1991, when members of the

Ninth Battalion kidnapped and murdered ten residents of the Acari slum. The police apparently were trying to find a large stash of money, perhaps as much as $3 million, hidden by a known criminal who hijacked trucks and sold the cargoes.

Even when investigations uncovered those responsible for such atrocities, few police officers were punished. Suspects were arrested in the Acari, Candelária, and Vigário Geral massacres, but intimidation and even murder of witnesses hampered prosecution. The military took charge of the cases involving on-duty police officers, and superior officers protected their accused men. Some officers wanted the military police to act even more ruthlessly against slum dwellers. In late 1993, the government transferred jurisdiction for crimes committed by military police against children to civil courts. Brazil's long experience with military dictatorship from 1964 to the 1980's made civilian investigators and courts hesitant to challenge the armed forces openly.

Consequences

Following the Candelária and Vigário Geral massacres, the Inter-American Development Bank made funding available to provide refuges, schools, and medical care for street children in Rio de Janeiro. Amnesty International and Americas Watch investigated violence against street children and the actions of the death squads. Brazilian activists created organizations to assist the children. Death squads threatened the activists, however, and Brazil lacked resources to provide much assistance. Many upper- and middle-class Brazilians had little sympathy for street children and slum dwellers, whom they viewed as criminals.

In October, 1994, President Franco sent federal troops into Rio de Janeiro to curb violent crime. According to *Brazil Report* (December 1, 1994), drug gangs operating out of the slums had effectively paid off many of the officers of the state-controlled police. Brazil's fragile democracy, political corruption, and economic woes continued to hinder its attempts to deal with the underlying causes of the slaughters at Candelária and Vigário Geral.

AP/Wide World Photos

Brazilian president Itamar Augusto Cautiero Franco.

Kendall W. Brown

Israel and PLO Sign Peace Agreement

In 1993, Israel and the Palestine Liberation Organization signed a peace agreement with the hope of ending decades of war between the Israeli and Arab peoples.

What: International relations
When: September 13, 1993
Where: Washington, D.C.
Who:

YASIR ARAFAT (1929-   ), chairman of the PLO

BILL CLINTON (1946-   ), president of the United States from 1993 to 2001

SHIMON PERES (1923-   ), foreign minister of Israel in 1977, from 1984 to 1986, and from 1995 to 1996

YITZHAK RABIN (1922-1995), prime minister of Israel from 1974 to 1977 and 1992 to 1995

Israel and the PLO Sign an Accord

On September 13, 1993, representatives from Israel and the Palestine Liberation Organization (PLO) attempted to end decades of strife by signing a peace agreement. In the document, Israel agreed to recognize limited self-rule for the 770,000 Palestinians living on the Gaza Strip and the more than one million Palestinians living on the West Bank, the region west of the Jordan River. In turn, the PLO, represented by its chairman, Yasir Arafat, agreed to end its conflict with Israel.

The agreement, actually a declaration of principles, consisted of seventeen articles and four annexes. Reflecting the wide areas of disagreement that remained between the two peoples, the declaration represented an outline for procedures to be carried out. There would be two significant steps in implementation of the agreement. First, a framework for self-rule would be worked out among the Palestinians themselves, with the role of the Israelis evolving into maintenance of external security. Second, an agreement would be developed in which the two sides would recognize each other's right to existence and develop means to permanently end the animosity between them.

Following introductory remarks by U.S. president Bill Clinton, host to the signing ceremony, Israeli foreign minister Shimon Peres and Arafat initialed the agreement. This was followed by an address by Yitzhak Rabin, prime minister of Israel, in which he called for an end to the hatred between Israel and the PLO.

Attempts to resolve further disputes continued. On February 9, 1994, Arafat and Peres signed an agreement in Cairo, Egypt, dealing with the safety of Jewish settlers in Palestinian territory. Another agreement dealing with the size of Palestinian enclaves, particularly around the city of Jericho on the West Bank, was signed on May 5, 1994. Further agreements were made over the course of the following year. As momentous as the signings themselves was the fact that the two enemies were willing to address, or argue, the issues face to face.

The Palestinian Issue

The background of the Palestinian issue dates from the partition of Palestine in 1947, but roots of the conflict can be traced through nearly two thousand years of Middle Eastern history. Israel (called Judea during the Roman occupation twenty centuries ago) historically has been the homeland of the Jewish people. It ceased to exist following its defeat in the Jewish War (so called by Josephus ben Mathius in the first century C.E.). The Romans renamed the land Palestine, and so it was called through a succession of rulers for the next nineteen hundred years.

Following World War I, Great Britian received a mandate to rule Palestine. Constant conflict involving Jews and Arabs, directed both against each other and against British troops occupying the region, finally persuaded the British to relin-

2539

AP/Wide World Photos

At the signing of the peace accord, Israel prime minister Yitzhak Rabin (left) shakes hands with PLO chairman Yasir Arafat on the White House lawn. President Bill Clinton stands behind them.

quish control. The United Nations voted on November 29, 1947, to partition the land into two separate states, Palestine for the Arabs and Israel for the Jews. The Arabs refused to accept partition and vowed to destroy the fledgling state of Israel. Nevertheless, on May 14, 1948, the state of Israel became a political reality for the first time in nineteen hundred years.

Immediately following independence and departure of the British troops, Israel was attacked by the Arab countries in the region. A bloody war of independence lasting more than seven months ended in a truce that did not include recognition of the Jewish state. A consequence of the war was the removal, both voluntary and under coercion, of many Palestinian families from their homes. Many of these people moved to refugee camps, with the thought of returning to their homes once Israel was eliminated. Most of these camps were on the western bank of the Jor-

dan River and within the Gaza Strip on the Mediterranean shore.

On June 2, 1964, the PLO was formed in Jerusalem, with the ultimate purpose of liberating Israel from the Jews. In 1969, Arafat became chairman of the PLO.

Israel continued to exist literally under the guns of its Arab neighbors. Following significant military threats by these neighbors, on June 5, 1967, Israel attacked and defeated Egypt, Jordan, and Syria in a period of six days. During the "lightning" war, Israel seized significant amounts of territory, including the West Bank and Gaza Strip, home to more than one million Palestinians, as well as the Golan Heights and Sinai Peninsula. Although the Sinai Peninsula eventually was returned to Egypt, Israel continued to occupy the Palestinian territories.

In December, 1987, Palestinians living in the occupied territories began an uprising, the *inti-*

fadah, against the Israelis. The PLO represented one fighting arm of the uprising. Tiring of constant fighting and catalyzed by events of the Persian Gulf War of 1991, both sides agreed to begin formal talks for the first time. On October 30, 1991, Israel and representatives of the PLO sat down together in Madrid, Spain. Talks continued in Norway and elsewhere. On August 29, 1993, it was announced that the two sides had reached an agreement on the issue of Palestinian self-rule in the occupied territories. This agreement was formally signed in Washington, D.C., on September 13, 1993.

Consequences

Although a peace agreement was signed, resulting in at least de facto recognition of both the Palestinian right to land and Israel's right to exist, peace itself remained elusive. Limited self-rule, in the sense that administration of a portion of the territories was now under Palestinian control, became a reality. Israeli troops and settlements, however, could still be found on the West Bank. Furthermore, economic realities were a significant disappointment to the Palestinians. Little industry and few trades were centered in these regions, and poverty remained a significant problem, sowing seeds for more conflict. Terrorism continued, fueled by extremists on both sides, though at least nominally condemned by Arafat. Fruition of the peace process remained elusive.

Richard Adler

Sihanouk Returns to Power in Cambodia

A central personality in the history of modern Cambodia, Prince Norodom Sihanouk resumed monarchical leadership of Cambodia to end his nation's civil strife and to augment the peace process.

What: National politics; Civil war
When: September 24, 1993
Where: Phnom Penh, Cambodia
Who:

NORODOM SIHANOUK (1922-   ), most
  enduring of the country's modern
  political leaders
POL POT (1928-1998), longtime leader of
  the Khmer Rouge
YASUSHI AKASHI (1931-   ), Japanese
  head of the United Nations
  Transitional Authority in Cambodia
NORODOM RANARIDDH (1946-   ),
  Sihanouk's son and prime minister

A Monarchy Restored

Popularly identified after 1953 as the father of Cambodian independence, Norodom Sihanouk served periodically over the next forty years as Cambodia's most influential and enduring politician. Amid ceremonies held in Phnom Penh, the Cambodian capital, on September 24, 1993, Prince Sihanouk once again occupied a throne he had abdicated in 1955 and revived a monarchy that he had abolished.

Sihanouk's restoration confirmed him as head of state in the context of a "liberal democracy." A few days earlier, on September 21, the Cambodian National Assembly had adopted a new constitution that sanctioned Sihanouk's authority "to rule but not to govern," a caveat that was more rhetorical than real. In addition to being head of state, Sihanouk was also the leader of Cambodia's major political party, the United National Front for an Independent, Neutral, Peaceful, and Cooperative Cambodia, the French acronym FUNCINPEC. A party of conservative royalists, FUNCINPEC shared power with the Cambodian People's Party (CCP), led by Hun

Sen. Sen and the CCP represented residual elements of the Vietnamese-backed government that had dominated the Phnom Penh region between 1979 and 1989 against opposition led by the left-wing, nationalist Khmer Rouge. Upon withdrawal of Vietnamese armed forces from Cambodia in 1989, however, the CCP had withered.

Sihanouk therefore presided over what was essentially a two-party coalition, with FUNCINPEC as the superior member. This was reflected by Sihanouk's appointment of Norodom Ranariddh, his son, as a "first" prime minister and of Hun Sen, the CCP leader, as a "second" prime minister. Khmer Rouge leaders and their still aggressively active guerrilla forces momentarily were excluded from government and refused to share power until they received 15 percent of government ministries and 20 percent of top military posts in exchange for relinquishing control over the territories held by their forces. Meanwhile, the signing of Cambodia's new constitution and the return of Prince Sihanouk to the throne automatically concluded the mission of the United Nations Transitional Authority in Cambodia (UNTAC).

Fragmented Cambodia

Since 1945 and the departure of a dominant French presence in the country, Cambodia, except for brief periods, had been a deeply divided nation seared by civil strife and political tumult. The independence formally gained in 1953 left the already well-known, flamboyant, and ambitious Sihanouk able to augment his political control.

In accord with the Geneva Convention of 1954, which was designed to bring an anticommunist peace to the remains of French Indochina, Sihanouk staged a referendum on independence that bore all the hallmarks of a plebiscite, amended the constitution so as to increase his executive authority, and in a political master stroke abdicated his throne in favor of his pliable father, Prince Norodom Suramarit. In addition, Sihanouk assumed leadership of a national political party (the People's Socialist Community). His handpicked candidates swiftly won 80 percent of the National Assembly's seats. Governing during these years, Sihanouk encountered only modest procommunist opposition throughout the country.

Cambodia and Sihanouk faced dramatically changed political circumstances after 1963 and the expansion of the Vietnam War. Sihanouk dealt with these circumstances by keeping communists from power while attacking the United States. He broke relations with the United States in 1965 and restored them in 1969. Massive Vietnamese violations of Cambodia's borders during the Vietnam War and the resultant conversion of large areas into North Vietnamese sanctuaries from American and South Vietnamese military forces soon caused Sihanouk's opposition to flourish.

Sihanouk had always banked on support from professional and bureaucratic elites but began encountering resistance from them. In addition, he suppressed a peasant uprising in 1967 after the massacre of more than ten thousand members of the rural population. Perhaps with Sihanouk's tacit approval, American bombing campaigns against Vietnamese Cambodian enclaves were launched in 1969. These drove hundreds of thousands of radicalized peasants into communist hands.

While on an overseas visit, Sihanouk was deposed as chief of state by his own National Assembly, and a new, U.S.-backed regime led by General Lon Nol assumed power. Lon Nol in turn was soon driven out by communist forces, and the Vietnamese-controlled government of Pol Pot was installed. That government's genocidal policies resulted in two million deaths and left Cambodia shattered by a many-sided and prolonged civil war.

During these years, Sihanouk nominally headed a Cambodian government-in-exile in Beijing, China. Only when Cambodia's warring factions, including the Khmer Rouge, agreed to negotiate their differences at the Paris Conference of 1991 and to do so under United Nations supervision did Sihanouk play a direct role in Cambodia's domestic politics. Two years later, he resumed the throne.

Consequences

Twenty thousand UNTAC personnel incurred a cost of more than $1 billion to create a neutral environment in which new Cambodian elections could be held. The UNTAC personnel then left. It was widely believed that the Cambodian economy would collapse and that the Khmer Rouge would continue armed opposition to Sihanouk's new government. Largely under the direction of Prime Minister Ranariddh and Finance Minister Sam Rainsy, the economy flourished as foreign investment increased. New wealth, in fact, accentuated corruption. The Khmer Rouge steadily lost both territory and supporters under government assaults on their guerrilla forces and on their lucrative smuggling and trading operations. Meanwhile, Sihanouk sought to induce Khmer Rouge leaders to join a national governing coalition.

By 1995, Cambodia generally was peaceful. Its major political problems, which seemed modest by earlier standards, revolved around the aging Sihanouk's attempts to become his own prime minister. As his disagreements with his son and other ministers grew, Ranariddh and Rainsy resigned from their posts.

For a time the Khmer Rouge appeared to be a genuine threat. In April, 1996, however, government troops launched an attack on the main Khmer Rouge base at the town of Phnom Veng and there were reports of splits within the leadership of the Khmer Rouge. Ieng Sary, one of the most powerful men in the radical guerrilla group, split with Pol Pot and applied to the Cambodian government for amnesty, which was granted by Sihanouk in September, 1996.

Throughout 1997 the remaining Khmer Rouge leaders turned against one another. After Pol Pot ordered the murder of his own defense minister along with the minister's entire family in June, Pol Pot himself was taken prisoner by the notorious Khmer Rouge general Ta Mok, nicknamed "the butcher." In April, 1998, while in captivity, Pol Pot died, reportedly of a heart attack.

2543

The splintering of the Khmer Rouge triggered a split between Hun Sen and Prince Ranariddh, who were competing for the loyalty of the defectors. Although Ranariddh was far more popular and would probably win any reasonably fair election, Hun Sen controlled more troops. During the summer of 1997 heavy fighting broke out in Phnom Penh between troops loyal to the two prime ministers. In July, a day after Ranariddh left the country for a visit to Paris, Hun Sen seized sole power and declared Ranariddh a traitor for having sought support among the Khmer Rouge forces.

Fighting between forces loyal to Ranariddh and Hun Sen's government troops continued, and the ongoing conflict in Cambodia appeared to be between the supporters of the royal family, who enjoyed popular support, and the former pro-Vietnamese leftists led by Hun Sen, who held most of the military power.

Clifton K. Yearley

Georgia's Abkhaz Rebels Capture Sukhumi

In a development indicating the strength of nationalism in the former Soviet Union, rebels in the Abkhazian region of Georgia captured the regional capital.

What: Civil war
When: September 27, 1993
Where: Abkhazia and Georgia
Who:
ZVIAD GAMSAKHURDIA (1939-1994), chairman of the Georgian Supreme Soviet from 1990 to 1992
VLADISLAV ARDZINBA (1945-), chairman of the Abkhaz Supreme Soviet (Parliament) from 1990
EDUARD SHEVARDNADZE (1928-), chairman of the State Council of Georgia in 1992, and chairman of the Georgian Supreme Soviet from 1992 to 1995
BORIS YELTSIN (1931-), president of the Russian Republic from 1991 to 1999

Abkhaz Separatists Seize Sukhumi

Demonstrations in early 1992 by ethnic Georgian supporters of ousted Georgian president Zviad Gamsakhurdia in Abkhazia prompted the new Georgian administration to send troops to the Abkhaz capital, Sukhumi. The troops clashed with the supporters of Gamsakhurdia, known as Zviadists. Tension in Abkhazia intensified after July 23, when the Abkhaz Supreme Soviet voted to substitute the constitution of 1925, which asserted that Abkhazia was a separate union republic, for that of 1978, which specified that Abkhazia was part of Georgia.

The Georgian State Council immediately declared the Abkhaz move invalid and sent an additional three thousand troops to the region. They were sent ostensibly to repress the Zviadists, who had taken Georgian officials hostage. Fighting broke out between Abkhaz troops and the Georgians, however, and on August 18 Georgian defense minister Tengiz Kitovani ordered an attack on the Abkhaz parliament. Abkhazian leader Vladislav Ardzinba and Abkhaz deputies fled to Gudauta and issued a call for armed resistance.

Although President Boris Yeltsin of Russia appealed to members of the Commonwealth of Independent States (CIS) not to interfere and asserted his dedication to Georgian integrity, Russian paratroops were dispatched, reputedly to protect Russian bases in the area. Clashes occurred between Russian and Georgian troops. The Georgians asserted that conservative elements in the Russian military were providing military assistance to the Abkhazians in order to tame Georgian independence. In October, 1992, an Abkhaz offensive overran northern Abkhazia. Hundreds of ethnic Georgians were killed in the process.

Fighting intensified between Georgians and Abkhazians in 1993. Although the Georgians continued to hold Sukhumi, their forces, simultaneously attempting to counter Zviadists in western Georgia, were nearing exhaustion. In mid-September, the Abkhazians launched a new offensive, and Eduard Shevardnadze, the new Georgian leader, flew to Sukhumi to direct its defense. The Georgians, who had withdrawn their artillery in accord with a provisional peace settlement brokered by the Russians in late July, were overrun in eleven days. Shevardnadze was forced to flee by air. Abkhaz rebels seized the capital on September 27, and by the end of the month, Georgian forces had been expelled from all of Abkhazia.

National Versus Ethnic Groups

The Abkhazians are a predominantly Muslim, Turkic-speaking people living on the Black Sea coast of the Caucasus. Russia asserted a protectorate over their region in 1810, then annexed it in 1864. In March, 1921, after the Red Army de-

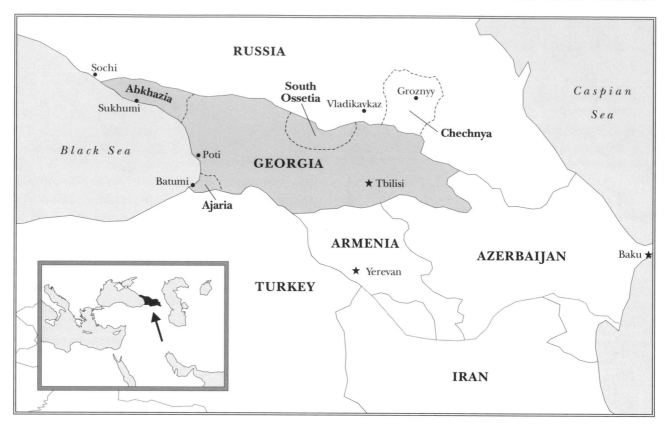

feated the independent Georgian state, Abkhazia became an independent union republic. In December, 1921, Abkhazia joined the Georgian Soviet Socialist Republic, but it retained its status as a union republic until April, 1930.

In 1930, Abkhazia was demoted to the status of an autonomous republic. Stalin ordered that western Georgians be resettled in the area. As a consequence, by 1989 ethnic Abkhazians constituted only 18 percent of the population and ethnic Georgians constituted 46 percent. Tensions existed between these two groups, both of which thought that they were being discriminated against to the advantage of the other. In 1978, the Abkhazians started a campaign to secede from Georgia and join the Russian Soviet Federated Socialist Republic. Although Moscow rejected this bid, it responded with economic and cultural concessions. These measures intensified the resentment of the Georgians.

On March 18, 1989, a mass meeting in the Abkhaz village of Lykhny demanded separation from Georgia and restoration of the former union republic status. In July, 1989, an effort by ethnic Georgians to transform the Georgian division of the Abkhaz State University into a branch of the Tiblisi State University provoked two weeks of ethnic violence in Sukhumi.

In August, 1990, the Georgians altered their election laws to exclude purely regional parties from the forthcoming Georgian Supreme Soviet elections. The exclusion applied to the Abkhazian Popular Front, Aydgylara. The Abkhaz delegates to the Abkhaz Supreme Soviet responded by declaring Abkhazia a completely independent republic. The Abkhazians rejected the authority of the new Georgian nationalist leader Gamsakhurdia, and in December they elected Ardzinba, an Abkhaz historian, as chairman of their Supreme Soviet. Ardzinba asserted the Abkhaz desire to remain in the Soviet Union as a union republic. The Abkhazians, rejecting a Georgian directive, participated in the March, 1991, referendum on preserving the Soviet Union. Of the 52.4 percent of eligible voters in Abkhazia who participated, more than 98 percent voted in the affirmative.

In the spring of 1991, Abkhazians and Georgians worked out a compromise electoral law for the region that would have guaranteed the

Abkhazians twenty-eight delegates, the Georgians twenty-six, and other groups a total of eleven. They also agreed to a two-thirds vote for important legislation. Gamsakhurdia, however, impeded the election. The ouster of Gamsakhurdia in January, 1992, did not improve the situation. When the new Abkhaz parliament met in early 1992, the Georgians viewed it as a body intent on secession. Georgians in Abkhazia launched a campaign of noncompliance and protests that eventually led to the rebel takeover.

Consequences

More than 200,000 Georgian civilians fled the vengeance wreaked by the conquering Abkhazians. Clashes continued despite peace talks sponsored by the United Nations and the signing of a memorandum by Georgian and Abkhaz representatives. Georgia claimed that the Abkhazians, who had succeeded in carrying out an almost totally successful campaign of ethnic cleansing, were refusing to allow ethnic Georgian refugees to return to the area.

As a result of the Abkhaz success, Russia was able to convince Georgia of its need for Russian assistance and hence of the necessity for Georgia to submit to Russian interests. Georgia, which had previously refused to join the CIS, signed a treaty of friendship with Russian in February, 1994, and joined the CIS. In return, Russia provided Georgia with needed military equipment.

Bernard A. Cook

Yeltsin Battles Russian Parliament

> *In what has been described as Russia's post-Soviet revolution, President Boris Yeltsin dissolved the parliament, sent troops to assault the legislature building, called new parliamentary elections, and replaced Russia's constitution.*

What: National politics
When: October 4, 1993
Where: Moscow, Russia
Who:
BORIS YELTSIN (1931-   ), president of the Russian Federation from 1991 to 1999
RUSLAN KHASBULATOV (1942-   ), chairman of the Russian parliament from 1991 to 1993
ALEKSANDR RUTSKOY (1947-   ), vice president of the Russian Federation from 1991 to 1993
VLADIMIR ZHIRINOVSKY (1946-   ), head of the right-wing, anti-Semitic Liberal Democratic Party

The Revolt and Elections

Through the spring and summer of 1993, confrontation developed between Russia's president, Boris Yeltsin, and Russia's parliament, led by Ruslan Khasbulatov. Although Yeltsin had chosen a hero of the Afghanistan war, Aleksandr Rutskoy, as his running mate in the 1991 presidential elections, the two men became estranged, and Rutskoy sided with the parliament. On September 21, 1993, Yeltsin announced his decision to dissolve the legislature, to dismiss Rutskoy, and to call new elections for December 12. The elections were to choose a new parliament and obtain popular approval for a new constitution that would strengthen presidential power.

News of Yeltsin's plan leaked prematurely, and Khasbulatov, Rutskoy, and a few hundred deputies and paramilitary supporters holed up in the Russian parliament building, a huge marble edifice known as the White House. Yeltsin ordered heat, light, hot water, telephone service, and other utilities cut off, and he tried to starve out the defiant legislators. On October 3, armed supporters of the parliamentary leadership broke through the police and military cordon surrounding the White House and joined its defenders. On Rutskoy's urging, irregular troops then sallied out from the White House and seized the lower floors of Moscow's city hall. They assaulted Russia's central television facilities, almost capturing them. Snipers on upper floors of the White House opened fire on government forces.

After some hesitation at Russia's Ministry of Defense, troops and tanks under Yeltsin's orders bombarded the White House, starting at 10:00 A.M. on October 4. They set fires in the building. Khasbulatov, Rutskoy, and the last of their supporters surrendered at 6:05 P.M., after which they were transported to Lefortovo prison. The official casualty figures were 144 killed and 878 wounded. This was the largest number of Russians slain in Moscow since the Bolshevik Revolution of 1917.

The December 12 elections resulted in Vladimir Zhirinovsky's ultranationalist Liberal Democratic Party gaining almost 25 percent of the vote. Opposition parties representing the former communists also did well. Parties supporting Yeltsin's reforms attracted only a minority of voters. A favorable vote by more than 50 percent of eligible voters in the country was required to approve Yeltsin's new constitution. The Yeltsin government declared after the election that the necessary majority had been obtained. The new constitution gave the president the right to dissolve parliament if it should obstruct presidential decrees and ministerial appointments.

Roots of Confrontation?

The issues in dispute in 1993 revolved around presidential power to direct Russia's government and name key officials. Many parliamentarians sided with the industrial managers of the previous communist era in opposing free market prices, competitive capitalism, and the privatization of factories and farms. Many legislators had also come to believe that President Yeltsin was trampling on the constitutional separation of the executive, the parliament, and the judiciary. The Constitutional Court took the same view, and Yeltsin suspended its work after the October showdown. Yeltsin himself admitted that he broke the law but argued that it was bad law that was pushing the country to the brink of collapse.

The parliament had become increasingly obstructionist, and it was using its powers to block Yeltsin's efforts to rescue the country from impending economic crises. The parliament had forced out Yeltsin's key reformist cabinet ministers, had supported the central bank president's wildly inflationary emission of paper currency, and had viciously attacked Yeltsin himself.

Yeltsin became increasingly confrontational. He was neither adept at nor inclined to make the political accommodations necessary to make a government of divided powers work. By September of 1993, it was clear that the country could not long endure the raging conflict of powers without a disaster.

The parliamentarians were the first to resort to military action; the armed forces vacillated. When fired upon by snipers from the White House, however, both soldiers and police officers entered the fray on the president's side.

The heavy voter support in December for extreme leftist and rightist parties can be explained in large part as a protest vote. Unemployment was rising, inflation was wiping out savings, and organized crime was gaining strength. Most independent observers, and even Russian official investigators, concluded after the elections that the new constitution had fallen short of receiving the necessary favorable votes from half of all eli-

Procommunist demonstrators and Russian nationalists denounce Boris Yeltsin and his proposed reforms in Lubyanka Square, in Moscow.

2549

gible voters. Apparently, election officials falsified the voting tallies. Yeltsin put the constitution into force anyway.

Consequences

As 1994 progressed, it became clear that the new parliament was not much easier for Yeltsin to manage than the old one had been. The confrontational mood took hold once again. Polls indicated that Yeltsin's popularity was plummeting. Political and economic conditions continued to drift or deteriorate. Nevertheless, the country did not fall back into the brutal dictatorship of the communist era. Freedom of speech, press, religion, and assembly continued, and the national security organs did not revert to their earlier practice of arresting dissidents and helpless citizens without legal sanction. Although the crackdown of October, 1993, may have ended Russia's dream of Western-style liberal democracy under the rule of law, the country did not revert to earlier conditions prevailing under communism.

Nathaniel Davis

Bhutto's Pakistan People's Party Returns to Power

> *For the first time since the creation of Pakistan in 1947, an election returned a prime minister who had been dismissed.*

What: National politics
When: October 6-19, 1993
Where: Pakistan
Who:
BENAZIR BHUTTO (1953-   ), prime minister of Pakistan from 1988 to 1990 and 1993 to 1996
GHULAM ISHAQ KHAN (1915-   ), president of Pakistan from 1988 to 1993
NAWAZ M. SHARIF (1949-   ), prime minister of Pakistan from 1990 to 1993

Benazir Bhutto's Party Wins Reelection

Since its creation in 1947, Pakistan has had a troubled political history. Early attempts to found a parliamentary democracy failed. Military regimes dominated the country from 1958 to 1971 and again from 1977 to 1988. Zulfiqar Ali Bhutto attempted to install a one-party system from 1972 to 1977. He was arrested and eventually hanged, and a military regime took over the country.

Benazir Bhutto, Zulfiqar Ali Bhutto's daughter, won election as prime minister in 1988 but was dismissed by President Ghulam Ishaq Khan in 1990. Khan himself was forced to resign in 1993 by order of the Supreme Court of Pakistan, setting the stage for the elections of October, 1993, in which Bhutto won election as prime minister. Her reelection was an exceptional event. On the surface, at least, it seemed to be a triumph for the democratic process.

Bhutto's first tenure as prime minister was highly controversial. Her first husband, Asif Zardari, was accused of corrupt financial dealings. Violence in Karachi and the Province of Sindh reached new heights. Critics held that because the power base of Bhutto's Pakistan People's Party (PPP) was located in Sindh, she was unwilling to suppress Sindhi gangs. In 1990, President Khan dismissed her on the grounds of corruption and ineptitude. Nawaz Sharif's Islamic Democratic Alliance—a coalition of nine political parties, including the Jamaat-i-Islami—came to power. The Alliance was weakened in 1992 when the Jamaat-i-Islami withdrew from the coalition, claiming that Sharif's government was not committed to turning Pakistan into a true Islamic state.

Charges of corruption and ineptitude began to be leveled at Sharif's government. Meanwhile, the courts cleared Bhutto's husband, Asif Zardari, of corruption charges. Bhutto organized a series of marches across the country that openly defied Sharif's order to halt all protests. President Khan dismissed Sharif's government. Apparently, he hoped that the army would step in once again. On this occasion, the army refused to take a hand in politics. Its leaders believed that the military's reputation had suffered too much during earlier periods of martial law.

When Pakistan's Supreme Court ruled that Khan had exceeded his powers as president, both he and Sharif resigned their offices. The country held elections judged to be fair and relatively free of violence. No party achieved an absolute majority in the National Assembly. Bhutto's PPP got a plurality of 86 of the total of 217 seats. Sharif's Pakistan Muslim League got 72, with the remaining seats going to a variety of smaller parties. Thus, Bhutto was able to form a government that was able to beat back votes of no confidence.

2551

Building Pakistan

When the British granted independence to their Indian empire in 1947, Pakistan was created out of the Muslim majority provinces of northwestern and eastern India. Much optimism accompanied the birth of Pakistan, a word that literally means "the land of the pure." Political leaders believed that Pakistan could be a model Islamic republic. Those high hopes quickly sank.

One of the most nagging political problems for Pakistan has been the persistence of regional and ethnic political identities. Although almost all of Pakistan's citizens adhere to Islam, religious unity has not so far covered over the distinctions among Punjabis, Sindhis, Baluchis, and Pakhtuns (Pathans). The Bengalis of East Pakistan declared their independence in 1971, creating the state of Bangladesh. Pakhtuns, Sindhis, and Baluchis have also established independence movements that sought to carve new nations of Sindh, Baluchistan, and Pakhtunistan out of Pakistan.

In the late 1980's, the city of Karachi became the scene of traumatic violence. It is Pakistan's biggest city, port, and commercial hub. Sindhis, natives of the area surrounding Karachi, began attacking immigrants (*muhajarin*) who had come from India after 1947. The *muhajars* responded in kind. Drive-by shootings became daily events. Annual death tolls from this type of violence in any given year were in the hundreds.

Pakistan's political turbulence began almost at the moment of its birth and continued into the last years of the twentieth century. Benazir Bhutto's father, Zulfiqar Ali Bhutto, founded the PPP during the crisis accompanying the separation of Bangladesh from East Pakistan. That crisis was marked by a war that brought down Yaya Khan's military regime. Though a brilliant politician, Zulfiqar Ali Bhutto resorted to violence to quiet his critics and political opponents. The elections of 1977 were fraudulent and brought the PPP a lopsided victory. In that year, General Zia al-Haq arrested Zulfiqar Ali Bhutto and created a military regime that lasted until 1988. Bhutto himself was tried for conspiracy to murder one of his opponents. He was convicted and hanged in July of 1979.

In 1988, General Zia al-Haq was killed in a plane crash. The collapse of Zia's military regime opened the way for the elections that brought the PPP to power. As leader of that party, Benazir Bhutto became prime minister.

Consequences

In the past, when the electoral process generated blatant fraud or excessive violence, Pakistan's army stepped in. The elections of 1993 were relatively free from corruption and violence. Civilian rule in Pakistan seemed to be affirmed. Nevertheless, Benazir Bhutto faced a number of political problems. Violence in Karachi and the province of Sindh continued. International tension between India and Pakistan persisted.

Although Pakistan's economy was growing, poorer Pakistanis suffered from the effects of rampant inflation. Bhutto was under enormous pressure to deal effectively with those problems, but her slender mandate made it difficult to achieve dramatic results, and she would hold office only another three years.

Gregory C. Kozlowski

Liberal Party's Chrétien Becomes Canada's Prime Minister

> *After nine years out of power, the Liberal Party under the leadership of veteran politician Jean Chrétien won a decisive victory in the 1993 federal election.*

What: Government; National politics
When: October 25, 1993
Where: Canada
Who:

JEAN CHRÉTIEN (1934-   ), leader of Canada's Liberal Party and winner of the 1993 election

KIM CAMPBELL (1947-   ), leader of the Progressive Conservative Party, who was defeated as prime minister in the 1993 election

LUCIEN BOUCHARD (1938-   ), leader of the Bloc Québecois, a party based in the province of Quebec

BRIAN MULRONEY (1939-   ), member of the Progressive Conservative Party and prime minister of Canada from 1984 to 1993

In with the New

On October 25, 1993, Canadians went to the polls and in the process elected a new government. The Liberal Party swept to power with its leader, Jean Chrétien, becoming the twentieth prime minister since the nation's birth in 1867. In the process, voters administered the worse defeat in Canadian history to the incumbent prime minister, Kim Campbell, and her Conservative Party. At the end of election night, the Liberals held 177 seats in the 295-member House of Commons, and the Conservatives were reduced to two seats, a substantial reduction from their preelection total of more than 150 seats. Campbell found herself defeated even in her own riding (electoral district) in the city of Vancouver.

Campbell had not been prime minister for long when she called the election. Her predecessor, Brian Mulroney, had served as prime minister since 1984, having been reelected in 1988. In February of 1993, he announced his decision to step down. That year was also the fifth since the previous election, and under Canadian law, an election had to be called. Mulroney left at a point when he and his party, after their lengthy stay in office, were extremely unpopular with the Canadian public. Campbell announced her decision to seek the party's leadership. Buoyed by the possibility of Canada having its first female prime minister, the delegates to the party convention elected her leader in June. After seeing the party's fortunes improve in opinion polls, she called an election for October 25, 1993. Her main opponent was the official opposition party, the Liberals, under Chrétien, who had become party leader in 1990.

The Campaign

Once the call was made, the five-week election campaign began. Canada's high level of unemployment quickly became a leading issue and a key divider between the two main contenders for prime minister. When asked by the news media what her government intended to do to create employment, Campbell replied that there was little it could do and that the cyclical nature of the economy meant high unemployment would continue for several more years. Her comments, although considered accurate, were viewed as insensitive to the plight of those Canadians without work. In contrast, the Liberals and Chrétien, in a party platform that they named the "Redbook," promised to create jobs by spending taxpayers' dollars through public works programs such as road construction.

Another significant point in the campaign was the televised debate between the respective party

AP/Wide World Photos

Jean Chrétien casting his ballot in a 1995 election.

leaders. Campbell was naturally the target of repeated criticism for her party's record while in office. Particularly damaging was an allegation by Lucien Bouchard, at one time a colleague of Campbell's in Mulroney's government, that the government was running a far larger budget deficit than it had admitted. During the debate, Bouchard repeatedly challenged Campbell to reveal the true figure. She stumbled in her response and in the process made herself appear at best ill prepared and at worst dishonest. The Conservative campaign never recovered.

The other aspect of the 1993 election that greatly aided the Chrétien campaign was that two of the parties, the Reform and the Bloc Québecois, were popular almost exclusively in regions, western Canada and Quebec respectively, that had traditionally been Conservative strongholds. The Liberal Party, on the other hand, was the only contender that had substantial voter loy-

alty in every part of Canada. As the election date neared, support for the Liberals surged at the expense of the Conservatives. Chrétien, who had been on the federal political scene since 1963, proved a popular figure, having long cultivated an image of being a regular citizen who cared about the ordinary working person.

By election day, the Conservative defeat appeared inevitable. The only questions remaining were how big the defeat would be and how significant the Liberal victory would be. The Liberals won easily, while the Conservatives found themselves barely surviving as a political presence with two seats. Campbell resigned as party leader shortly afterward. The Bloc Québecois, a party that had as its main goal the separation of Quebec from Canada, finished second and became the official opposition party in the Canadian parliament.

Consequences

For several reasons, the 1993 Canadian election was one of the most remarkable in the nation's history. First, the Liberal Party, frequently nicknamed "Canada's governing party," resumed its traditional position of power. Next, the Conservatives, under the leadership of Canada's first female prime minister, Campbell, found itself virtually destroyed as a national political entity. Such a defeat was unprecedented in Canadian political history. The Liberals had won in part because of the popularity of Chrétien. More important, however, was the unpopularity of the Conservative government.

Also significant to the Liberal victory was the regionally divided result of the election. The Bloc Québecois captured the most votes and seats, many of which in previous elections had gone to the Conservatives in Quebec. On the other hand, the newly created Reform Party, which had as its primary goal a better deal for western Canada, finished first in Canada's two most western provinces, Alberta and British Columbia, which were also former Conservative strongholds. In the end, only the Liberals could credibly make the claim of being a national party.

Steve Hewitt

Maastricht Treaty Formally Creates European Union

> *The European Union was created on November 1, 1993, with the ratification of the Maastricht Treaty by twelve member nations.*

What: International relations
When: November 1, 1993
Where: Maastricht, The Netherlands
Who:
JACQUES DELORS (1925-　　　), president of the European Commission
JEAN MONNET (1888-1979), French statesman
ROBERT SCHUMAN (1886-1963), French foreign minister from 1948 to 1953

The Maastricht Treaty Is Ratified

On November 1, 1993, the Maastricht Treaty on European Union formally went into effect, following its ratification by Germany. Germany was the last of the twelve member nations to ratify the treaty. The twelve charter member nations of the European Union were Belgium, Denmark, France, Germany, Greece, Ireland, Italy, Luxembourg, the Netherlands, Portugal, Spain, and the United Kingdom.

Prior to the Maastricht Treaty, the organization of European nations was known as the European Community (EC). Earlier, it was known as the European Economic Community (EEC). The goal of the organization's founders was to construct a united Europe through peaceful means and create conditions promoting economic growth, social cohesion among the European people, and greater political integration and cooperation among governments. The Maastricht Treaty's objectives include a common foreign security policy, common defense, close cooperation on justice and domestic affairs, and economic monetary union.

The Maastricht Treaty's idea of economic and monetary union, along with creation of a single currency, became the focus of much of the debate over the treaty. The goal of a single currency for use by all European countries was set to be reached in 1999.

One major advantage of a single currency is the complete elimination of exchange rate risk. Traders would not have to worry that the rate of exchange between currencies would change against their favor. This would encourage more trade and capital flows across European borders. In addition, there would be no need to exchange currencies to make trades in goods, so the costs of currency exchanges would be eliminated, making trade less costly. Thus, European countries would find trading more profitable. Increased trade would allow them to specialize in producing particular goods that they could sell for export, using the proceeds to buy goods that they no longer had to produce for themselves.

Further implications of a single European currency would be a single money supply throughout Europe, rather than a separate money supply for each country. Thus, European monetary policy would be consolidated. A major concern in adopting a single European currency was the idea of a single European monetary policy. Each country's government might prefer to implement its own monetary policy. Another potential hindrance to implementation of a single currency was the uncertainty as to what exchange rate would be used for conversion of each local currency to the single currency. Countries naturally would want the most favorable rate for their currencies.

The Road to Unity

A plan for a unified Europe was first put forward by French statesman Jean Monnet after World War II. The idea for the union of European countries originated in 1950 with a plan proposed by French foreign minister Robert Schuman. In 1950, Schuman proposed a plan that resulted in the creation of the European Coal and Steel Community in 1952. This was the first step in joining Europe and was aimed at pooling the coal and steel industries in the wake of World War II. Schuman's long-term goal was to provide a formal structure for the political unification of Europe through economic integration.

With final ratification by all twelve countries on November 1, 1993, the Maastricht Treaty went into effect. Technically, it is an international treaty and therefore supersedes all national law.

The first step toward the ultimate goal of monetary union was to keep exchange rates among the European currencies within a specified range around "normal" levels for two years. Originally, this range was 2.25 percent, but that goal changed to plus or minus 15 percent. Germany's Bundesbank did not want to accept that wide a variation and wanted to adhere to the original 2.25 percent. The planned single currency was called the European Currency Unit (ECU). The ECU was an artificial construction with a value based on an average of the values of various European currencies.

The European Commission, in recognition of Germany's concerns, planned to set out a monetary union timetable that would delay the introduction of a single currency well into the twenty-first century. It was believed that once exchange rates were set, it would take another two to four years for currency and coins of the new monetary system to achieve widespread circulation. However, this timetable was accelerated when the European nations agreed to create a single unit of currency, the "euro," in 1998.

Consequences

One goal of the European Union is monetary union. The overall objective is to raise Europe to the economic and fiscal level of the United States. The Maastricht Treaty was designed to carry out this goal. Many observers agree that the advantages to monetary union outweigh the disadvantages, but there has yet to be any tangible proof of this. In addition, at various world trade talks European countries appeared unable to agree on methods to achieve free trade in various commodities, either among themselves or between Europe and the rest of the world. Therefore, achieving free trade within the European Union seemed to be a formidable task.

Anita B. Pasmantier

GREAT EVENTS

1900-2001

CATEGORY INDEX

LIST OF CATEGORIES

AGRICULTURE

Congress Reduces Federal Farm Subsidies and Price Supports, **7**-2760

Gericke Reveals Significance of Hydroponics, **2**-569

Insecticide Use Increases in American South, **1**-359

Ivanov Develops Artificial Insemination, **1**-44

Morel Multiplies Plants in Vitro, Revolutionizing Agriculture, **3**-1114

Müller Develops Potent Insecticide DDT, **2**-824

ANTHROPOLOGY

Anthropologists Find Earliest Evidence of Modern Humans, **6**-2220

Benedict Publishes *Patterns of Culture*, **2**-698

Boas Lays Foundations of Cultural Anthropology, **1**-216

Boule Reconstructs Neanderthal Man Skeleton, **1**-183

Dart Finds Fossil Linking Apes and Humans, **2**-484

Humans and Chimpanzees Are Found to Be Genetically Linked, **5**-2129

Johanson Discovers "Lucy," Three-Million-Year-Old Hominid Skeleton, **5**-1808

Last Common Ancestor of Humans and Neanderthals Found in Spain, **7**-2841

Leakeys Find 1.75-Million-Year-Old Hominid Fossil, **4**-1328

Mead Publishes *Coming of Age in Samoa*, **2**-556

117,000-Year-Old Human Footprints Are Found Near South African Lagoon, **7**-2860

Pottery Suggests Early Contact Between Asia and South America, **3**-1230

Simons Identifies Thirty-Million-Year-Old Primate Skull, **4**-1553

Weidenreich Reconstructs Face of Peking Man, **2**-786

Zdansky Discovers Peking Man, **2**-476

VI

CIVIL STRIFE

CIVIL WAR

COMMUNICATIONS

COMPUTER SCIENCE

CONSUMER ISSUES

COUPS

CRIME

DISASTERS

EDUCATION

ENERGY

ENGINEERING

XIV

INTERNATIONAL LAW

INTERNATIONAL RELATIONS

LABOR

XVIII

POLITICS

XXV

XXVI

WEAPONS TECHNOLOGY